Encyclopedia of Quantum Mechanics: Recent Advances

Volume V

Encyclopedia of Quantum Mechanics: Recent Advances Volume V

Edited by **Ian Plummer**

NY RESEARCH PRESS

New York

Published by NY Research Press,
23 West, 55th Street, Suite 816,
New York, NY 10019, USA
www.nyresearchpress.com

Encyclopedia of Quantum Mechanics: Recent Advances
Volume V
Edited by Ian Plummer

International Standard Book Number: 978-1-63238-160-6 (Hardback)

Printed in the United States of America.

Contents

Preface

Over the recent decade, advancements and applications have progressed exponentially. This has led to the increased interest in this field and projects are being conducted to enhance knowledge. The main objective of this book is to present some of the critical challenges and provide insights into possible solutions. This book will answer the varied questions that arise in the field and also provide an increased scope for furthering studies.

The advancement of quantum mechanics has given physics a completely new direction from that of classical physics in the early days. In fact, there is a constant development in this subject of a very fundamental nature, such as implications for the foundations of physics, physics of entanglement, geometric phases, gravity and cosmology and elementary particles as well. This book will be an important resource for researchers with respect to present topics of research in this developing area. The book addresses important subjects grouped under three sections: Classical-Quantum Correspondence, Schrödinger Equation and Path Integrals.

I hope that this book, with its visionary approach, will be a valuable addition and will promote interest among readers. Each of the authors has provided their extraordinary competence in their specific fields by providing different perspectives as they come from diverse nations and regions. I thank them for their contributions.

Editor

Section 1

The Classical-Quantum Correspondence

Classical and Quantum Conjugate Dynamics – The Interplay Between Conjugate Variables

Gabino Torres-Vega

Additional information is available at the end of the chapter

1. Introduction

There are many proposals for writing Classical and Quantum Mechanics in the same language. Some approaches use complex functions for classical probability densities [1] and other define functions of two variables from single variable quantum wave functions [2,3]. Our approach is to use the same concepts in both types of dynamics but in their own realms, not using foreign unnatural objects. In this chapter, we derive many inter relationships between conjugate variables.

1.1. Conjugate variables

An important object in Quantum Mechanics is the eigenfunctions set $\{| n>\}_{n=0}^{\infty}$ of a Hermitian operator $\hat{F}$. These eigenfunctions belong to a Hilbert space and can have several representations, like the coordinate representation $\psi_n(q)=\langle q \mid n\rangle$. The basis vector used to provide the coordinate representation, $| q>$, of the wave function are themselves eigenfunctions of the coordinate operator $\hat{Q}$ We proceed to define the classical analogue of both objects, the eigenfunction and its support.

Classical motion takes place on the associated cotangent space T^*Q with variable $z=(q, p)$, where q and p are n dimensional vectors representing the coordinate and momentum of point particles. We can associate to a dynamical variable $F(z)$ its eigensurface, i.e. the level set

$$\Sigma_F(f)=\{z \in T^*Q \mid F(z)=f\} \tag{1}$$

Where f is a constant, one of the values that $F(z)$ can take. This is the set of points in phase space such that when we evaluate $F(z)$, we obtain the value f. Examples of these eigensurfaces are the constant coordinate surface, $q=X$, and the energy shell, $H(z)=E$, the surface on which the evolution of classical systems take place. These level sets are the classical analogues of the support of quantum eigenfunctions in coordinate or momentum representations.

Many dynamical variables come in pairs. These pairs of dynamical variables are related through the Poisson bracket. For a pair of conjugate variables, the Poisson bracket is equal to one. This is the case for coordinate and momentum variables, as well as for energy and time. In fact, according to Hamilton's equations of motion, and the chain rule, we have that

$$\{t,\ H\}=\sum_i\left(\frac{\partial t}{\partial q^i}\frac{\partial H}{\partial p_i}-\frac{\partial H}{\partial q^i}\frac{\partial t}{\partial p_i}\right)=\sum_i\left(\frac{dt}{dq^i}\frac{dq^i}{dt}+\frac{dp_i}{dt}\frac{dt}{dp_i}\right)=\frac{dt}{dt}=1 \tag{2}$$

Now, a point in cotangent space can be specified as the intersection of $2n$ hypersurfaces. A set of $2n$ independent, intersecting, hypersurfaces can be seen as a coordinate system in cotangent space, as is the case for the hyper surfaces obtained by fixing values of coordinate and momentum, i.e. the phase space coordinate system with an intersection at $z=(q,\ p)$. We can think of alternative coordinate systems by considering another set of conjugate dynamical variables, as is the case of energy and time.

Thus, in general, the T^*Q points can be represented as the intersection of the eigensurfaces of the pair of conjugate variables F and G,

$$\Sigma_{FG}(f,\ g)=\{z\in T^*Q \mid F(z)=f,\ G(z)=g\}. \tag{3}$$

A point in this set will be denoted as an abstract bra $(f,\ g\,|$, such that $(f,\ g\,|\,u)$ means the function $u(f,\ g)$.

We can also have marginal representations of functions in phase space by using the eigensurfaces of only one of the functions,

$\Sigma_F(f)=\{z\in T^*Q \mid F(z)=f\}$, and $\Sigma_G(g)=\{z\in T^*Q \mid G(z)=g\}$.

A point in the set $\Sigma_F(f)[\Sigma_G(g)]$ will be denoted by the bra $(f\,|[(g\,|]$ and an object like $(f\,|\,u)[(g\,|\,u)]$ will mean the $f[g]$ dependent function $u(f)[u(g)]$.

1.2. Conjugate coordinate systems

It is usual that the origin of one of the variables of a pair of conjugate variables is not well defined. This happens, for instance, with the pair of conjugate variables q and p. Even though the momentum can be well defined, the origin of the coordinate is arbitrary on the trajectory of a point particle, and it can be different for each trajectory. A coordinate system fixes the origin of coordinates for all of the momentum eigensurfaces.

A similar situation is found with the conjugate pair energy-time. Usually the energy is well defined in phase space but time is not. In a previous work, we have developed a method for defining a time coordinate in phase space [4]. The method takes the hypersurface $q^1 = X$, where X is fixed, as the zero time eigensurface and propagates it forward and backward in time generating that way a coordinate system for time in phase space.

Now, recall that any phase space function $G(z)$ generates a motion in phase space through a set of symplectic system of equations, a dynamical system,

$$\frac{dz}{df} = X_G, \qquad X_G = \left(\frac{\partial G}{\partial p}, -\frac{\partial G}{\partial q}\right), \tag{4}$$

where f is a variable with the same units as the conjugate variable $F(z)$. You can think of $G(z)$ as the Hamiltonian for a mechanical system and that f is the time. For classical systems, we are considering conjugate pairs leading to conjugate motions associated to each variable with the conjugate variable serving as the evolution parameter (see below). This will be applied to the energy-time conjugate pair. Let us derive some properties in which the two conjugate variables participate.

1.3. The interplay between conjugate variables

Some relationships between a pair of conjugate variables are derived in this section. We will deal with general $F(z)$ and $G(z)$ conjugate variables, but the results can be applied to coordinate and momentum or energy and time or to any other conjugate pair.

The magnitude of the vector field $|X_G|$ is the change of length along the f direction

$$|X_G| = \sqrt{\frac{dq^i}{df}\frac{dq^i}{df} + \frac{dp_i}{df}\frac{dp_i}{df}} = \frac{dl_F}{df}, \tag{5}$$

where $dl_F = \sqrt{(dq^i)^2 + (dp^i)^2}$ is the length element.

A unit density with the eigensurface $\Sigma_G(g)$ as support

$$(z|g) = \delta(z - v), \qquad v \in \Sigma_G(g) \tag{6}$$

is the classical analogue of the corresponding quantum eigenstate in coordinate $\langle q \mid g\rangle$ and momentum $\langle p \mid g\rangle$ representations. When $G(z)$ is evaluated at the points of the support of $(z \mid g)$, we get the value g. We use a bra-ket like notation to emphasise the similarity with the quantum concepts.

The overlap between a probability density with an eigenfunction of $\hat{F}$ or $\hat{G}$ provides marginal representations of a probability density,

$$\varrho(f) := (f \mid \rho) := \int (f \mid z)(z \mid \rho) dz = \int \delta(z - f)\rho(z) dz, \quad f \in \sum_F (f). \tag{7}$$

$$\varrho(g) := (g \mid \rho) := \int (g \mid z)(z \mid \rho) dz = \int \delta(z - g)\rho(z) dz, \quad g \in \sum_G (g). \tag{8}$$

But, a complete description of a function in T^*Q is obtained by using the two dimensions unit density $(z \mid f, g) = \delta(z - (f, g))$, the eigenfunction of a location in phase space,

$$\varrho(f, g) := (f, g \mid \rho) := \int (f, g \mid z)(z \mid \rho) dz = \int \delta(z - (f, g))\rho(z) dz, \quad (f, g) \in \sum_{FG} (f, g). \tag{9}$$

In this way, we have the classical analogue of the quantum concepts of eigenfunctions of operators and the projection of vectors on them.

1.4. Conjugate motions

Two dynamical variables with a constant Poisson bracket between them induce two types of complementary motions in phase space. Let us consider two real functions $F(z)$ and $G(z)$ of points in cotangent space $z \in T^*Q$ of a mechanical system, and a unit Poisson bracket between them,

$$\{F, G\} = \frac{\partial F}{\partial q^i}\frac{\partial G}{\partial p_i} - \frac{\partial G}{\partial q^i}\frac{\partial F}{\partial p_i} = 1, \tag{10}$$

valid on some domain $\mathrm{D} = \mathrm{D}\left(\frac{\partial F}{\partial q^i}\right) \cap \mathrm{D}\left(\frac{\partial F}{\partial q^i}\right) \cap \mathrm{D}\left(\frac{\partial F}{\partial q^i}\right) \cap \mathrm{D}\left(\frac{\partial F}{\partial q^i}\right)$, according to the considered functions F and G. The application of the chain rule to functions of p and q, and Eq. (10), suggests two ways of defining dynamical systems for functions F and G that comply with the unit Poisson bracket. One of these dynamical systems is

$$\frac{dp_i}{dF} = -\frac{\partial G}{\partial q^i}, \qquad \frac{dq^i}{dF} = \frac{\partial G}{\partial p_i}. \tag{11}$$

With these replacements, the Poisson bracket becomes the derivative of a function with respect to itself

$$\{F, G\} = \frac{\partial F}{\partial q^i}\frac{\partial q^i}{\partial F} + \frac{\partial p_i}{\partial F}\frac{\partial F}{\partial p_i} = \frac{dF}{dF} = 1. \tag{12}$$

Note that F is at the same time a parameter in terms of which the motion of points in phase space is written, and also the conjugate variable to G.

We can also define other dynamical system as

$$\frac{dp_i}{dG}=\frac{\partial F}{\partial q^i}, \qquad \frac{dq^i}{dG}=-\frac{\partial F}{\partial p_i}. \tag{13}$$

Now, G is the shift parameter besides of being the conjugate variable to F. This also renders the Poisson bracket to the identity

$$\{F,G\}=\frac{dp_i}{dG}\frac{\partial G}{\partial p_i}+\frac{dq^i}{dG}\frac{\partial G}{\partial q^i}=\frac{dG}{dG}=1. \tag{14}$$

The dynamical systems and vector fields for the motions just defined are

$$\frac{dz}{dG}=X_F, \quad X_F=\left(-\frac{\partial F}{\partial p_i},\frac{\partial F}{\partial q^i}\right), \quad \text{and} \quad \frac{dz}{dF}=X_G, \quad X_G=\left(\frac{\partial G}{\partial p_i},-\frac{\partial G}{\partial q^i}\right) \tag{15}$$

Then, the motion along one of the F or G directions is determined by the corresponding conjugate variable. These vector fields in general are not orthogonal, nor parallel.

If the motion of phase space points is governed by the vector field (15), F remains constant because

$$\frac{dF}{dG}=\frac{\partial F}{\partial q^i}\frac{\partial q^i}{\partial G}+\frac{\partial F}{\partial p_i}\frac{\partial p_i}{\partial G}=\frac{\partial p_i}{\partial G}\frac{\partial q^i}{\partial G}-\frac{\partial q^i}{\partial G}\frac{\partial p_i}{\partial G}=0. \tag{16}$$

In contrast, when motion occurs in the F direction, by means of Eq. (16), it is the G variable the one that remains constant because

$$\frac{dG}{dF}=\frac{\partial G}{\partial q^i}\frac{\partial q^i}{\partial F}+\frac{\partial G}{\partial p_i}\frac{\partial p_i}{\partial F}=-\frac{\partial p_i}{\partial F}\frac{\partial q^i}{\partial F}+\frac{\partial q^i}{\partial F}\frac{\partial p_i}{\partial F}=0. \tag{17}$$

Hence, motion originated by the conjugate variables $F(z)$ and $G(z)$ occurs on the shells of constant $F(z)$ or of constant $G(z)$, respectively.

The divergence of these vector fields is zero,

$$\nabla\bullet X_F=-\frac{\partial}{\partial q^i}\frac{\partial F}{\partial p_i}+\frac{\partial}{\partial p_i}\frac{\partial F}{\partial q^i}=0, \qquad \nabla\bullet X_G=\frac{\partial}{\partial q^i}\frac{\partial G}{\partial p_i}-\frac{\partial}{\partial p_i}\frac{\partial G}{\partial q^i}=0. \tag{18}$$

Thus, the motions associated to each of these conjugate variables preserve the phase space area.

A constant Poisson bracket is related to the constancy of a cross product because

$$X_G \wedge X_F = \frac{dz}{dF} \wedge \frac{dz}{dG} = \begin{vmatrix} \hat{q} & \hat{p} & \hat{n} \\ \frac{\partial G}{\partial p} & -\frac{\partial G}{\partial q} & 0 \\ -\frac{\partial F}{\partial p} & \frac{\partial F}{\partial q} & 0 \end{vmatrix} = \hat{n}\left(\frac{\partial G}{\partial p}\frac{\partial F}{\partial q} - \frac{\partial G}{\partial q}\frac{\partial F}{\partial p}\right) = \hat{n}\{F, G\}. \tag{19}$$

where $\hat{n}$ is the unit vector normal to the phase space plane. Then, the magnitudes of the vector fields and the angle between them changes in such a way that the cross product remains constant when the Poisson bracket is equal to one, i.e. the cross product between conjugate vector fields is a conserved quantity.

The Jacobian for transformations from phase space coordinates to (f, g) variables is one for each type of motion:

$$J = \begin{vmatrix} \frac{\partial q}{\partial f} & \frac{\partial p}{\partial f} \\ \frac{\partial q}{\partial g} & \frac{\partial p}{\partial g} \end{vmatrix} = \begin{vmatrix} \frac{\partial G}{\partial p} & -\frac{\partial G}{\partial q} \\ \frac{\partial q}{\partial g} & \frac{\partial p}{\partial g} \end{vmatrix} = \frac{\partial G}{\partial p}\frac{\partial p}{\partial g} + \frac{\partial G}{\partial q}\frac{\partial q}{\partial g} = \frac{dG}{dg} = 1, \tag{20}$$

and

$$J = \begin{vmatrix} \frac{\partial q}{\partial f} & \frac{\partial p}{\partial f} \\ \frac{\partial q}{\partial g} & \frac{\partial p}{\partial g} \end{vmatrix} = \begin{vmatrix} \frac{\partial q}{\partial f} & \frac{\partial p}{\partial f} \\ -\frac{\partial F}{\partial p} & \frac{\partial F}{\partial q} \end{vmatrix} = \frac{\partial F}{\partial q}\frac{\partial q}{\partial f} + \frac{\partial F}{\partial p}\frac{\partial p}{\partial f} = \frac{dF}{df} = 1. \tag{21}$$

We have seen some properties related to the motion of phase space points caused by conjugate variables.

1.5. Poisson brackets and commutators

We now consider the use of commutators in the classical realm.

The Poisson bracket can also be written in two ways involving a *commutator*. One form is

$$\{F, G\} = \left(\frac{\partial G}{\partial p}\frac{\partial}{\partial q} - \frac{\partial G}{\partial q}\frac{\partial}{\partial p}\right)F = [L_G, F] = 1, \tag{22}$$

and the other is

$$\{F, G\}=\left(\frac{\partial F}{\partial q}\frac{\partial}{\partial p}-\frac{\partial F}{\partial p}\frac{\partial}{\partial q}\right)G=[\mathrm{L}_F, G]=1 . \tag{23}$$

With these, we have introduced the Liouville type operators

$$\mathrm{L}_F=\frac{\partial F}{\partial q}\frac{\partial}{\partial p}-\frac{\partial F}{\partial p}\frac{\partial}{\partial q}=X_F\bullet\nabla, \quad \text{and } \mathrm{L}_G=\frac{\partial G}{\partial p}\frac{\partial}{\partial q}-\frac{\partial G}{\partial q}\frac{\partial}{\partial p}=X_G\bullet\nabla . \tag{24}$$

These are Lie derivatives in the directions of X_F and X_G, respectively. These operators generate complementary motion of functions in phase space. Note that now, we also have operators and commutators as in Quantum Mechanics.

Conserved motion of phase space functions moving along the f or g directions can be achieved with the above Liouvillian operators as

$$\frac{\partial}{\partial f}=-\mathrm{L}_G, \text{ and } \quad \frac{\partial}{\partial g}=-\mathrm{L}_F . \tag{25}$$

Indeed, with the help these definitions and of the chain rule, we have that the total derivative of functions vanishes, i.e. the total amount of a function is conserved,

$$\frac{d}{df}=\frac{dq}{df}\frac{\partial}{\partial q}+\frac{dp}{df}\frac{\partial}{\partial p}+\frac{\partial}{\partial f}=\frac{dz}{df}\bullet\nabla+\frac{\partial}{\partial f}=X_G\bullet\nabla+\frac{\partial}{\partial f}=\mathrm{L}_G+\frac{\partial}{\partial f}=-\frac{\partial}{\partial f}+\frac{\partial}{\partial f}=0, \tag{26}$$

and

$$\frac{d}{dg}=\frac{dq}{dg}\frac{\partial}{\partial q}+\frac{dp}{dg}\frac{\partial}{\partial p}+\frac{\partial}{\partial g}=\frac{dz}{dg}\bullet\nabla+\frac{\partial}{\partial g}=X_F\bullet\nabla+\frac{\partial}{\partial g}=\mathrm{L}_F+\frac{\partial}{\partial g}=-\frac{\partial}{\partial g}+\frac{\partial}{\partial g}=0 . \tag{27}$$

Also, note that for any function $u(z)$ of a phase space point z, we have that

$$[\mathrm{L}_F, u(z)]=\mathrm{L}_F u(z)=X_F\bullet\nabla u(z)=\frac{\mathrm{dz}}{\mathrm{dG}}\bullet\nabla \mathrm{u}(z)=-\frac{\partial}{\partial g}u(z), \tag{28}$$

and

$$[\mathrm{L}_G, u(z)]=\mathrm{L}_G u(z)=X_G\bullet\nabla u(z)=\frac{dz}{dF}\bullet\nabla u(z)=-\frac{\partial}{\partial f}u(z) , \tag{29}$$

which are the evolution equations for functions along the conjugate directions f and g. These are the classical analogues of the quantum evolution equation $\frac{d}{dt}=\frac{1}{i\hbar}[\ ,\hat{H}]$ for time dependent operators. The formal solutions to these equations are

$$u(z;g)=e^{-g\mathsf{L}_F}u(z), \quad \text{and} \quad u(z;f)=e^{-f\mathsf{L}_G}u(z). \tag{30}$$

With these equations, we can now move a function $u(z)$ on T^*Q in such a way that the points of their support move according to the dynamical systems Eqs. (15) and the total amount of u is conserved.

1.6. The commutator as a derivation and its consequences

As in quantum theory, we have found commutators and there are many properties based on them, taking advantage of the fact that a commutator is a derivation.

Since the commutator is a derivation, for conjugate variables $F(z)$ and $G(z)$ we have that, for integer n,

$$[\mathsf{L}_G^n, F]=n\,\mathsf{L}_G^{n-1}, \qquad [\mathsf{L}_G, F^n]=n\,F^{n-1}, \qquad [\mathsf{L}_F^n, G]=n\,\mathsf{L}_F^{n-1}, \qquad [\mathsf{L}_F, G^n]=n\,G^{n-1}. \tag{31}$$

Based on the above equalities, we can get translation relationships for functions on T^*Q. We first note that, for a holomorphic function $u(x)=\sum_{n=0}^{\infty} u_n x^n$,

$$[u(\mathsf{L}_G), F]=\left[\sum_{n=0}^{\infty} u_n \mathsf{L}_G^n, F\right]=\sum_{n=0}^{\infty} n u_n \mathsf{L}_G^{n-1}=u'(\mathsf{L}_G). \tag{32}$$

In particular, we have that

$$[e^{f\mathsf{L}_G}, F]=f e^{f\mathsf{L}_G}. \tag{33}$$

Then, $e^{f\mathsf{L}_G}$ is the eigenfunction of the commutator $[\bullet, F]$ with eigenvalue f.

From Eq. (32), we find that

$$u(\mathsf{L}_G)F - Fu(\mathsf{L}_G)=u'(\mathsf{L}_G). \tag{34}$$

But, if we multiply by $u^{-1}(\mathsf{L}_G)$ from the right, we arrive to

$$u(\mathsf{L}_G)F u^{-1}(\mathsf{L}_G)=F + u'(\mathsf{L}_G)u^{-1}(\mathsf{L}_G). \tag{35}$$

This is a generalized version of a shift of F, and the classical analogue of a generalization of the quantum Weyl relationship. A simple form of the above equality, a familiar form, is obtained with the exponential function, i.e.

$$e^{f\mathsf{L}_G}F e^{-f\mathsf{L}_G}=F + f. \tag{36}$$

This is a relationship that indicates how to translate the function $F(z)$ as an operator. When this equality is acting on the number one, we arrive at the translation property for F as a function

$$F(z;f)=\left(e^{f\mathcal{L}_G}F(z)\right)=F(z)\,e^{f\mathcal{L}_G}1+f\,e^{f\mathcal{L}_G}1=F(z)+f\ . \tag{37}$$

This implies that

$$\frac{d}{df}F(z;f)=1\,, \tag{38}$$

i.e., up to an additive constant, f is the value of $F(z)$ itself, one can be replaced by the other and actually they are the same object, with f the classical analogue of the spectrum of a quantum operator.

Continuing in a similar way, we can obtain the relationships shown in the following diagram

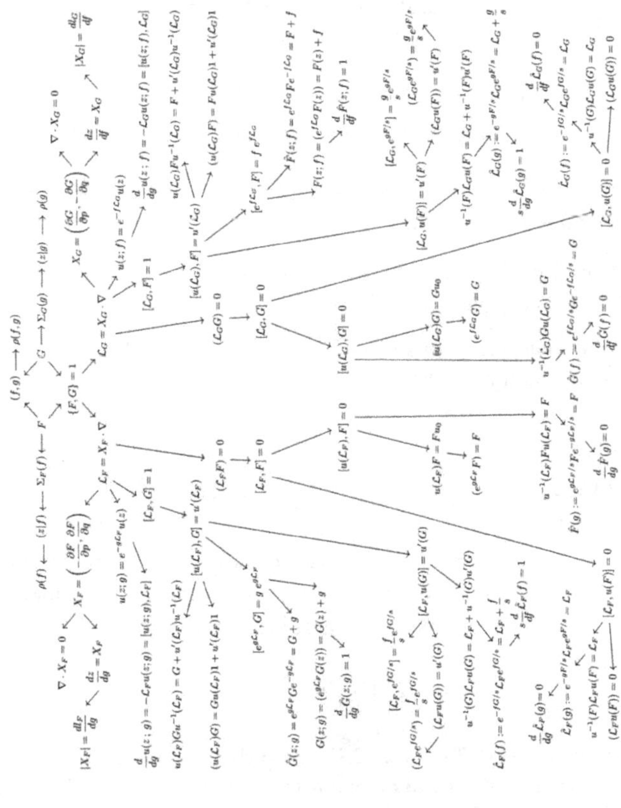

Diagram 1.

where the constant s has units of action, length times momentum, the same units as the quantum constant $\hbar$.

Some of the things to note are:

The operator $e^{g\mathsf{L}_F}$ is the eigenoperator of the commutator $[\,\bullet,\,G]$ and can be used to generate translations of $G(z)$ as an operator or as a function. This operator is also the propagator for the evolution of functions along the g direction. The variable g is more than just a shift parameter; it actually labels the values that $G(z)$ takes, the classical analogue of the spectrum of a quantum operator.

The operators L_F and $G(z)$ are also a pair of conjugate operators, as well as the pair L_G and $F(z)$.

But L_F commutes with $F(z)$ and then it cannot be used to translate functions of $F(z)$, $F(z)$ is a conserved quantity when motion occurs along the $G(z)$ direction.

The eigenfunction of $[\mathsf{L}_F,\ \bullet\,]$ and of $s\mathsf{L}_F$ is $e^{fG(z)/s}$ and this function can be used to shift L_F as an operator or as a function.

The variable f is more than just a parameter in the shift of $s\mathsf{L}_F$, it actually is the value that $s\mathsf{L}_F$ can take, the classical analogue of the spectra of a quantum operator.

The steady state of L_F is a function of $F(z)$, but $e^{gF(z)/s}$ is an eigenfunction of L_G and of $[\mathsf{L}_G,\ \bullet\,]$ and it can be used to translate L_G.

These comments involve the left hand side of the above diagram. There are similar conclusions that can be drawn by considering the right hand side of the diagram.

Remember that the above are results valid for classical systems. Below we derive the corresponding results for quantum systems.

2. Quantum systems

We now derive the quantum analogues of the relationships found in previous section. We start with a Hilbert space H of wave functions and two conjugate operators $\hat{F}$ and $\hat{G}$ acting on vectors in H, and with a constant commutator between them

$$[\hat{F},\hat{G}]=i\hbar\,, \tag{39}$$

together with the domain $\mathsf{D}=\mathsf{D}(\hat{F}\hat{G})\cap\mathsf{D}(\hat{G}\hat{F})$ in which the commutator holds. Examples of these operators are coordinate $\hat{Q}$ and momentum $\hat{P}$ operators, energy $\hat{H}$ and time $\hat{T}$ operators, creation $\hat{a}^{\dagger}$ and annihilation $\hat{a}$ operators

The eigenvectors of the position, momentum and energy operators have been used to provide a representation of wave functions and of operators. So, in general, the eigenvectors $| f\rangle$ and $| g\rangle$ of the conjugate operators $\hat{F}$ and $\hat{G}$ provide with a set of vectors for a representation of dynamical quantities like the wave functions $\langle f | \psi\rangle$ and $\langle g | \psi\rangle$.

With the help of the properties of commutators between operators, we can see that

$$[\hat{F}^n, \hat{G}]=i\hbar\hat{F}^{n-1}, \qquad [\hat{F}, \hat{G}^n]=i\hbar\hat{G}^{n-1}. \tag{40}$$

Hence, for a holomorphic function $u(z)=\sum_{n=0}^{\infty} u_n z^n$ we have that

$$[\hat{u}(\hat{F}), \hat{G}]=i\hbar\hat{u}'(\hat{F}), \qquad [\hat{F}, \hat{u}(\hat{G})]=i\hbar\hat{u}'(\hat{G}), \tag{41}$$

i.e., the commutators behave as derivations with respect to operators. In an abuse of notation, we have that

$$\frac{1}{i\hbar}[\bullet, \hat{G}]=\frac{d}{dF}, \qquad \frac{1}{i\hbar}[\hat{F}, \bullet]=\frac{d}{dG}. \tag{42}$$

We can take advantage of this fact and derive the quantum versions of the equalities found in the classical realm.

A set of equalities is obtained from Eq. (43) by first writing them in expanded form as

$$\hat{u}(\hat{F})\hat{G} - \hat{G}\hat{u}(\hat{F})=i\hbar\hat{u}'(\hat{F}), \quad \text{and} \quad \hat{F}\hat{u}(\hat{G}) - \hat{u}(\hat{G})\hat{F}=i\hbar\hat{u}'(\hat{G}). \tag{43}$$

Next, we multiply these equalities by the inverse operator to the right or to the left in order to obtain

$$\hat{u}(\hat{F})\hat{G}\hat{u}^{-1}(\hat{F})=\hat{G}+i\hbar\hat{u}'(\hat{F})\hat{u}^{-1}(\hat{F}), \quad \text{and} \quad \hat{u}^{-1}(\hat{G})\hat{F}\hat{u}(\hat{G})=\hat{F}+i\hbar\hat{u}^{-1}(\hat{G})\hat{u}'(\hat{G}). \tag{44}$$

These are a set of generalized shift relationships for the operators $\hat{G}$ and $\hat{F}$. The usual shift relationships are obtained when $u(x)$ is the exponential function, i.e.

$$\hat{G}(g):=e^{-ig\hat{F}/\hbar}\hat{G}e^{ig\hat{F}/\hbar}=\hat{G}+g, \quad \text{and} \quad \hat{F}(f):=e^{if\hat{G}/\hbar}\hat{F}e^{-if\hat{G}/\hbar}=\hat{F}+f. \tag{45}$$

Now, as in Classical Mechanics, the commutator between two operators can be seen as two different derivatives introducing quantum dynamical system as

$$\frac{d\hat{P}(f)}{df} = -\frac{\partial \hat{G}(\hat{Q}, \hat{P})}{\partial \hat{Q}} = \frac{1}{i\hbar}[\hat{P}(f), \hat{G}(\hat{Q}, \hat{P})], \qquad \frac{d\hat{Q}(f)}{df} = \frac{\partial \hat{G}(\hat{Q}, \hat{P})}{\partial \hat{P}} = \frac{1}{i\hbar}[\hat{Q}(f), \hat{G}(\hat{Q}, \hat{P})], \tag{46}$$

$$\frac{d\hat{P}(g)}{dg} = \frac{\partial \hat{F}(\hat{Q}, \hat{P})}{\partial \hat{Q}} = \frac{1}{i\hbar}[\hat{F}(\hat{Q}, \hat{P}), \hat{P}(g)], \quad \text{and} \quad \frac{d\hat{Q}(g)}{dg} = -\frac{\partial \hat{F}(\hat{Q}, \hat{P})}{\partial \hat{P}} = \frac{1}{i\hbar}[\hat{F}(\hat{Q}, \hat{P}), \hat{Q}(g)], \tag{47}$$

where

$$\hat{P}(f) = e^{if\hat{G}/\hbar}\hat{P}e^{-if\hat{G}/\hbar}, \qquad \hat{Q}(f) = e^{if\hat{G}/\hbar}\hat{Q}e^{-if\hat{G}/\hbar}, \tag{48}$$

$$\hat{P}(g) = e^{-ig\hat{F}/\hbar}\hat{P}e^{if\hat{F}/\hbar}, \quad \text{and} \quad \hat{Q}(g) = e^{-ig\hat{F}/\hbar}\hat{Q}e^{ig\hat{F}/\hbar}. \tag{49}$$

These equations can be written in the form of a set of quantum dynamical systems

$$\frac{d\hat{z}}{df} = \hat{X}_G, \qquad \hat{X}_G = \left(\frac{\partial \hat{G}}{\partial \hat{P}}, -\frac{\partial \hat{G}}{\partial \hat{Q}}\right), \qquad \frac{d\hat{z}}{dg} = \hat{X}_F, \qquad \hat{X}_F = \left(-\frac{\partial \hat{F}}{\partial \hat{P}}, \frac{\partial \hat{F}}{\partial \hat{Q}}\right), \tag{50}$$

where $\hat{z} = (\hat{Q}, \hat{P})$.

The inner product between the operator vector fields is

$$\hat{X}_F^{\dagger} \bullet \hat{X}_F = \left(-\frac{\partial \hat{F}}{\partial \hat{P}}, \frac{\partial \hat{F}}{\partial \hat{Q}}\right)^{\dagger} \bullet \left(-\frac{\partial \hat{F}}{\partial \hat{P}}, \frac{\partial \hat{F}}{\partial \hat{Q}}\right) = \left(\frac{d\hat{Q}}{dg}\right)^2 + \left(\frac{d\hat{P}}{dg}\right)^2 := \left(\frac{d\hat{l}_F}{dg}\right)^2, \tag{51}$$

where $(d\hat{l}_F)^2 := (d\hat{Q})^2 + (d\hat{P})^2$, evaluated along the g direction, is the quantum analogue of the square of the line element $(dl_F)^2 = (dq)^2 + (dp)^2$.

We can define many of the classical quantities but now in the quantum realm. Liouville type operators are

$$\hat{\hat{L}}_F := \frac{1}{i\hbar}[\hat{F}, \bullet], \quad \text{and} \quad \hat{\hat{L}}_G := \frac{1}{i\hbar}[\bullet, \hat{G}]. \tag{52}$$

These operators will move functions of operators along the conjugate directions $\hat{G}$ or $\hat{F}$, respectively. This is the case when $\hat{G}$ is the Hamiltonian $\hat{H}$ of a physical system, a case in which we get the usual time evolution of operator.

There are many equalities that can be obtained as in the classical case. The following diagram shows some of them:

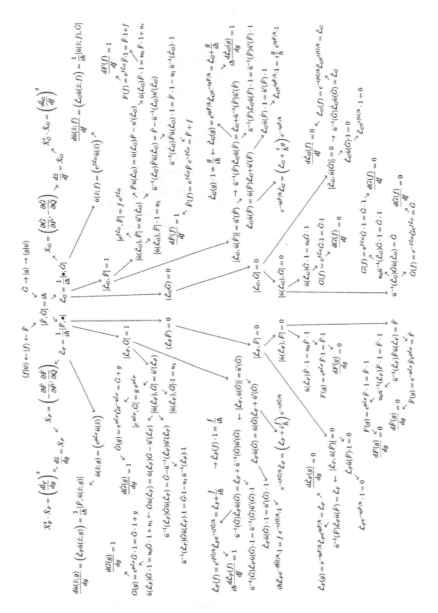

Diagram 2.

Note that the conclusions mentioned at the end of the previous section for classical systems also hold in the quantum realm.

Next, we illustrate the use of these ideas with a simple system.

3. Time evolution using energy and time eigenstates

As a brief application of the abovee ideas, we show how to use the energy-time coordinates and eigenfunctions in the reversible evolution of probability densities.

Earlier, there was an interest on the classical and semi classical analysis of energy transfer in molecules. Those studies were based on the quantum procedure of expanding wave functions in terms of energy eigenstates, after the fact that the evolution of energy eigenstates is quite simple in Quantum Mechanics because the evolution equation for a wave function $i\hbar\frac{\partial}{\partial t}|\psi>=\hat{H}|\psi>$ is linear and contains the Hamiltonian operator. In those earlier calculations, an attempt to use the eigenfunctions of a complex classical Liouville operator was made [5-8]. The results in this chapter show that the eigenfunction of the Liouville operator L_H is $e^{gT(z)}$ and that it do not seems to be a good set of functions in terms of which any other function can be written, as is the case for the eigenfunctions of the Hamiltonian operator in Quantum Mechanics. In this section, we use the time eigenstates instead.

With energy-time eigenstates the propagation of classical densities is quite simple. In order to illustrate our procedure, we will apply it to the harmonic oscillator with Hamiltonian given by (we will use dimensionless units)

$$H(z)=\frac{p^2}{2}+\frac{q^2}{2}. \quad (53)$$

Given and energy scaling parameter E_s and the frequency ω of the harmonic oscillator, the remaining scaling parameters are

$$p_s=\sqrt{mE_s}, \quad q_s=\sqrt{\frac{E_s}{m\omega^2}}, \quad t_s=\frac{1}{\omega}. \quad (54)$$

We need to define time eigensurfaces for our calculations. The procedure to obtain them is to take the curve $q=0$ as the zero time curve. The forward and backward propagation of the zero time curve generates the time coordinate system in phase space. The trajectory generated with the harmonic oscillator Hamiltonian is

$$q(t)=\sqrt{2E}\cos\left(t+\frac{\pi}{2}\right), \quad p(t)=\sqrt{2E}\sin\left(t+\frac{\pi}{2}\right). \quad (55)$$

With the choice of phase we have made, $q=0$ when $t=0$, which is the requirement for an initial time curve. Then, the equation for the time curve is

$$p = q\tan\left(t+\frac{\pi}{2}\right), \quad \text{or} \quad q = p\cot\left(t+\frac{\pi}{2}\right). \tag{56}$$

These are just straight lines passing through the origin, equivalent to the polar coordinates. The value of time on these points is t, precisely. In Fig. 1, we show both coordinate systems, the phase space coordinates (q, p), and the energy time coordinates (E, t) on the plane. This is a periodic system, so we will only consider one period in time.

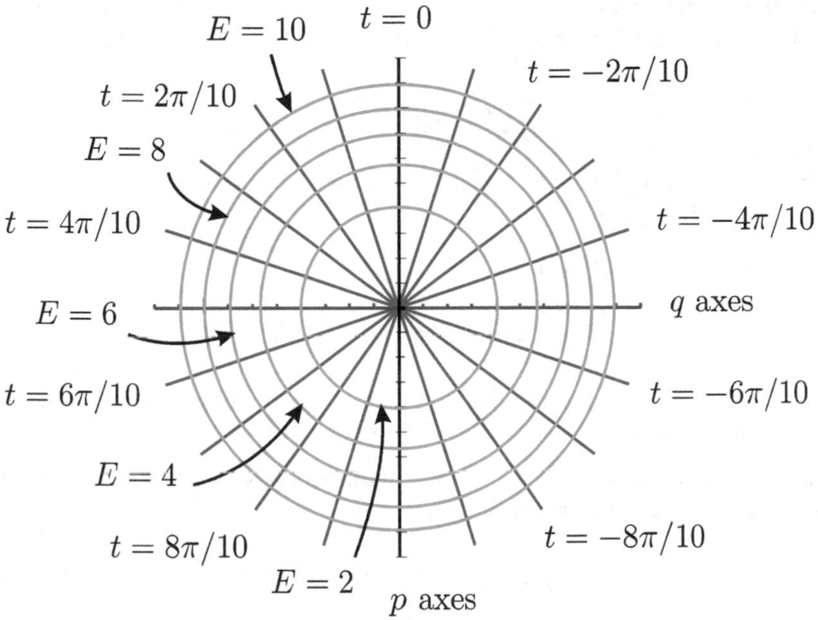

Figure 1. Two conjugate coordinate systems for the classical harmonic oscillator in dimensionless units. Blue and black lines correspond to the (q, p) coordinates and the red and green curves to the (E, t) coordinates.

At this point, there are two options for time curves. Both options will cover the plane and we can distinguish between the regions of phase space with negative or positive momentum. One is to use half lines and t in the range from $-\pi$ to π, with the curve $t=0$ coinciding with the positive p axes. The other option is to use the complete curve including positive and negative momentum values and with $t \in (-\pi/2, \pi/2)$. In the first option, the positive momentum part of a probability density will correspond to the range $t \in (-\pi/2, \pi/2)$, and the negative values will correspond to $t \in \left(-\pi, -\frac{\pi}{2}\right) \cup \left(\pi/2, \pi\right)$. We take this option.

Now, based on the equalities derived in this chapter, we find the following relationship for a marginal density dependent only upon $H(z)$, assuming that the function $\rho(H)$ can be written as a power series of H, $\rho(H) = \sum_i \rho_i H^i$,

$$e^{-\tau \mathsf{L}_H} \rho(H) = \sum_n \frac{(-\tau)^n}{n!} \mathsf{L}_H^n \sum_i \rho_i H^i = \sum_i \rho_i H^i = \rho(H), \tag{57}$$

where we have made use of the equality $\mathsf{L}_H H = 0$. Then, a function of H does not evolve in time, it is a steady state. For a marginal function dependent upon t, we also have that

$$e^{-\tau \mathsf{L}_H} \rho(t) = e^{\tau d/dt} \rho(t) = \rho(t+\tau). \tag{58}$$

where we have made use of the result that $\frac{d}{dt} = -\mathsf{L}_H$. Therefore, a function of t is only shifted in time without changing its shape.

For a function of H and t we find that

$$e^{-\tau \mathsf{L}_H} \rho(H, t) = e^{-\tau d/dt} \rho(H, t) = \rho(H, t+\tau). \tag{59}$$

This means that evolution in energy-time space also is quite simple, it is only a shift of the function along the t axes without a change of shape.

So, let us take a concrete probability density and let us evolve it in time. The probability density, in phase space, that we will consider is

$$\rho(z) = H(z) e^{-((q-q_0)^2 + (p-p_0)^2)/2\sigma^2}, \tag{60}$$

with $(q_0, p_0) = (1,2)$ and $\sigma = 1$. A contour plot of this density in phase-space is shown in (a) of Fig. 2. The energy-time components of this density are shown in (b) of the same figure. Time evolution by an amount τ correspond to a translation along the t axes, from t to $t+\tau$, without changing the energy values. This translation is illustrated in (d) of Fig. 2 in energy-time space and in (c) of the same figure in phase-space.

Recall that the whole function $\rho(z)$ is translated in time with the propagator $e^{-\tau \mathsf{L}_H}$. Then, there are two times involved here, the variable t as a coordinate and the shift in time τ. The latter is the time variable that appears in the Liouville equation of motion $\frac{d\rho(z;\tau)}{d\tau} = -\mathsf{L}_H \rho(z;\tau)$.

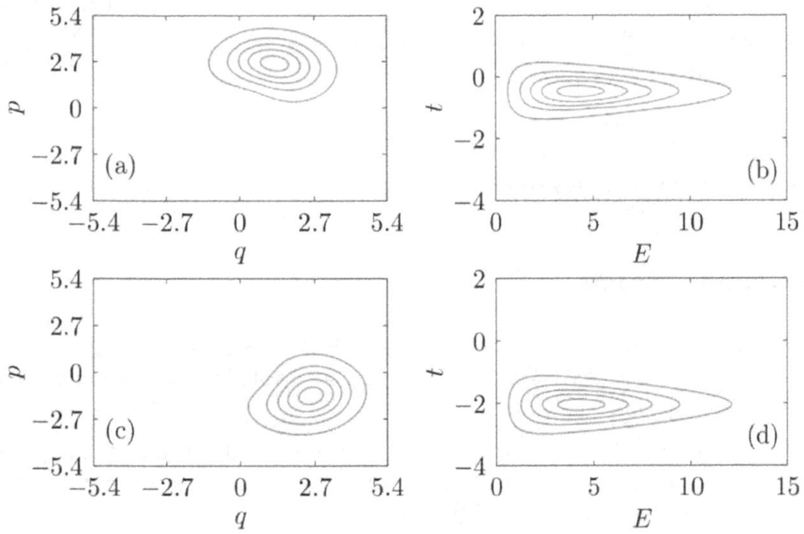

Figure 2. Contour plots of the time evolution of a probability density on phase-space and on energy-time space. Initial densities (a) in phase space, and (b) in energy-time space. (d) Evolution in energy-time space is accomplished by a shift along the *t*axes. (c) In phase space, the density is also translated to the corresponding time eigensurfaces.

This behaviour is also observed in quantum systems. Time eigenfunctions can be defined in a similar way as for classical systems. We start with a coordinate eigenfunction $|q>$ for the eigenvalue $q=0$ and propagate it in time. This will be our time eigenstate

$$|t> = e^{\frac{it\hat{H}}{\hbar}} |q=0> . \tag{61}$$

The projection of a wave function onto this vector is

$$<t|\psi> = <q=0| e^{-\frac{it\hat{H}}{\hbar}} |\psi> = \psi(q=0;t) , \tag{62}$$

Which is the time dependent wave function, in the coordinate representation, and evaluated at $q=0$. This function is the time component of the wave function.

The time component of a propagated wave function for a time τ is

$$<t|\psi(\tau)> = <q=0| e^{-\frac{it\hat{H}}{\hbar}} e^{-\frac{i\tau\hat{H}}{\hbar}} |\psi> = <t+\tau|\psi> . \tag{63}$$

Then, time evolution is the translation in time representation, without a change in shape. Note that the variable τ is the time variable that appears in the Schrödinger equation for the wave function.

Now, assuming a discrete energy spectrum with energy eigenvalue E_n and corresponding eigenfunction $|n>$, in the energy representation we have that

$$<n|\psi(\tau)> = <n|e^{-\frac{i\tau\hat{H}}{\hbar}}|\psi> = e^{-\frac{i\tau E_n}{\hbar}}<n|\psi>, \tag{64}$$

i.e. the wave function in energy space only changes its phase after evolution for a time τ.

4. Concluding remarks

Once that we have made use of the same concepts in both classical and quantum mechanics, it is more easy to understand quantum theory since many objects then are present in both theories.

Actually, there are many things in common for both classical and quantum systems, as is the case of the eigensurfaces and the eigenfunctions of conjugate variables, which can be used as coordinates for representing dynamical quantities.

Another benefit of knowing the influence of conjugate dynamical variables on themselves and of using the same language for both theories lies in that some puzzling things that are found in one of the theories can be analysed in the other and this helps in the understanding of the original puzzle. This is the case of the Pauli theorem [9-14] that prevents the existence of a hermitian time operator in Quantum Mechanics. The classical analogue of this puzzle is found in Reference [15].

These were some of the properties and their consequences in which both conjugate variables participate, influencing each other.

Author details

Gabino Torres-Vega

Physics Department, Cinvestav, México

References

[1] Woodhouse NMJ. Geometric Quantization. Oxford: Osford University Press; 1991.

[2] Wigner E. Phys Rev A 1932; 40 749

[3] Husimi K. Proc Phys Math Soc Jpn 1940; 22 264

[4] Torres-Vega G, Theoretical concepts of quantum mechanics. Rijeka: InTech; 2012.

[5] Jaffé C. Classical Liouville mechanics and intramolecular relaxation dynamics. The Journal of Physical Chemistry 1984; 88 4829.

[6] Jaffé C and Brumer C. Classical-quantum correspondence in the distribution dynamics of integrable systems. Journal of Chemical Physics 1985; 82 2330.

[7] Jaffé C. Semiclassical quantization of the Liouville formulation of classical mechanics. Journal of Chemical Physics 1988; 88 7603.

[8] Jaffé C. Sheldon Kanfer and Paul Brumer, Classical analog of pure-state quantum dynamics. Physical Review Letters 1985; 54 8.

[9] Pauli W. Handbuch der Physics. Berlin: Springer-Verlag; 1926

[10] Galapon EA, Proc R Soc Lond A 2002; 458 451

[11] Galapon EA, Proc R Soc Lond A 2002; 458 2671

[12] Galapon EA, quant-ph/0303106

[13] Galindo A, Lett Math Phys 1984; 8 495

[14] Garrison JC and Wong J, J Math Phys 1970; 11 2242

[15] Torres-Vega G, J Phys A: Math Theor 45, 215302 (2012)

Chapter 2

Charathéodory's "Royal Road" to the Calculus of Variations: A Possible Bridge Between Classical and Quantum Physics

Francisco De Zela

Additional information is available at the end of the chapter

1. Introduction

Constantin Carathéodory, a Greek-born, well-known German mathematician, is rarely mentioned in connection to physics. One of his most remarkable contributions to mathematics is his approach to the calculus of variations, the so-called Carathéodory's "royal road" [1]. Among physicists, Carathéodory's name is most frequently related to his contributions to the foundations of thermodynamics [2] and to topics of classical optics, though, as a pupil of Hermann Minkowski, he also worked on the development of especial relativity. In our opinion, however, Caratheodory's formulation of the variational problem deserves to be better known among physicists. For mathematicians, features like rigor and non-redundancy of basic postulates are of utmost importance. Among physicists, a more pragmatic attitude is usually behind efforts towards a theoretical construction, whose principal merit should be to offer an adequate description of Nature. Such a construction must provide us with predictive power. Rigor of the theoretical construction is necessary but not sufficient. Elegance – which is how non-redundancy and simplicity usually manifest themselves – can be sometimes just a welcome feature. Some other times, however, elegance has become a guiding principle when guessing at how Nature works. Nevertheless, once the basic principles of a theoretical construction have been identified, elegance may recede in favor of clarity, and redundancy might become acceptable. Such differences between the perspectives adopted by mathematicians and physicists have been presumably behind the different weight they have assigned to Caratheodory's achievements in the calculus of variations. To be sure, variational calculus does play a central role in physics, nowadays even more than ever before. It is by seeking for the appropriate Lagrangian that we hope to find out the most basic principles ruling physical behavior. Concepts like Feynman's path integral have become basic tools for the calculation of probability amplitudes of different processes, as well as for suggesting new developments in quantum field theory. Symmetry, such a basic concept underlying those aspects of Nature which appear to us in the form of interactions among fundamental particles, is best accounted for within the framework of a variational principle.

Within the domain of classical physics, only two fundamental interactions have been addressed: the gravitational and the electromagnetic interactions. The theoretical construction may correspondingly rest on two variational principles, one for gravitation and the other for electromagnetism. These principles lead to so-called "equations of motion": the Maxwell and the Lorentz equations for electromagnetism, and the Einstein and the geodesic equations for gravitation. All these differential equations can be derived as Euler-Lagrange equations from the appropriate Lagrangian or Lagrangian density.

The usual approach to variational calculus in physics starts by considering small variations of a curve which renders extremal the action integral $\int Ldt$, with L being the Lagrangian. This leads to the Euler-Lagrange equations of motion. By submitting L to a Legendre transformation one obtains the corresponding Hamiltonian, in terms of which the Hamilton equations of motion can be established. By considering canonical transformations of these equations, one arrives at the Hamilton-Jacobi equation for a scalar function $S(t,x)$. It is last one that has been used to connect the classical approach with the quantum one, e.g., in Madelung's hydrodynamic model [3] or in Bohm's "hidden variables" approach [4]. This appears natural, because both the Hamilton-Jacobi and the Schrödinger equation rule the dynamics of quantities like $S(t,x)$ and $\psi(t,x)$, respectively, which are scalar fields. Their scalar nature is in fact irrelevant; they could be tensors and spinors. The relevant issue is that while the Euler-Lagrange and the Hamilton equations refer to a single path, quantum equations address a field. The quantum-classical connection thus requires making a field out of single paths, something which occurs by going to the Hamilton-Jacobi equation, or else by establishing a path-integral formulation, as Feynman did. The latter considers a family of trajectories and assigns a probability to each of them. Now, Carathéodory's approach has the advantage of addressing right from the start a field of extremals. In fact, as the calculus of variations shows, a solution of the extremal problem exists only when the sought-after extremal curve can be embedded in a field of similar extremals. Carathéodory exploited this fact by introducing the concepts of "equivalent variational problems" and the "complete figure". It is then possible to elegantly derive from a single statement the Euler-Lagrange and the Hamilton equations, as well as the Hamilton-Jacobi equation, all of them as field equations. The familiar Euler-Lagrange and Hamilton equations can be obtained afterwards by singling out a particular extremal of the field. But – as already stressed – it is not the inherent elegance of the formulation what drives our interest towards Carathéodory's approach. It is rather its potentiality as a bridge between classical and quantum formulations what should be brought to the fore. Indeed, Carathéodory's approach can provide new insights into the connection between classical and quantum formulations. These insights could go beyond those already known, which were obtained by extending the Hamilton-Jacobi equation with the inclusion of additional terms. By dealing with the other field equations that appear within Carathéodory's approach, one may hope to gain additional insight.

The present chapter, after discussing Carathéodory's approach, shows how one can classically explain two phenomena that have been understood as being exclusively quantum mechanical: superconductivity and the response of a sample of charged particles to an external magnetic field. The London equations of superconductivity were originally understood as an *ad-hoc* assumption, with quantum mechanics lying at its roots. On the other hand, according to classical mechanics there can be no diamagnetism and no paramagnetism at all. We will deal with these two issues, showing how it is possible to classically derive the London equations and the existence of magnetic moments. This is not to say that there is a classical explanation of these phenomena. What is meant is that, specifically, the London equations of superconductivity can be derived from a classical Lagrangian. It is worth noting that a previous attempt in this direction, due to W. F. Edwards [5], proved false [6–8]. The failure was due to an improper application of the principle of least action. The approach presented here is free from any shortcomings. It leads to the London equations both in the relativistic and in the nonrelativistic domains. It should be stressed that this does not explain the appearance of the superconducting phase. It only shows how the London equations follow from a purely classical approach. Also the expulsion of a magnetic field from the interior of a superconductor, i.e., the Meissner effect, follows. That is, perfect diamagnetism can be explained classically, as has been shown recently [9] but under restricted conditions. This is in contradiction with the Bohr-van Leeuwen theorem, according to which there can be no classical magnetism [10]. This point has been recently discussed (see, e.g., [11]) and it has been shown that the Bohr-van Leeuwen theorem does not hold when one uses the Darwin Hamiltonian, which was proposed back in 1920. The Darwin Hamiltonian contains additional terms with respect to the standard one that is used to describe a charged particle interacting with an electromagnetic field. Applying Carathéodory's approach it can be shown that it is unnecessary to go beyond the standard Hamiltonian or Lagrangian to conclude that a magnetic response may be explained classically. The main point is that the Bohr-van Leeuwen theorem did not consider a constant of the motion which in Carathéodory's approach naturally arises. By considering this constant of the motion, the possibility of magnetic response in a sample of charged particles automatically appears.

After dealing with the above two cases, the rest of the chapter will be devoted to show how gauge invariance can be considered within Carathéodory's framework. This may have some inspiring effect for future work aiming at exploring the quantum-classical correspondence.

2. Carathéodory's royal road

2.1. Preliminaries

Let us begin by recapitulating the approach usually employed in physics. For the sake of describing a particle's motion we use a variational principle based on a Lagrangian L. When describing the dynamics of a field we use instead a variational principle based on a Lagrangian density $\mathcal{L}$. The Euler-Lagrange equations are, respectively,

$$\frac{d}{dt}\left(\frac{\partial L}{\partial \dot{x}^i}\right) - \frac{\partial L}{\partial x^i} = 0, \tag{1}$$

for a Lagrangian $L(t, x^i, \dot{x}^i)$, with $i = 1, \ldots, n$, and

$$\frac{\partial}{\partial x^\mu}\left(\frac{\partial \mathcal{L}}{\partial(\partial_\mu \psi^i)}\right) - \frac{\partial \mathcal{L}}{\partial \psi^i} = 0, \tag{2}$$

for a Lagrangian density $\mathcal{L}(\psi^i, \partial_\mu \psi^i)$ that depends on n fields ψ^i and their derivatives $\partial_\mu \psi^i$ with respect to space-time coordinates x^μ. The convention of summing over repeated indices has been used in Eq.(2), as we will do henceforth. The above equations are necessary conditions that are derivable from the action principle

$$\delta I = 0, \tag{3}$$

with the action given by $I = \int L dt$ for the particle motion and $I = \int \mathcal{L} d^4x$ for the field dynamics. The variation δ means that we consider different paths joining some fixed initial and end points – hypersurfaces in the case of $\int \mathcal{L} d^4x$ – and seek for the path that affords I an extremal value. Curves which are solutions of the Euler-Lagrange equations are called *extremals*.

Let us concentrate on the case $I = \int L dt$ in what follows and sketch how the standard derivation of Eq.(1) is usually obtained: one takes the variation $\delta \int L dt = \int dt \left[(\partial L/\partial x^i)\delta x^i + (\partial L/\partial \dot{x}^i)\delta \dot{x}^i\right]$, and observing that $\delta \dot{x}^i = d(\delta x^i)/dt$, integration by parts gives $\delta I = \int dt \left(\partial L/\partial x^i - d(\partial L/\partial \dot{x}^i)/dt\right) \delta x^i = 0$, where we have considered that $\delta x^i = 0$ at the common endpoints of all the paths involved in the variation. The arbitrariness of δx^i leads to Eq.(1) as a necessary condition for δI to be zero.

The important case of a time-independent Lagrangian ($\partial L/\partial t = 0$) leads to the conservation of the quantity

$$\frac{\partial L}{\partial \dot{x}^i} \dot{x}^i - L \tag{4}$$

along an extremal, as can be seen by taking its time-derivative and using Eq.(1). By introducing the canonical momenta $p_i = \partial L(t, x, \dot{x})/\partial \dot{x}^i$ and assuming that we can solve these equations for the $\dot{x}^i$ as functions of the new set of independent variables, $\dot{x}^i = \dot{x}^i(t, x^i, p_i)$, we can define a Hamiltonian $H(t, x, p)$ through the expression given by Eq.(4), written in terms of the new variables (t, x^i, p_i):

$$H(t, x, p) = p_i \dot{x}^i(t, x, p) - L(t, x, \dot{x}(t, x, p)). \tag{5}$$

The Euler-Lagrange equations are then replaced by the *Hamilton equations*:

$$\frac{dx^i}{dt} = \frac{\partial H}{\partial p_i}, \; \frac{dp_i}{dt} = -\frac{\partial H}{\partial x^i}. \tag{6}$$

Eq.(5) can be seen as a Legendre transformation leading from the set $(x^i, \dot{x}^i, t)$ to the set (x^i, p_i, t) by means of the function $H(t, x, p)$. Taking the differential of $H(t, x, p)$ on the left-hand side of Eq.(5),

$$dH = \frac{\partial H}{\partial t} dt + \frac{\partial H}{\partial x^i} dx^i + \frac{\partial H}{\partial p_i} dp_i, \tag{7}$$

and on the right-hand side,

$$dH = p_i d\dot{x}^i + \dot{x}^i dp_i - \frac{\partial L}{\partial x^i} dx^i - \frac{\partial L}{\partial \dot{x}^i} d\dot{x}^i - \frac{\partial L}{\partial t} dt, \tag{8}$$

and replacing $\partial L/\partial \dot{x}^i$ by p_i, after equating both sides we see that Eqs.(6) must hold true, together with

$$\frac{\partial H}{\partial t} = -\frac{\partial L}{\partial t}. \tag{9}$$

A third way to deal with the motion problem is given by the *Hamilton-Jacobi* equation. In order to introduce it, one usually starts by considering canonical transformations, i.e., those being of the form $\{x^i, p_i\} \to \{X^i(x,p,t), P_i(x,p,t)\}$ and leaving the action I invariant. They lead to equations for $\{X^i, P_i\}$ that are similar to Eqs.(6) but with a new Hamiltonian, $K(t, X, P)$. From the set of canonical variables $\{x^i, p_i, X^i, P_i\}$ only $2n$ of them are independent. One considers then four types of transformations, in accordance to the chosen set of independent variables: $\{x^i, X^i\}$, $\{x^i, P_i\}$, $\{p_i, X^i\}$, and $\{p_i, P_i\}$. The transformation from the old to the new canonical variables can be afforded by a so-called "generating function" S, which depends on the chosen set of independent variables and the time t. The old and new Hamiltonians are related by $K = H + \partial S/\partial t$. If we succeed in finding a transformation such that $K = 0$, the Hamilton equations for K can be trivially solved. One is thus led to seek for a transformation whose generating function is such that $K = 0$. When the set of independent variables is $\{x^i, P_i\}$ the p_i are given by $p_i = \partial S/\partial x^i$, while the new momenta have constant values $P_i = \alpha_i$ in virtue of $K = 0$. This last equation reads, in terms of the original Hamiltonian,

$$\partial S/\partial t + H(t, x^i, \partial S/\partial x^i) = 0. \tag{10}$$

This is the Hamilton-Jacobi equation. It has played an important role beyond the context in which it originally arose, becoming a sort of bridge that links classical and quantum mechanics. As a first attempt to obtain a quantum-mechanical formalism it was Sommerfeld who, following Bohr, considered action-angle variables for the case of a conservative Hamiltonian, $\partial H/\partial t = 0$. This Hamiltonian was furthermore assumed to allow the splitting of $S(t, x^i, \alpha_i)$ as $S = Et - \sum_j W_j(x^i; \alpha_i)$. Restricting the treatment to cases where the relationship between the p_i and the x^i, given through $p_i = \partial S/\partial x^i$, is such that the orbits $p_i = p_i(x^j, \alpha_j)$ are either closed (*libration*-like) or else periodic (*rotation*-like), action-angle variables can be introduced as new canonical variables [12]. By imposing that the action variables are integer-multiples of a fundamental action, i.e., Planck's h, it was possible to obtain a first formulation of quantum mechanics. This version is known as "old quantum mechanics". A second attempt went along Schrödinger's reinterpretation of the left-hand-side of Eq.(10) as a Lagrangian density of a new variational principle. Schrödinger considered first the case $\partial H/\partial t = 0$, with $H = \sum_{i=1,3} p_i^2/2m + V$, and introduced ψ through $S = k \ln \psi$, with k a constant. From the left-hand side of Eq.(10), after multiplying it by ψ^2, Schrödinger obtained an expression that he took as a Lagrangian density: $\mathcal{L} = \sum_{i=1,3} k^2 (\partial\psi/\partial x^i)^2/2m + (V - E)\psi^2$. Inserting this $\mathcal{L}$ into the Euler-Lagrange equations (2) one readily obtains the (time-independent) Schrödinger equation. The constant k could be identified with $\hbar$ by comparison with Bohr's energy levels in the case of the hydrogen atom ($V \sim 1/r$). We recall that parallel to this approach, another one, due to Heisenberg, Born, Jordan and Dirac, was constructed out of a reformulation of the action-angle formalism applied to multiple periodic motions. This reformulation led to a formalism in which the Poisson brackets were replaced by commutators, and the canonical variables by operators.

Coming back to the general action principle, we have so far followed the road usually employed by physicists. This road was build out of manifold contributions, made at different times and with different purposes. As a consequence, it lacks the unity and compactness that a mathematical theory usually has. At the beginning of the 20th century mathematicians were concerned with the construction and extension of a sound theory for the calculus of variations. It is in this context that Carathéodory made his contributions to the subject. They were thus naturally conceived from a mathematical viewpoint. Apparently, they added nothing new that could be of use for physicists, and so passed almost unnoticed to them. Our purpose here is to show how Carathéodory's formulation can provide physical insight and inspire new approaches. In the following, we give a short account of Carathéodory's approach. We will try to show the conceptual unity and potential usefulness that Carathéodory's formulation entails. Such a unity roots on the so-called *complete figure* that Carathéodory introduces as a central concept of his approach. It serves as the basis of a formulation in which the Euler-Lagrange, the Hamilton and the Hamilton-Jacobi equations appear as three alternative expressions of one and the same underlying concept.

2.2. The non-homogeneous case

Let us first consider the so-called *non-homogeneous* case, i.e., one in which the action principle – and with it the Euler-Lagrange equation – is not invariant under a change of the curve parameter. In physics, this parameter usually corresponds to time. By solving the equations of motion one obtains not only the geometrical path traced by the particle – or group of particles – being described, but also how, i.e., the rate at which this path is traveled. The *non-homogeneous* case applies to non-relativistic formulations.

The equation of motion follows from the variational principle $\delta \int L(t, x^i, \dot{x}^i)dt = 0$. As physicists, we usually visualize the variational principle as expressing how Nature works: among all possible paths joining two given points, Nature chooses the one which affords $\int Ldt$ an extremal value. In some sense, this presupposes a non-local behavior, as two distant points determine the extremal curve that should join them. This is reminiscent of the action-at-a-distance invoked by earlier formulations, in whose context the variational principle originally arose. The approach proposed by Carathéodory is more in accordance with our modern view of local interactions. He replaced the problem of finding an extremum for the action integral by one of finding a local extremal value for a function. Thus, the field concept is at the forefront, playing a major role.

Let us recall some important assumptions [1, 13–15] concerning the central problem of variational calculus:

a) To find an extremal curve $x^i = x^i(t)$ that satisfies $\delta \int Ldt = 0$ requires that we restrict ourselves to a simply-connected domain. Though apparently technical, this point might entail a profound physical significance.

b) An extremal curve exists only in case that it can be embedded in a whole set of extremals, a so-called "Mayer field".

Now, having a field of curves is equivalent to defining a vector field $v^i(t,x^j)$: at each point x^j we just define $v^i(t,x^j)$ to be tangent to the unique curve which goes through x^j. In other words, the curves that constitute the field are integral curves of $v^i(t,x^j)$:

$$\frac{dx^i(t)}{dt} = v^i(t,x(t)). \tag{11}$$

Finding all the extremals $x^i(t)$ is thus equivalent to fixing $v^i(t,x)$. Once we have $v^i(t,x)$, the extremals can be obtained by integration of Eq.(11). The task of finding $v^i(t,x)$ can be approached locally. To this end, observe that the extremals we are seeking, for which $\delta \int Ldt = 0$, are also extremals of the modified, *"equivalent variational problem"*, $\delta \int (L - dS/dt)dt = 0$. This can be written as

$$\delta \int (L - \partial_t S - \dot{x}^i \partial_i S)dt = 0. \tag{12}$$

Now, assume that we are dealing with a particular Lagrangian $L^*(t,x,v)$, for which the following requirements are met: First, it is possible to find a vector field v such that $L^*(t,x,v) = 0$. Second, $L^*(t,x,w) > 0$ for any other field $w \neq v$. It is then easy to show that the integral curves of v are extremals of the variational problem $\delta \int L^* dt = 0$. Of course, not every Lagrangian will satisfy the requirements we have put on L^*; but by making use of the freedom we have to change our original problem into an *"equivalent variational problem"*, we let $L^* = L - dS/dt$ and seek for a vector field v such that

$$L(t,x,v) - \partial_t S - v^i \partial_i S = 0 \tag{13}$$

identically, the value zero being an extremal one with respect to variations of v. This happens for a *suitably* chosen $S(t,x)$ that remains fixed in this context. The function $S(t,x)$ must be just the one for which the value of $\int (\partial_t S + \dot{x}^i \partial_i S)dt = \int L(t,x,\dot{x})dt$, this last integral being calculated along an extremal curve. In other words, among all equivalent variational problems we seek the one for which the conditions put upon L^* are fulfilled. Thus, for the extremal value being, e.g., a minimum, it must hold Eq.(13), while $L(t,x^i,w^i) - \partial_t S - w^i \partial_i S > 0$ for any other field $w \neq v$. In this way our variational problem becomes a local one: v has to be determined so as to afford an extremal value to the expression at the left-hand side of Eq.(13). Thus, taking the partial derivative of this expression with respect to v and equating it to zero we obtain

$$\frac{\partial S}{\partial x^i} = \frac{\partial L(t,x,v(t,x))}{\partial v^i}. \tag{14}$$

Eqs.(13) and (14) are referred to as the *fundamental equations* in Carathéodory's approach. From these two equations we can derive all known results of the calculus of variations. We see, for instance, that defining $p_i = \partial L(t,x,v)/\partial v^i$, Eq.(14) gives $p_i = \partial S(t,x)/\partial x^i$. If we now introduce, by means of a Legendre transformation, the Hamiltonian $H(t,x,p) = v^i(t,x,p)p_i - L(t,x,v(t,x,p))$, Eq.(13) reads

$$\partial_t S + H\left(t, x^i, \partial_i S\right) = 0, \tag{15}$$

which is the Hamilton-Jacobi equation. In this way we obtain an equation for S, the auxiliary function that was so far undetermined. It is also straightforward to deduce the Euler-Lagrange and the Hamilton equations within the present approach. For the sake of brevity, we will show how to derive the Euler-Lagrange equations in the homogeneous case only. The non-homogeneous case can be treated along similar lines.

2.3. The homogeneous case

Let us turn into the so-called homogeneous problem, the one appropriate for a relativistic formulation. In relativity, we consider a space-time continuum described by four variables x^μ. Our variational principle is of the same form as before, i.e., $\delta \int L d\tau = 0$; but we require it to be invariant under Lorentz transformations and under parameter changes. Indeed, all we need in order to fix the motion is the geometrical shape of the extremal curve $x^\mu(\tau)$ in space-time, so that the parameter τ has no physical meaning and the theory must be invariant under arbitrary changes of it. This is achieved when L does not depend explicitly on τ and, furthermore, it is homogeneous of first degree in the generalized velocities $\dot{x}^\mu$: $L(x^\mu, \alpha\dot{x}^\mu) = \alpha L(x^\mu, \dot{x}^\mu)$, for $\alpha \geq 0$. From this requirement, it follows the identity

$$\dot{x}^\mu \frac{\partial L(x,\dot{x})}{\partial \dot{x}^\mu} = L, \tag{16}$$

which holds true for homogeneous Lagrangians. This property, however, precludes us from introducing a Hamiltonian in a similar manner as we did in the non-homogeneous case. We come back to this point later on.

As before, we seek also now for a velocity field $v(x)$ and a function $S(x^\mu)$, such that

$$L(x,v) - v^\mu \partial_\mu S = 0, \tag{17}$$

the value zero being an extremal one with respect to v, for a suitably chosen $S(x)$ that remains fixed in this context. For a maximum, for example, it must hold $L(x^\mu, w^\mu) - w^\mu \partial_\mu S < 0$ for any other field[1] $w \neq v$. Differentiating the left-hand side of Eq.(17) with respect to v and equating the result to zero we get

$$\frac{\partial S}{\partial x^\mu} = \frac{\partial L(x,v)}{\partial v^\mu}. \tag{18}$$

[1] The considered fields w^α are essentially different from v^α. A field $w^\alpha = \phi v^\alpha$, with ϕ a scalar function, is essentially the same as v^α.

From the *fundamental equations*, (17) and (18), we can derive all known results also in this case. In particular, we see that $S(x)$ must satisfy the integrability conditions

$$\frac{\partial^2 S}{\partial x^\mu \partial x^\nu} = \frac{\partial^2 S}{\partial x^\nu \partial x^\mu}, \tag{19}$$

which are, as we will shortly see, at the very basis of the Euler-Lagrange equations. Indeed, from Eq.(17) we obtain, by taking the derivative with respect to x^μ,

$$\frac{\partial L}{\partial x^\mu} + \frac{\partial L}{\partial v^\sigma}\frac{\partial v^\sigma}{\partial x^\mu} = \frac{\partial v^\sigma}{\partial x^\mu}\frac{\partial S}{\partial x^\sigma} + v^\sigma \frac{\partial^2 S}{\partial x^\mu \partial x^\sigma}. \tag{20}$$

On using Eq.(18), Eq.(20) reduces to

$$\frac{\partial L}{\partial x^\mu} = v^\sigma \frac{\partial^2 S}{\partial x^\mu \partial x^\sigma}. \tag{21}$$

From Eqs.(18) and (19) we thus obtain

$$\frac{\partial^2 S}{\partial x^\mu \partial x^\sigma} = \frac{\partial^2 S}{\partial x^\sigma \partial x^\mu} = \frac{\partial^2 L}{\partial x^\sigma \partial v^\mu} + \frac{\partial^2 L}{\partial v^\tau \partial v^\mu}\frac{\partial v^\tau}{\partial x^\sigma}, \tag{22}$$

so that

$$\frac{\partial L}{\partial x^\mu} = v^\sigma \frac{\partial^2 L}{\partial x^\sigma \partial v^\mu} + \frac{\partial^2 L}{\partial v^\tau \partial v^\mu}\frac{\partial v^\tau}{\partial x^\sigma} v^\sigma. \tag{23}$$

If we now evaluate this last relation along a single extremal, $dx^\mu/d\tau = v^\mu(x(\tau))$, we obtain, after recognizing the right hand side of Eq.(23) as $d(\partial L/\partial v^\mu)/d\tau$, the Euler-Lagrange equation:

$$\frac{\partial L}{\partial x^\mu} = \frac{d}{d\tau}\left(\frac{\partial L}{\partial \dot{x}^\mu}\right). \tag{24}$$

Eq.(23) is therefore more general than the Euler-Lagrange equation. The latter follows from Eq.(23); but not the other way around.

For the non-homogeneous case we obtain a similar result

$$\frac{d}{dt}\left(\frac{\partial L}{\partial \dot{x}^i}\right) - \frac{\partial L}{\partial x^i} = 0, \tag{25}$$

but with the important difference that now the curve-parameter t is fixed: the solution of Eq.(25) provides us not only with the geometrical shape of the extremal curve, but also with the rate at which this curve is traced back.

2.4. The arbitrariness of the curve parameter

Let us see how the arbitrariness of the curve parameter τ manifests itself when dealing with fields of extremals. It is usual to take advantage of such an arbitrariness in order to simplify the equations of motion. It is well known that in the cases of electromagnetism, for which $L(x,\dot{x}) = mc(\eta_{\mu\nu}\dot{x}^\mu\dot{x}^\nu)^{1/2} + \frac{e}{c}A_\mu(x)\dot{x}^\mu$, and gravitation, for which $L(x,\dot{x}) = (g_{\mu\nu}(x)\dot{x}^\mu\dot{x}^\nu)^{1/2}$, by choosing τ such that $(\eta_{\mu\nu}\dot{x}^\mu\dot{x}^\nu)^{1/2} = 1$, and $(g_{\mu\nu}(x)\dot{x}^\mu\dot{x}^\nu)^{1/2} = 1$, respectively, the equations of motion acquire a simple form. We are so led to ask whether the field v satisfying the fundamental Eqs.(17) and (18) has a corresponding arbitrariness. That this is indeed the case can be seen as follows. We wish to prove that in case v^μ satisfies Eqs.(17) and (18), so does $w^\mu = \phi v^\mu$, with $\phi(x) > 0$ an arbitrary, scalar function. From the homogeneity of the Lagrangian we have $L(x^\mu, \phi v^\mu) = \phi L(x^\mu, v^\mu)$, so that it is seen at once that w^μ satisfies Eq.(17) if v^μ does. Indeed, multiplying Eq.(17) by $\phi(x) > 0$ leads to

$$\phi(x)\left(L(x,v) - v^\mu\partial_\mu S\right) = L(x,\phi v) - (\phi v^\mu)\,\partial_\mu S = L(x,w) - w^\mu\partial_\mu S. \tag{26}$$

The Lagrangian of the "equivalent variational problem" is $L^* = L - v^\mu\partial_\mu S$. Clearly, $L^*(x,\phi v) = \phi L^*(x,v)$, and hence it follows that

$$\frac{\partial L^*(x,w)}{\partial v^\mu} = \frac{\partial L^*(x,w)}{\partial w^\nu}\frac{\partial w^\nu}{\partial v^\mu} = \frac{\partial L^*(x,w)}{\partial w^\mu}\phi. \tag{27}$$

On the other hand,

$$\frac{\partial L^*(x,w)}{\partial v^\mu} = \frac{\partial}{\partial v^\mu}\left(\phi L^*(x,v)\right) = \phi\frac{\partial L^*(x,v)}{\partial v^\mu} = \phi\left(\frac{\partial L}{\partial v^\mu} - \frac{\partial S}{\partial x^\mu}\right) = 0, \tag{28}$$

on account of Eq.(17). In view of Eq.(27) we have then that $\partial L^*(x,w)/\partial w^\mu = 0$. In summary, Eqs.(17,18) hold with v being replaced by w, so that both velocity fields solve our variational problem for the same $S(x)$. We have thus the freedom to choose ϕ according to our convenience. The integral curves of $v^\mu(x)$ and $w^\mu(x)$ coincide with each other, differing only in their parametrization.

2.5. Hamiltonians

The introduction of a Hamiltonian offers no problem in the non-homogeneous case, where it was defined as $H(x^i, p_i) \equiv \dot{x}^i(t,x,p)p_i - L(t,x,\dot{x}(t,x,p))$, with $p_i \equiv \partial L/\partial \dot{x}^i$; the condition for solving $\dot{x}^i$ in terms of (x^j, p_j) being assumed to be fulfilled: $\det(\partial^2 L/\partial\dot{x}^i\partial\dot{x}^j) \neq 0$. It is then straightforward [1, 13] to obtain

$$\frac{\partial H}{\partial p_i} = \dot{x}^i = \frac{dx^i}{dt}, \tag{29}$$

which constitute half of the Hamilton equations. It is also easy to sow that $\partial H/\partial t = -\partial L/\partial t$ and $\partial H/\partial x^i = -\partial L/\partial x^i$. Using this last result together with $p_i = \partial L/\partial \dot{x}^i$ in the

Euler-Lagrange equation, Eq.(25), one gets

$$\frac{dp_i}{dt} = -\frac{\partial H}{\partial x^i}, \tag{30}$$

the other half of the Hamilton equations.

In the homogeneous case, as already mentioned, the corresponding expression for H, i.e., $\dot{x}^\mu \partial L/\partial \dot{x}^\mu - L$, vanishes identically by virtue of Eq.(16) . It is nonetheless generally possible to introduce a Hamiltonian in a number of ways. Carathéodory's approach leads to an infinite set of Hamiltonians, from which we can choose the most suitable one for the problem at hand. We will not go into the details here, but refer the reader to the standard literature [1, 13] in which this material is discussed at length.

3. Electromagnetism: The London equations and the Bohr-van Leeuwen theorem

3.1. The London equations of superconductivity

As mentioned before, there are only two interactions relevant to classical physics: electromagnetism and gravitation. In electromagnetism, the Lagrangian is given by

$$L(x, \dot{x}) = mc(\eta_{\mu\nu}\dot{x}^\mu \dot{x}^\nu)^{1/2} + \frac{e}{c}A_\mu(x)\dot{x}^\mu. \tag{31}$$

Here, $\eta_{\mu\nu} = diag(+1, -1, -1, -1)$ is the Minkowski metric tensor and summation over repeated indices from 0 to 3 is understood. The electromagnetic field is given by the four-potential A_μ, whose components are $\phi(t, x^i)$ and $\mathbf{A}(t, x^i)$.

We are now in a position to show how the London equations follow as a *logical* consequence of the relations presented above, when we use Eq.(31). From Eqs.(18) and (19) we obtain, in general,

$$\frac{\partial}{\partial x^\mu}\left(\frac{L(x, v(x))}{\partial v^\nu}\right) - \frac{\partial}{\partial x^\nu}\left(\frac{L(x, v(x))}{\partial v^\mu}\right) = 0. \tag{32}$$

This equation can be used to obtain the relativistic version of the London equations: As stated before, because the Lagrangian is homogeneous of first order in v, this vector field can be chosen so as to satisfy $(v_\mu v^\mu)^{1/2} = 1$ in the region of interest. From Eq.(32) and Eq.(31) we get

$$\frac{\partial v_\nu}{\partial x^\mu} - \frac{\partial v_\mu}{\partial x^\nu} + \frac{e}{mc^2}\left(\frac{\partial A_\nu}{\partial x^\mu} - \frac{\partial A_\mu}{\partial x^\nu}\right) = 0. \tag{33}$$

This condition leads to the London equations, if we go to the non-relativistic limit, $v^2/c^2 \ll 1$. Indeed, after multiplication by ne, with n meaning a uniform particle density, Eq.(33) can be brought into the form:

$$\frac{\partial j_\nu}{\partial x^\mu} - \frac{\partial j_\mu}{\partial x^\nu} + \frac{ne^2}{mc^2}\left(\frac{\partial A_\nu}{\partial x^\mu} - \frac{\partial A_\mu}{\partial x^\nu}\right) = 0, \tag{34}$$

where $j_\mu \equiv nev_\mu$. In the non-relativistic limit Eq.(34) reduces, for $\mu, \nu = i, k = 1, 2, 3$, to

$$\frac{\partial j_k}{\partial x^i} - \frac{\partial j_i}{\partial x^k} = -\frac{ne^2}{mc}\left(\frac{\partial A_k}{\partial x^i} - \frac{\partial A_i}{\partial x^k}\right), \tag{35}$$

where we have used $\dot{x}^\mu(\tau) = \gamma(1, \mathbf{v}(t)/c)$ with $\gamma \equiv (1 - \mathbf{v}^2/c^2)^{-1/2} \approx 1$. In three-vector notation this equation reads

$$\nabla \times \mathbf{j} = -\frac{ne^2}{mc}\nabla \times \mathbf{A} = -\frac{ne^2}{mc}\mathbf{B}, \tag{36}$$

which is the London equation [16]. Eq. (36), together with the steady-state Maxwell equation, $\nabla \times \mathbf{B} = (4\pi/c)\mathbf{j}$, lead to $\nabla^2\mathbf{B} = \left(4\pi ne^2/mc^2\right)\mathbf{B}$, from which the Meissner effect follows. By considering now the case $\mu = 0, \nu = k = 1, 2, 3$ in Eq.(34), we obtain

$$\frac{\partial j_k}{\partial x^0} - \frac{\partial j_0}{\partial x^k} = \frac{ne^2}{mc^2}\left(\frac{\partial A_0}{\partial x^k} - \frac{\partial A_k}{\partial x^0}\right). \tag{37}$$

Multiplying this equation by $-c^2$ and using three-vector notation it reads, with $j_0 = c\rho$,

$$\frac{\partial \mathbf{j}}{\partial t} + c^2\nabla\rho = \frac{ne^2}{m}\mathbf{E}. \tag{38}$$

This equation was also postulated by the London brothers as part of the phenomenological description of superconductors. It was guessed as a relativistic generalization of the equation that should hold for a perfect conductor. Without the ρ-term (which in our case vanishes due to the assumed uniformity of n) it is nothing but the Newton, or "acceleration" equation for charges moving under the force $e\mathbf{E}$. The ρ-term was originally conceived as a relativistic "time-like supplement " to the current $\mathbf{j}$ [16]. We see that the London equations are in fact the non-relativistic limit of an integrability condition, Eq.(33), which follows from the variational principle $\delta \int L ds = 0$ alone. The physical content of this procedure appears when we interpret the integral curves of $v(x)$ as streamlines of an ideal fluid. By contracting Eq.(33) with v^μ and using $v^\mu \partial_\nu v_\mu = 0$ (which follows from $v^\mu v_\mu = 1$) we obtain

$$v^\mu \frac{\partial v_\nu}{\partial x^\mu} = \frac{e}{mc^2} F_{\nu\mu} v^\mu, \tag{39}$$

with $F_{\nu\mu} = \partial_\nu A_\mu - \partial_\mu A_\nu$, which relates to **E** and **B** by $E_i = F_{0i}$ and $B_i = -\epsilon_{ijk}F_{jk}/2$, with ϵ_{ijk} the totally antisymmetric symbol and latin indices running from 1 to 3. The nonrelativistic limit of Eq.(39) reads

$$\frac{\partial \mathbf{v}}{\partial t} + (\mathbf{v} \cdot \nabla)\, \mathbf{v} = \frac{e}{m}\left(\mathbf{E} + \frac{\mathbf{v}}{c} \times \mathbf{B}\right). \tag{40}$$

The left-hand side of this equation is the *convective derivative*, which reduces to $d\mathbf{v}/dt$ by restriction to a single extremal. Analogously, Eq.(39) becomes the well-known Lorentz equation when evaluated along a single extremal: $dx^\mu/ds = v^\mu(x(s))$. In this case, $v^\mu(x(s))\partial v_\nu(x(s))/\partial x^\mu = dv_\nu(s)/ds$. Thus, we see that the Lorentz equation for a single particle follows from the more general Eq.(39). For $\mu = 0$ Eq.(39) gives an equation which can be derived from Eq.(40) by scalar multiplication with **v**. This is the energy equation. It is worth mentioning that this last fact is a particular manifestation of a well-known result valid for *homogeneous* Lagrangians: only $n-1$ out of the n Euler-Lagrange equations are independent from each other in this case, due to the identity $\dot{x}^\mu E_\mu = 0$, with $E_\mu \equiv d(\partial L/\partial \dot{x}^\mu)/d\tau - \partial L/\partial x^\mu$ being the Euler vector[13]. Such a result follows from Eq.(16).

Some remarks are in place here. Our derivation of the London equations brings into evidence that they have a validity that goes beyond their original scope. They cannot be seen by themselves as characterizing the phenomenon of superconductivity. Instead, they describe a "dust" of charged particles moving along the extremals of the Lagrangian given by Eq.(31). The field A^μ under which these particles move could be produced by external sources, or else be the field resulting from the superposition of some external fields with those produced by the charges themselves. In this last case, the Maxwell and London equations constitute a self-consistent system. Only under special circumstances, the system of charges can be in a state of collective motion that may be described by the field of extremals obeying Eq.(39). This is the superconducting phase, for which quantum aspects are known to play a fundamental role[17]. However, *once* the phase transition from the normal to the superconducting state has occurred, it becomes possible to describe some aspects of the superconducting state by classical means. This is a case analogous to the one encountered in laser theory. Indeed, several features of a lasing system can be understood within a semi-classical laser theory, whereby the electromagnetic field is treated as a classical, non-quantized field. Perhaps some plasmas could reach the limit of perfect conductivity. However, in order to produce a Meissner-like effect some conditions should be met. It is necessary, for instance, that the available free-energy of the plasma is sufficient to overcome the magnetic field energy, so that the magnetic field can be driven out of the plasma [5–8]. The so-called helicity of the system should also play a role, attaining the value zero for the superconducting state to be reached [9].

In any case, we see that Carathéodory's approach can be a fruitful one in physics. In the case of superconductivity, from the sole assumption that the Lagrangian be given by Eq.(31) one can derive all the equations that were more or less guessed, in the course of almost twenty five years, since Kamerlingh Onnes discovered superconductivity in 1911, until the London model was proposed, in 1935. But beyond this, there are other aspects that can be illuminated by following Carathéodory's approach, as we shall see next.

3.2. Beyond the London equations

Let us address the case when the charge density ρ is not constant, as previously assumed. There is a close relationship between the norm of our velocity field, i.e., $\phi(x) = \left(v_\mu v^\mu\right)^{1/2}$, and $\rho(x)$. It can be shown that it is always possible to choose v^μ so that the continuity equation $\partial_\mu j^\mu = 0$ holds. Here, $j^\mu := \rho v^\mu$ and $\rho = nec\phi^{-1}$, n being a free parameter whose dimensions are 1/volume. Indeed, in view of the aforementioned possibility of changing the field v^μ by $w^\mu = \tilde{\phi} v^\mu$, we can always satisfy the continuity equation. For, if $\partial_\mu j^\mu = -f$, we may choose $\tilde{j}^\mu = \tilde{\phi} j^\mu$ such that $\partial_\mu \tilde{j}^\mu = \tilde{\phi}\partial_\mu j^\mu + j^\mu \partial_\mu \tilde{\phi} = 0$. Putting $\psi = \log \tilde{\phi}$, we need to solve $j^\mu \partial_\mu \psi = f$, which is always possible.

Coming back to our Lagrangian of Eq.(31), by replacing it in Eq.(18), we obtain

$$v_\mu = \frac{\phi}{mc}\left(\partial_\mu S - \frac{e}{c}A_\mu\right), \tag{41}$$

with $\phi := \left(v_\mu v^\mu\right)^{1/2}$. Using the gauge freedom of A_μ we may replace this field by

$$A'_\mu = A_\mu - \frac{c}{e}\partial_\mu S, \tag{42}$$

in which case Eq.(41) reads

$$v_\mu = -\phi\left(\frac{e}{mc^2}\right)A'_\mu. \tag{43}$$

From this equation and $v_\mu v^\mu = \phi^2$ we get

$$A'_\mu A'^\mu = \left(\frac{mc^2}{e}\right)^2. \tag{44}$$

Eq.(32) applied to the present case gives

$$\frac{\partial}{\partial x^\mu}\left(\frac{v_\nu}{\phi}\right) - \frac{\partial}{\partial x^\nu}\left(\frac{v_\mu}{\phi}\right) + \frac{e}{mc^2}\left(\frac{\partial A_\nu}{\partial x^\mu} - \frac{\partial A_\mu}{\partial x^\nu}\right) = 0. \tag{45}$$

It is clear that this equation holds for A'_μ as well. Eq.(43) is a particular solution of this equation. By Fourier-transforming Eq.(45) we obtain, with $w^\mu := v^\mu/\phi$,

$$k^\mu w^\nu - k^\nu w^\mu = -\frac{e}{mc^2}\left(k^\mu A^\nu - k^\nu A^\mu\right). \tag{46}$$

As for the Fourier-transformed version of Eq.(43), it is given by

$$w_\mu(k) = -\frac{e}{mc^2}A'_\mu(k). \tag{47}$$

As we saw before, v^μ can be chosen so that $j_\mu = necw_\mu = (nec/\phi)v_\mu \equiv \rho(x)v_\mu$ satisfies the continuity equation $\partial_\mu j^\mu = 0$. The factor nec is included for dimensional purposes: c/ϕ has no dimensions and n is a free parameter such that ne has dimensions of charge per unit volume. While n is a constant, $\rho(x)$ is a non-uniform charge density. Thus, the scalar field $\phi = \left(v_\mu v^\mu\right)^{1/2}$, the norm of the velocity field, is related to the density $\rho(x)$ by $\rho = nec\phi^{-1}$.

Note that $\partial_\mu j^\mu = 0$ implies a restriction on $\partial_\mu v^\mu$. To see this, observe that $\partial_\mu j^\mu = \rho\partial_\mu v^\mu + v^\mu\partial_\mu\rho = 0$. This can be rewritten as

$$\frac{v^\mu}{\phi}\partial_\mu\phi = f, \tag{48}$$

with $\partial_\mu v^\mu = f$. On the other hand, from $v_\mu v^\mu = \phi^2$ it follows that $\phi\partial_\mu\phi = v^\nu\partial_\mu v_\nu$. Eq.(48) then implies that

$$\frac{v^\mu v^\nu}{2\phi^2}(\partial_\mu v_\nu + \partial_\nu v_\mu) = \partial_\sigma v^\sigma. \tag{49}$$

It is also worth noting that instead of Eq.(39) we have now

$$v^\mu\frac{\partial v_\nu}{\partial x^\mu} = \frac{e\phi}{mc^2}F_{\nu\mu}v^\mu + \left(v^\mu\frac{\partial}{\partial x^\mu}\ln\phi\right)v_\nu = \frac{e\phi}{mc^2}F_{\nu\mu}v^\mu + fv_\nu. \tag{50}$$

We could argue that the second term on the right hand side is not physical, because we could choose $\phi = 1$, as we did before, getting Eq.(39). However, such a choice is not available any longer when we invoke charge (or matter) conservation. In such a case, $\partial_\mu j^\mu = 0$, and we must relate j^μ with v^μ by $j^\mu = \rho v^\mu$, so that the above considerations apply.

Coming back to Eq.(47), we see that it implies that $\partial_\mu A'^\mu = 0$, i.e., A'^μ is in the Lorentz gauge. Because of $A'_\mu = A_\mu - (c/e)\,\partial_\mu S$, the scalar function S must satisfy

$$\partial_\mu\partial^\mu S \equiv \Box S = \frac{e}{c}\partial_\mu A^\mu. \tag{51}$$

Let us consider now Maxwell equations, $\partial_\mu F^{\mu\nu} = (4\pi/c)j^\nu$. If we take j^ν to be the same as before, we are assuming that $F^{\mu\nu}$ is generated by the same currents upon which this field is acting. That is, we are considering a closed system of charges and fields. We have then, using $F^{\mu\nu} = \partial^\mu A'^\nu - \partial^\nu A'^\mu$ and $\partial_\mu A'^\mu = 0$,

$$\partial_\mu F^{\mu\nu} = \Box A'^\nu = \frac{4\pi}{c}j^\nu, \tag{52}$$

while from Eq. (47) we get

$$A'^\nu = -\frac{mc}{ne^2}j^\nu, \tag{53}$$

so that we can write Eq. (52) as

$$\Box j^{\nu} = -\frac{4\pi n e^2}{mc^2} j^{\nu} \equiv -\frac{1}{\lambda_L^2} j^{\nu}, \tag{54}$$

in which we have identified the London penetration length λ_L. This equation can be rewritten in the form of the Klein-Gordon equation:

$$\left(\Box + \lambda_L^{-2}\right) j^{\nu} = 0, \tag{55}$$

with λ_L replacing $\lambda_C = \hbar/mc$, the Compton wavelength that appears in the Klein-Gordon equation. For the steady-state ($\partial_0 j^{\nu} = 0$), Eq.(54) reads

$$\nabla^2 j^{\nu} = +\frac{1}{\lambda_L^2} j^{\nu}. \tag{56}$$

Taking the usual configuration of a superconductor filling half the space ($z > 0$), the solution of this equation (satisfying appropriate boundary conditions: $\lim_{z\to\infty} j^{\nu}(z) = 0$) is

$$j^{\nu}(z) = \exp\left(-\frac{z}{\lambda_L}\right) j^{\nu}(0). \tag{57}$$

In general, however, Eq.(55) admits several other solutions that depend on the assumed boundary conditions. Note that Eq.(55) corresponds to a field-free case of the Klein-Gordon equation. This is because $j^{\nu} \sim A'^{\nu}$, so that electromagnetic fields and current share the same dynamics. This is a consequence of having assumed that the Euler-Lagrange equations (written as field equations) and Maxwell equations conform a closed system. Notably, A'^{ν} behaves like a source-free Proca field [18] whose mass (in units of inverse length) is fixed by λ_L.

3.3. The Bohr-van Leeuwen theorem

Dropping the prime, Eq.(53) gives $j^{\nu} = -(ne^2/mc)A^{\nu}$, which can be rewritten as $v^{\nu} = -\left(e/mc^2\right) A^{\nu}$, with $v_{\nu}v^{\nu} = 1$. We get thus $v_{\nu}A^{\nu} = -mc^2/e$, which in the nonrelativistic limit reads

$$\mathbf{v} \cdot \mathbf{A} = \frac{mc^3}{e}. \tag{58}$$

This condition is important for the following reason. Our considerations have confirmed the possibility of classical diamagnetism, in contradiction with the Bohr-van Leeuwen theorem. Therefore, this theorem should be modified. Eq.(58) represents a constant of the motion that must be taken into account when constructing the phase density for a system of charged particles. The original version of the Bohr-van Leeuwen theorem did not consider condition (58). We will show next how this condition modifies the theorem.

The Bohr-van Leeuwen theorem addresses a sample of charged particles subjected to a uniform magnetic field $\mathbf{B}$. The nonrelativistic Lagrangian of the system is $L = \sum_{i=1}^{N}(m_i/2)\mathbf{v}_i^2 + (e_i/c)\mathbf{v}_i \cdot \mathbf{A}$. We can take $\mathbf{A} = \mathbf{B} \times \mathbf{r}/2$. The partition function is given by

$$Z = \int_{-\infty}^{+\infty} \dots \int_{-\infty}^{+\infty} \exp\left[-\beta\left(\sum_i \dot{x}^i \frac{\partial L}{\partial \dot{x}^i} - L\right)\right] \left|\frac{\partial^2 L}{\partial \dot{x}^i \partial \dot{x}^j}\right| d^N\tau, \tag{59}$$

with $d^N\tau$ a properly normalized volume element in configuration space. We see that the terms in L that depend on magnetic potentials are linear in the velocities, so that the integrand in Z turns out to be independent of magnetic potentials. The Bohr-van Leeuwen theorem then follows: because Z is independent of magnetic potentials, there is no effect on the system in response to $\mathbf{B}$. This prediction changes when we take into account the constant of motion, Eq.(58), or equivalently, $G := (e\mathbf{B}/2mc) \cdot (\mathbf{r} \times \mathbf{v}) = c^2$. For a sample of identical particles we define $G_s := \sum_i (e\mathbf{B}/2mc) \cdot (\mathbf{r}_i \times \mathbf{v}_i) \equiv \sum_i \boldsymbol{\omega}_L \cdot (\mathbf{r}_i \times \mathbf{v}_i)$, with $\boldsymbol{\omega}_L$ the Larmor frequency. The phase density D for the corresponding Hamiltonian $H = \sum_i (2m)^{-1}(\mathbf{p}_i - (e_i/c)\mathbf{A}_i)^2$ is given by $D = Z^{-1}\exp(-\beta H - \lambda G_s)$, with Z being the partition function that normalizes D and $\beta = (k_B T)^{-1}$. Both λ and β are Lagrange multipliers, introduced to take account of the restrictions imposed by Eq. (58) and the fixed mean energy, respectively. Thus,

$$D = \frac{1}{Z}\exp\left[\sum_i\left(\frac{-\beta(\mathbf{p}_i - (e/c)\mathbf{A}_i)^2}{2m} + \lambda(\boldsymbol{\omega}_L \cdot \mathbf{r}_i \times \mathbf{v}_i)\right)\right]. \tag{60}$$

The single-particle velocity distribution that can be obtained from D is proportional to

$$\exp\left[-\frac{\beta m}{2}\left(\mathbf{v} - \frac{\lambda}{\beta m}\boldsymbol{\omega}_L \times \mathbf{r}\right)^2 + \frac{\lambda^2}{2\beta m}(\boldsymbol{\omega}_L \times \mathbf{r})^2\right]. \tag{61}$$

This gives the mean velocity at $\mathbf{r}$. We have thus $\langle \mathbf{v} \rangle = (\lambda/\beta m)\boldsymbol{\omega}_L \times \mathbf{r}$, which determines the value of the Lagrange multiplier as $\lambda = \beta m$. The phase density can finally be written as

$$D = \frac{1}{Z}\exp\left[-\beta\left(\sum_i \frac{m}{2}\mathbf{v}_i^2 + \mathbf{B} \cdot \mathbf{M}\right)\right], \tag{62}$$

with $\mathbf{M} \equiv \sum_i (e/2c)\mathbf{r}_i \times \mathbf{v}_i$ naturally arising as the magnetic moment of the system. A magnetic response shows up therefore also classically, contrary to what the original version of the Bohr-van Leeuwen theorem stated. It has been shown before that this theorem does not hold whenever the magnetic field produced by the moving charges is taken into account. Such a field is included in the Darwin Lagrangian [11], which is correct to order $(v/c)^2$. In our case, we do not need to modify the standard Lagrangian.

4. Hamilton-Jacobi equations without Hamiltonian

We have already mentioned that for homogeneous Lagrangians the definition of a Hamiltonian is precluded by the vanishing of $\dot{x}^\mu \partial L/\partial \dot{x}^\mu - L$. It is nonetheless possible to introduce a Hamiltonian in a number of ways. Carathéodory's approach leads to an infinite set of Hamiltonians, from which we can choose the most suitable one for dealing with the problem at hand. Here, we focus on the two Lagrangians of interest to us, given by $L = mc(\eta_{\mu\nu}\dot{x}^\mu \dot{x}^\nu)^{1/2} + eA_\mu(x)\dot{x}^\mu/c$ for electromagnetism and

$$L(x, \dot{x}) = \left(g_{\mu\nu}(x)\dot{x}^\mu \dot{x}^\nu\right)^{1/2} \tag{63}$$

for gravitation. We will prove that in these two particular cases it is possible to derive the equation which the function $S(x)$ has to satisfy, *without* having to introduce a Hamiltonian.

Let us start with gravitation. From Eq.(63) with v^μ replacing $\dot{x}^\mu$, it follows that

$$\frac{\partial L}{\partial v^\mu} = \frac{1}{L} g_{\mu\nu} v^\nu. \tag{64}$$

Using $g_{\mu\nu}g^{\nu\sigma} = \delta^\sigma_\mu$ this equation leads to $v^\nu = Lg^{\mu\nu}p_\mu$, with $p_\mu \equiv \partial L/\partial v^\mu$. Considering that $L^2 = g_{\mu\nu}v^\mu v^\nu = L^2 g^{\mu\nu}p_\mu p_\nu$, it follows $g^{\mu\nu}p_\mu p_\nu = 1$. And because $\partial S/\partial x^\mu = \partial L/\partial v^\mu = p_\mu$, we obtain the Hamilton-Jacobi equation for S:

$$g^{\mu\nu}(x)\frac{\partial S}{\partial x^\mu}\frac{\partial S}{\partial x^\nu} = 1. \tag{65}$$

In the electromagnetic case the corresponding Lagrangian leads, by the same token, to $v_\mu = (\partial_\mu S - \frac{e}{c}A_\mu)\phi/mc$ with $\phi \equiv (\eta_{\mu\nu}v^\mu v^\nu)^{1/2}$, again as a consequence of $\partial S/\partial x^\mu = \partial L/\partial v^\mu$. By replacing the above expression of v_μ in $\phi^2 = \eta^{\mu\nu}v_\mu v_\nu$, it follows the Hamilton-Jacobi equation

$$\eta^{\mu\nu}\left(\frac{\partial S}{\partial x^\mu} - \frac{e}{c}A_\mu\right)\left(\frac{\partial S}{\partial x^\nu} - \frac{e}{c}A_\nu\right) = m^2c^2. \tag{66}$$

We remark that there was no need to choose v so as to satisfy either $\phi = const.$ in the electromagnetic case, or $L = const.$ in the gravitational case, as it is usually done for obtaining the respective Euler-Lagrange equations in their simplest forms. As a consequence, the constants appearing on the right-hand sides of Eqs.(65) and (66) are independent of the way by which we decide to fix the parameter τ of the extremal curves. Let us remark that it is not unusual to find in textbooks Eq.(65) written with m^2c^2 instead of the 1 on the right-hand side (see, e.g. [19]). This occurs because Eq.(65) is usually introduced as a generalization of Eq.(66), with $A_\mu = 0$ (field-free case). Invoking the equivalence principle, one replaces $\eta^{\mu\nu}$ by $g^{\mu\nu}$ and so arrives at the equation which is supposed to describe a "free" particle moving in a curved space-time region. Now, the metric tensor $g^{\mu\nu}$ embodies all the information that determines how a test particle moves under gravity, irrespective of its inertial mass m. There is therefore no physical reason to put a term like m^2c^2 on the right-hand side of Eq.(65). To be sure, for all practical purposes it is irrelevant that we set *any* constant on the right-hand side of Eq.(65), as this constant will drop afterwards in the equations describing the motion. But, as a matter of principle, the mass of a test particle should not appear in an equation which describes how it moves under the sole action of gravity.

5. Gauge invariance in electromagnetism and gravitation

Gauge invariance is presently understood as a key principle that lies at the root of fundamental interactions. An equation like Schrödinger's (or Dirac's) for a free electron is invariant under the transformation $\psi \to \exp(i\alpha)\psi$, for constant α. This is in accordance with the physical meaning of the wave-function and the way it enters in all expressions related to measurable quantities. However, one expects that Nature should respect such an invariance not only *globally*, i.e., with constant α, but also *locally*, with α a function of time and position. It is, so to say, by recourse to the appropriate interaction that Nature manages to reach this goal. For achieving invariance under the $U(1)$ transformation $\psi \to \exp(i\alpha)\psi$, it is necessary to introduce a *gauge field*, in this case a field represented by $A_\mu(x)$, which couples to the particle. The equation for a free particle is correspondingly changed into one in which A_μ appears. In this context, gauge invariance means invariance under the simultaneous change $\psi \to \exp(i\alpha)\psi$ and an appropriate one for A_μ. This last one must be so designed that the equation now containing A_μ remains invariant. The change of A_μ turns out to be $A_\mu \to A_\mu - (\hbar c/e)\partial_\mu\alpha$, which is the one corresponding to a gauge transformation of the *electromagnetic* field. Hence, one is led to interpret electromagnetic interactions as a consequence of local $U(1)$-invariance. Other fundamental interactions stem from similar gauge invariances: $SU(2) \times U(1)$ gives rise to electroweak interactions, $SU(3)$ to the strong interaction [20], and local Lorentz invariance to gravitation [21, 22].

In this Section we want to show how gauge invariance leads, within the classical context, to considerations paralleling those of quantum mechanics. Carathéodory's formulation will be particularly useful to this end. Let us start with the electromagnetic case. Replacing the Lagrangian of Eq.(31) in the fundamental Eq.(17), we get

$$mc(v_\mu v^\mu)^{1/2} + \frac{e}{c}A_\mu v^\mu - v^\mu\partial_\mu S = 0. \tag{67}$$

Now, the observable predictions we can make concern the integral curves of the velocity field v^μ. This field remains invariant under the replacement

$$A^*_\mu = A_\mu - \frac{c}{e}\partial_\mu W, \tag{68}$$

whenever a simultaneous change in S is undertaken. This change is given by $S \to S^* = S + W$. It leaves Eq.(67) unchanged, for a fixed $v^\mu(x)$. Eq.(66), to which the velocity field $v^\mu(x)$ belongs, is also fulfilled with S^* and A^*_μ. The quantum-mechanical counterpart of this result could have suggested such a conclusion, in view of the relationship $\psi \sim \exp(iS/\hbar)$. Indeed, a change $\psi \to \psi^* = \exp(i\alpha)\psi$ means that $\psi^* \sim \exp(iS^*/\hbar)$, with $S^* = S + W$, where $W = \hbar\alpha$.

Now, we are naturally led to ask about a similar invariance in the gravitational case. Here, Eq.(17) reads

$$(g_{\mu\nu}v^{\mu}v^{\nu})^{1/2} - v^{\mu}\partial_{\mu}S = 0, \tag{69}$$

and we ask how a simultaneous change of $g_{\mu\nu}$ and S might be, in order that v^{μ} remains fixed and with it the field of extremals. In the present case, it is better to start with Eq.(23) instead of Eq.(69). The reasons will become clear in what follows. Working out Eq.(23) for the present Lagrangian we obtain, after some manipulations,

$$v^{\tau}\frac{\partial v^{\nu}}{\partial x^{\tau}} + g^{\mu\nu}(\partial_{\tau}g_{\mu\sigma} - \frac{1}{2}\partial_{\mu}g_{\sigma\tau})v^{\sigma}v^{\tau} = \frac{\partial(\ln L(x, v(x)))}{\partial x^{\tau}}v^{\nu}v^{\tau} = \frac{\partial\Phi(x)}{\partial x^{\tau}}v^{\nu}v^{\tau}, \tag{70}$$

where $\Phi(x) = \ln L(x, v(x))$. The right-hand side of Eq.(70) can be written in the form $\frac{1}{2}(\delta^{\nu}_{\sigma}\partial_{\tau}\Phi + \delta^{\nu}_{\tau}\partial_{\sigma}\Phi)v^{\sigma}v^{\tau}$. This suggests us to symmetrize the coefficient of $v^{\sigma}v^{\tau}$ on the left-hand side, thereby obtaining

$$g^{\mu\nu}(\partial_{\tau}g_{\mu\sigma} - \frac{1}{2}\partial_{\mu}g_{\sigma\tau})v^{\sigma}v^{\tau} = \frac{1}{2}g^{\mu\nu}(\partial_{\tau}g_{\mu\sigma} + \partial_{\sigma}g_{\mu\tau} - \partial_{\mu}g_{\sigma\tau})v^{\sigma}v^{\tau} \equiv \Gamma^{\nu}_{\sigma\tau}v^{\sigma}v^{\tau}, \tag{71}$$

with $\Gamma^{\nu}_{\sigma\tau}$ the Christoffel symbols. Eq.(70) then reads

$$v^{\tau}\frac{\partial v^{\nu}}{\partial x^{\tau}} + \Gamma^{\nu}_{\sigma\tau}v^{\sigma}v^{\tau} = \frac{1}{2}(\delta^{\nu}_{\sigma}\partial_{\tau}\Phi + \delta^{\nu}_{\tau}\partial_{\sigma}\Phi)v^{\sigma}v^{\tau}. \tag{72}$$

Note that if we choose v such that $L = const.$, then Eq.(72) becomes the usual geodesic equation, when it is calculated along an extremal curve, $dx^{\mu}/d\tau = v^{\mu}(x(\tau))$:

$$\frac{d\dot{x}^{\mu}}{d\tau} + \Gamma^{\mu}_{\sigma\rho}\frac{dx^{\sigma}}{d\tau}\frac{dx^{\rho}}{d\tau} = 0. \tag{73}$$

If $L \neq const.$, we obtain a geodesic equation with a right-hand side of the form $(df/d\tau)\dot{x}^{\nu}$. In both cases we obtain the same curves – geodesics – but with a different parametrization.

Now, assume that a change, $g_{\mu\sigma} \to g^{*}_{\mu\sigma}$ can be found, so that the corresponding $\Gamma^{*\nu}_{\tau\sigma}$ satisfy

$$\Gamma^{*\nu}_{\tau\sigma} - \Gamma^{\nu}_{\tau\sigma} = \frac{1}{2}\left(\delta^{\nu}_{\sigma}\partial_{\tau}\Lambda(x) + \delta^{\nu}_{\tau}\partial_{\sigma}\Lambda(x)\right), \tag{74}$$

with $\Lambda(x)$ being arbitrary. Such a change leads to an equation equivalent to Eq.(70), with Φ being replaced by $\Phi^{*} = \Phi + \Lambda$, and hence to the same extremals. In this way we recover an old result due to Weyl: if Christoffel symbols are related to each other by Eq.(74), then they have the same geodesics [23]. Given $g_{\mu\nu}$ and Λ, it is always possible to find a $g^{*}_{\mu\nu}$ satisfying Eq.(74). This is because this equation can be put in the form

$$\partial_{\lambda}g^{*\mu\nu} = -g^{*\nu\sigma}\Omega^{\mu}_{\lambda\sigma} - g^{*\mu\sigma}\Omega^{\nu}_{\lambda\sigma}, \tag{75}$$

with $\Omega^{\mu}_{\lambda\sigma} := \Gamma^{\mu}_{\lambda\sigma} + \frac{1}{2}(\delta^{\mu}_{\sigma}\partial_{\lambda}\Lambda + \delta^{\mu}_{\lambda}\partial_{\sigma}\Lambda)$, and it can be straightforwardly proved that the integrability conditions for the above equation are identically satisfied.

In fact, Weyl arrived at a relation like Eq.(74) but having the expression $w_{\tau}\delta^{\nu}_{\sigma} + w_{\sigma}\delta^{\nu}_{\tau}$ on the right hand side, with w_{σ} taken to be a covariant vector. Now, it is easy to see that w_{σ} must be a gradient. Indeed, after writing Eq.(74) in Weyl's form, with w_{μ} replacing $\frac{1}{2}\partial_{\mu}\Lambda$, we contract both sides of this equation with respect to ν and τ, thereby obtaining $w_{\sigma} = \frac{1}{5}(\Gamma^{*\nu}_{\nu\sigma} - \Gamma^{\nu}_{\nu\sigma})$. Using $\Gamma^{\nu}_{\nu\sigma} = \frac{1}{2}\partial_{\sigma}\ln g$, with $g = \mid \det(g_{\mu\nu}) \mid$, we get $w_{\sigma} = \partial_{\sigma}(\ln(g^*/g))/10$.

If we take geodesics as the only observable objects, then it is natural to seek transformations that leave them invariant. Such transformations are given by Eq.(74). However, a transformation of the metric tensor that fulfils Eq.(74) does not leave invariant Einstein's field equations:

$$R_{\mu\nu} - \frac{1}{2}g_{\mu\nu} = \varkappa T_{\mu\nu}, \tag{76}$$

where, we recall, $R_{\mu\nu} = R^{\sigma}_{\mu\nu\sigma}$ is the Ricci tensor stemming from the Riemann tensor $R^{\lambda}_{\mu\nu\sigma}$ by contraction of λ and σ, and $T_{\mu\nu}$ means the energy-momentum tensor.

If our transformations do not leave Eq.(76) invariant but we insist in viewing geodesic invariance as a fundamental requirement, then we are led to ask for alternative equations for the gravitational field. These equations should be invariant under Eq.(74). Weyl found a tensor that is invariant under Eq.(74), i.e., a candidate for replacing $R^{\lambda}_{\mu\nu\sigma}$ as the starting point of the sought-after equations. It is given by

$$W^{\lambda}_{\mu\nu\sigma} = R^{\lambda}_{\mu\nu\sigma} - \frac{1}{4}\left(\delta^{\lambda}_{\sigma}R_{\mu\nu} - \delta^{\lambda}_{\nu}R_{\mu\sigma}\right). \tag{77}$$

Unfortunately, any contraction of $W^{\lambda}_{\mu\nu\sigma}$ vanishes identically, thereby precluding an alternative setting of equations analogous to those of Einstein.

One could argue that it remains still open the possibility of changing our very starting point, so that we should look for a Lagrangian which does not depend on a metric tensor. A natural candidate for this would be an affine connection (the Christoffel symbols being a special case). However, we can show that, even if we start from very general assumptions, we will end up with a Lagrangian like that of Eq. (63). That is, if we take our variational principle in the general form $\delta \int L(x,v)d\tau = 0$, and require that L is invariant under local Lorentz transformations, then L must be of the form $(g_{\mu\nu}(x)v^{\mu}v^{\nu})^{1/2}$. The requirement of invariance under local Lorentz transformations follows from the principle of equivalence: at any given point we can choose our coordinate system so that a body subjected only to gravity appears to move freely in a small neighborhood of the given point. This requirement leads to the particular form of L just given, as can be seen as follows [24]: From the homogeneity of L with respect to v it follows that we can write L in the form $L = (g_{\mu\nu}(x,v)v^{\mu}v^{\nu})^{1/2}$, with $g_{\mu\nu}(x,v) := \frac{1}{2}\partial^2 L^2(x,v)/\partial v^{\mu}\partial v^{\nu}$. This puts our variational problem within the framework of Finsler spaces [25]. But local Lorentz invariance implies that $g_{\mu\nu}$ is independent of v, as we shall see, so that we end up within the framework of Riemann spaces, a special case of Finsler spaces.

A transformation in the tangent space, $v \to w$, defined through $w^\mu = \overset{\sim\mu}{\Lambda}_\nu v^\nu$, is a local Lorentz transformation if it satisfies $g_{\mu\nu}(x,v) = g_{\lambda\sigma}(x,v)\Lambda^\lambda_\nu(x)\Lambda^\sigma_\nu(x)$ at any *fixed* point x. Here, Λ^μ_ν means the inverse of $\overset{\sim\mu}{\Lambda}_\nu$. Invariance of L under local Lorentz transformations means that $L(x^\mu, w^\mu) = L(x^\mu, \overset{\sim\mu}{\Lambda}_\nu v^\nu) = L(x^\mu, v^\mu)$. From this equality, by taking partial derivatives with respect to v, we obtain the two following equations:

$$\frac{\partial L(x,w)}{\partial v^\mu}\Lambda^\mu_\nu = \frac{\partial L(x,v)}{\partial v^\nu} \tag{78}$$

$$\frac{\partial^2 L(x,w)}{\partial v^\mu \partial v^\nu}\Lambda^\mu_\sigma \Lambda^\nu_\tau = \frac{\partial^2 L(x,v)}{\partial v^\sigma \partial v^\tau} \tag{79}$$

When these equations are substituted into the identity

$$g_{\mu\nu}(x,v) = \frac{1}{2}\frac{\partial^2 L^2}{\partial v^\mu \partial v^\nu} = \frac{\partial L}{\partial v^\mu}\frac{\partial L}{\partial v^\nu} + L\frac{\partial^2 L}{\partial v^\mu \partial v^\nu}, \tag{80}$$

one obtains

$$g_{\mu\nu}(x,w) = g_{\lambda\sigma}(x,v)\Lambda^\lambda_\mu(x)\Lambda^\sigma_\nu(x). \tag{81}$$

We conclude therefore, in view of this last equation and the definition of the Lorentz transformation given above, that the equality $g_{\mu\nu}(x,w) = g_{\mu\nu}(x,v)$ holds true for any w and v that are connected to each other by a Lorentz transformation. Thus, setting $w = v + \delta v$, we obtain

$$\frac{\partial g_{\mu\nu}(x,v)}{\partial v^\lambda} = \lim_{\delta v \to 0}\left(\frac{g_{\mu\nu}(x,v+\delta v) - g_{\mu\nu}(x,v)}{\delta v^\lambda}\right) = 0. \tag{82}$$

Thus, L must be of the form $(g_{\mu\nu}(x)v^\mu v^\nu)^{1/2}$. As we have seen, this result follows from the requirement of local Lorentz invariance. Such an assumption is the counterpart of the condition put by Helmholtz on a general metric space, in order to geometrically characterize Riemann spaces [23]. In this last case, local rotations played the role that is assigned to local Lorentz transformations in the physical case.

6. Summary and conclusions

Carathéodory's approach to the calculus of variations appears to be an appropriate tool for uncovering some aspects of the quantum-classical relationship. Because it describes a whole field of extremals rather than a single one, Carathéodory's approach is, by its very nature, more akin to the quantal formulation. It remains still open how to introduce in this framework the second basic element of the quantal formulation, namely probability. By blending field and probability issues, it is likely that the ensuing result shed some light on questions concerning the quantum-classical correspondence. Here, by way of illustration of the capabilities of Carathéodory's approach, we have dealt with the two fundamental interactions of classical physics: electromagnetism and gravitation. We have seen that the London equations of superconductivity can be formally derived from the standard Lagrangian of a particle interacting with a prescribed electromagnetic field. The London equations have therefore not a distinctive quantum-mechanical origin, as it is often assumed. This does not mean, however, that we can explain superconductivity by recourse to classical physics alone. The conditions under which a system of charged particles behaves as described by the standard, classical Lagrangian, might be explainable only through quantum mechanics.

In the gravitational case, we recovered Weyl's results about the invariance of geodesics under some special transformation of the Christoffel symbols. Carathéodory's fundamental equations led us to formulate Weyl's result without having to resort to the tools of differential geometry. Furthermore, we have seen that the Lagrangian $L = (g_{\mu\nu}(x)v^{\mu}v^{\nu})^{1/2}$ is a direct consequence of the assumption of local Lorentz invariance. The underlying principle that led us to state the appropriate questions was the principle of gauge invariance, something usually tied to a quantal approach.

In summary, Carathéodory's approach to variational calculus represents an alternative way to introduce some of the most basic principles of classical physics. It unifies different aspects that otherwise appear to be independent from one another, and it can help us in our quest for delimiting the quantum-classical correspondence.

Acknowledgements

Partial financial support from DGI-PUCP and from the Science Department (PUCP) is gratefully acknowledged.

Author details

Francisco De Zela

Departamento de Ciencias, Sección Física, Pontificia Universidad Católica del Perú, Lima, Peru

References

[1] C. Carathéodory, *Calculus of Variations and Partial Differential Equations of the First Order. Part II: Calculus of Variations*, Holden-Day, San Francisco, 1967.

[2] O. Redlich, "Fundamental Thermodynamics since Carathéodory", *Rev. Mod. Phys.* vol. 40 pp. 556-563, 1968.

[3] P. R. Holland, *The quantum theory of motion: an account of the de Broglie-Bohm causal interpretation of quantum mechanics*, Cambridge University Press, Cambridge 1993.

[4] D. Bohm, "A suggested interpretation of the quantum theory in terms of 'hidden' variables", *Phys. Rev.* vol. 85, pp. 166-179, 1952.

[5] W. Farrell Edwards, "Classical Derivation of the London Equations", *Phys. Rev. Lett.* vol. 47, pp. 1863-1866, 1981.

[6] F. S. Henyey, "Distinction between a Perfect Conductor and a Superconductor", *Phys. Rev. Lett.* vol. 49, pp. 416, 1982.

[7] B. Segall, L. L. Foldy, and R. W. Brown, "Comment on 'Classical Derivation of the London Equations"', *Phys. Rev. Lett.* vol. 49, p. 417, 1982.

[8] J. B. Taylor, "A classical derivation of the Meissner effect?", *Nature* vol. 299, pp. 681-682, 1982.

[9] S. M. Mahajan, "Classical Perfect Diamagnetism: Expulsion of Current from the Plasma Interior", *Phys. Rev. Lett.*, vol. 100, pp. 075001-1-075001-4, 2008.

[10] R. Balian, *From Microphysics to Macrophysics*, Vol. I, Springer-Verlag, Berlin, Heidelberg, New York, 1991.

[11] H. Essén and M. C. N. Fiolhais, "Meissner effeect, diamagnetism, and classical physics – a review", *Am. J. Phys.*, vol. 80, pp. 164-169, 2011.

[12] H. Goldstein, *Classical Mechanics*, 2nd. Ed., Addison-Wesley, Reading, Massachusetts, 1980.

[13] H. Rund, *The Hamilton-Jacobi Theory in the Calculus of Variations*, D.van Nostrand Comp., London, 1966.

[14] D. Lovelock and H. Rund, *Tensors, Differential Forms, and Variational Principles* Wiley, New York, 1975.

[15] G. A. Bliss, *Lectures on the Calculus of Variations*, University of Chicago Press, Chicago, 1946.

[16] F. London and H. London, "The Electromagnetic Equations of the Supraconductor", *Proc. Roy. Soc.(London)* vol. A149, pp. 71-88, 1935.

[17] J. Bardeen, L. N. Cooper, J. R. Schrieffer, "Theory of Superconductivity", *Phys. Rev.* vol. 108, pp. 1175-1204, 1957.

[18] J. D. Jackson, *Classical Electrodynamics*, 3rd edition, Wiley, New York, 1999.

[19] L. D. Landau, E. M. Lifshitz, *The Classical Theory of Fields*, Pergamon Press, Oxford, 1962.

[20] F. Halzen and A. D. Martin, *Quarks and Leptons: An Introductory Course in Modern Particle Physics*, Wiley, New York, 1984.

[21] R. Utiyama, "Invariant Theoretical Interpretation of Interaction", *Phys. Rev.* vol. 101, pp. 1597-1607, 1956.

[22] T. W. B. Kibble, "Lorentz Invariance and the Gravitational Field", *J. Math. Phys.* vol. 2, pp. 212-221, 1961.

[23] D. Laugwitz, *Differential and Riemannian Geometry* , Academic Press, New York, 1965.

[24] F. De Zela, "Über mögliche Grenzen einer Gravitationstheorie", *Ann. der Physik*, vol. 48, pp. 269-282, 1991.

[25] H. Rund, *The differential geometry of Finsler spaces*, Springer, Berlin, 1959.

Chapter 3

Classical and Quantum Correspondence in Anisotropic Kepler Problem

Keita Sumiya, Hisakazu Uchiyama,
Kazuhiro Kubo and Tokuzo Shimada

Additional information is available at the end of the chapter

1. Introduction

If the classical behavior of a given quantum system is chaotic, how is it reflected in the quantum properties of the system? To elucidate this correspondence is the main theme of the quantum chaos study. With the advent of nanophysics techniques, this has become also of experimental importance. With the advent of new technology, various quantum systems are now challenging us. These include nano-scale devices, laser trapping of atoms, the Bose-Einstein condensate, Rydberg atoms, and even web of chaos is observed in superlattices.

In this note we devote ourselves to the investigation of the quantum scars which occurs in the Anisotropic Kepler Problem (AKP) – the classical and quantum physics of an electron trapped around a proton in semiconductors. The merit of AKP is that its chaotic property can be controlled by changing the anisotropy from integrable Kepler limit down ergodic limit where the tori are completely collapsed and isolated unstable periodic orbits occupy the classical phase space. Thus in AKP we are able to investigate the classical quantum correspondence at varying chaoticity. Furthermore each unstable periodic orbit (PO) can be coded in a Bernoulli code which is a large merit in the formulation of quantum chaos in term of the periodic orbit theory (POT) [1, 2, 6].

The AKP is an old home ground of the quantum chaos study. Its low energy levels were used as a test of the periodic orbit theory in the seminal work of Gutzwiller [3-7]. Then an efficient matrix diagonalization scheme was devised by Wintgen et al. (WMB method) [8]. With this method, the statistics of up to nearly 8000 AKP quantum levels were examined and it was found that the quantum level statistics of AKP change from Poisson to Wigner distribution with the increase of mass anisotropy [9]. Furthermore, an intriguing classical Poincaré surface of section (POS) was found at medium anisotropy ($\gamma \equiv m(\text{light})/m(\text{heavy}) = 0.8$), which indicates remnants of tori (cantori) in the classical phase space [9]. Thus, over two decades from the early 70th, AKP was a good testing ground of theories (along with billiards) as well as a constant source of important information to quantum chaos studies. However, there has not been much recent theory investigation on AKP. Especially, to our knowledge, the quantum scar of the classical periodic orbits in AKP has not been directly examined, even though intriguing phenomena was discovered by Heller [10] in 1984. On the other hand, for an analogous system – the hydrogen under a magnetic field (diamagnetic Kepler problem (DKP)), the scars of periodic orbits were extensively studied using highly efficient tool called as scar strength functions [11]. We note that AKP is by far simpler; for DKP it is necessary to code the POs by a sequence of symbols consisting of three letters.

Recently the level statistics of AKP was examined from the random matrix theory view [12]. It was considered that the AKP level statistic in the transitive region from Poisson to Wigner distribution correspond to the critical level statistics of an extended GOE random matrix theory and it was conjectured that the wave functions should exhibit characteristic multifractality. This aspect has been further developed in [13, 14]; it is considered that *Anderson transition* occurs in the quantum physics of a class of physical systems such as AKP and periodically driven kicked rotator in their critical parameter regions. Further very recently a well devised new solid state experiment has been conducted for AKP and ADKP [15, 16]. We also refer [17] for a recent overview including this interesting conjecture.

Such is the case we have recently conducted AKP high accuracy matrix diagonalization based on the WMB method. This is not a perturbation calculation; the anisotropy term is not regarded as a perturbation and the full Hamiltonian matrix is diagonalized. Thus the approximation comes only from the size of the matrix. But, as a trade-off, a scaling parameter is unavoidably included; it is crucial to choose a correct parameter value at every anisotropy parameter. We have derived a simple rule of thumb to choose a suitable value [17]. After comparing with original WMB result in Sturmian basis, we have also worked with tensored-harmonic-wavefunction basis (THWFB) [11], which is more suitable for the Husimi function calculation to investigate the quantum scars. Our contribution here is the calculation of anisotropy term in the AKP Hamiltonian in THWFB [17], which is harder than the diamagnetic case. Comparing the results from two independent bases we have verified that both results agree completely thus the choices of scaling parameters (in both bases) are validated.

Aimed by these numerical data, we report in section 2 salient evidences of quantum scars in AKP for the first time. We compare the features of various known observables; thus this section will serve as a comparative test of methods and fulfills the gap in the literature pointed out above. Most interesting is the test using the scar strength function. We show that even in the ergodic regime ($\gamma = 0.2$), we can quantitatively observe that prominent periodic orbits systematically contribute to the quantum theory endowed with random energy spectrum.

In section 3 we investigate that how the scaring phenomena are affected by the variation of the anisotropy parameter. It is well known that the energy levels show successive avoiding crossings. On the other hand, in the periodic orbit formula, each term in the series for the density of states (DOS) consists of a contribution of an unstable PO with a pole (with an imaginary part given by the Lyapunov exponent of the PO) at the Bohr-Sommerfeld-type energy; thus each term smoothly varies with the anisotropy. We show that how these two seemingly contradicting features intriguingly compromise. The localization patterns in the wave functions or Husimi functions are swapped between two eigenstates of energy at every avoiding crossing. Repeating successively this swap process characteristic scarring patterns follow the POs responsible to them. In this sense the quantum scarring phenomena are robust. We conclude in section 4.

2. Manifestation of Scars in AKP

We first explain how we have prepared the energy levels and wave functions. Then we introduce the indispensable ingredients to study the scars in AKP. After briefly explaining Husimi functions, we explain periodic orbit theory. The quantum scars will be observed along the classical unstable periodic orbits.

2.1. Matrix diagonalization

2.1.1. AKP Hamiltonian

The AKP Hamiltonian in the dimensionless form is given by

$$H_G = \frac{1}{2\mu}p_x^2 + \frac{1}{2\nu}(p_y^2 + p_z^2) - \frac{1}{r} \tag{1}$$

where $r = \sqrt{x^2+y^2+z^2}$ and $\mu > \nu$[1,4] with which POT was formulated in the history, or equivalently it may be also written as [9,11] (Harmonic basis)

$$H_W^{88} = \frac{1}{2}(p_x^2 + p_y^2) + \frac{\gamma}{2}p_z^2 - \frac{1}{r} \tag{2}$$

with $\gamma = \nu/\mu = 1/\mu^2$ or

$$H_W^{87} = p_x^2 + p_y^2 + \gamma p_z^2 - \frac{2}{r} \tag{3}$$

as used in WMB (Sturmian basis) [8]. We recapitulate POT predictions in terms of (1), our formula for AKP eigenvalue calculation in tensored harmonic wave function basis in terms of (2), and we discuss quantum scars using energy values in (3) in order to facilitate comparison with literature.

2.1.2. Matrix diagonalization in Sturmian basis

We here summarize WMB method for efficient matrix diagonalization.

Firstly, in the Sturmian basis

$$\langle\vec{r}|n\ell m\rangle = \frac{1}{r}\sqrt{\frac{n!}{(2\ell+n+1)!}}e^{-\frac{\lambda r}{2}}(\lambda r)^{\ell+1}L_n^{2\ell+1}(\lambda r)Y_{\ell m}(\theta,\varphi) \tag{4}$$

with a scaling parameter λ, the Schrödinger equation of the AKP becomes a matrix equation:

$$\left[-\lambda\overleftrightarrow{\Delta^{(3)}} + (1-\gamma)\lambda\overleftrightarrow{\frac{\partial^2}{\partial z^2}} - 2\overleftrightarrow{\frac{1}{r}}\right]\mathbf{\Psi} = \frac{E}{\lambda}\overleftrightarrow{\mathrm{Id}}\mathbf{\Psi} \tag{5}$$

Dividing the whole equation by λ and packing E/λ^2 into a parameter ε, one obtains

$$\overleftrightarrow{M}\mathbf{\Psi} \equiv \left[-\overleftrightarrow{\Delta^{(3)}} + (1-\gamma)\overleftrightarrow{\frac{\partial^2}{\partial z^2}} - \epsilon\overleftrightarrow{\mathrm{Id}}\right]\mathbf{\Psi} = \frac{2}{\lambda}\mathbf{\Psi}. \tag{6}$$

This ϵ is to be fixed at some constant value. In principle any value will do, but for finite size of Hamiltonian matrix, the best choice is given [17] approximately

$$\varepsilon^* \simeq -\frac{1}{4}\gamma. \tag{7}$$

With this choice, we can get the largest number of reliable energy levels at a given matrix size. The ratio of reliable levels to the matrix size can be estimated as

$$R_{eff} \simeq \sqrt{\gamma}. \tag{8}$$

After fixing ε, the diagonalization of (6) is performed for $2/\lambda_i$ s and finally we obtain the energy eigenvalues by

$$E_i = \varepsilon\lambda_i^2. \tag{9}$$

2.1.3. Matrix diagonalization in Sturmian basis

For the (tensored) harmonic wave function basis (THWFB) [11] we convert the Hamiltonian of AKP into the Hamiltonian of two of two-dimensional harmonic oscillators.

For this purpose semi-parabolic coordinates are introduced

$$\mu\nu = \rho = \sqrt{x^2+y^2},\ \ \frac{1}{2}(\mu^2-\nu^2) = z,\ \ \phi = \tan^{-1}\left(\frac{y}{x}\right) \tag{10}$$

and the AKP Schrödinger equation becomes

$$\left[-\frac{1}{2\left(\mu^2+\nu^2\right)}\left(\Delta_\mu^{(2)}+\Delta_\nu^{(2)}\right)+\frac{1-\gamma}{2}\frac{\partial^2}{\partial z^2}-\frac{2}{\mu^2+\nu^2}\right]|\Psi\rangle=E\,|\Psi\rangle. \tag{11}$$

Multiplying by $\mu^2+\nu^2$and swapping the Coulombic interaction term and the E term one obtains

$$\left[-\frac{1}{2}\left(\Delta_\mu^{(2)}+\Delta_\nu^{(2)}\right)+|E|\left(\mu^2+\nu^2\right)+\frac{1-\gamma}{2}\left(\mu^2+\nu^2\right)\frac{\partial^2}{\partial z^2}\right]|\Psi\rangle=2\,|\Psi\rangle. \tag{12}$$

Thanks to the semi-parabolic coordinates, the Coulombic singularity has removed [19] for $\gamma=1$. Corresponding to the Sturmian basis with a scaling parameter λ in (4), we introduce the harmonic wave function basis

$$\langle\mu,\nu\mid i,j,\kappa\rangle=\frac{\kappa}{\pi}L_i(\kappa\mu^2)L_j(\kappa\nu^2)\exp\left(-\frac{\mu^2+\nu^2}{2}\right)$$

with a scaling parameter κ and, corresponding to ε in (6), we introduce a parameter

$$\tilde{\varepsilon}=2\frac{|E|}{\kappa^2} \tag{13}$$

and we solve (12) after transforming it into the matrix equation of WMB form with eigenvalues $\Lambda_n=2/\kappa_n$. The matrix element calculation of the mass anisotropy term in (12) is somewhat involved and we refer to [17] for detail. Energy levels are then determined by

$$E_n=-\frac{\kappa_n^2}{2}\tilde{\varepsilon}=-\frac{2}{\Lambda_n^2}\tilde{\varepsilon} \tag{14}$$

The best value of $\tilde{\varepsilon}$ is given by

$$\tilde{\varepsilon}^*\approx\gamma \tag{15}$$

which is similar to (7).

We have found precise agreement between our calculations by the Sturmian basis and by the harmonic oscillator basis which in turn validates our choices of scaling parameter ε and $\tilde{\varepsilon}$.

For the calculation of Husimi functions and scar strength function which uses Husimi functions, we use the THWFB since the projection of the basis functions to the Gaussian packets are easy to calculate [17].

2.2. Husimi function

Husimi function is defined via the scalar product of the wave function $|\psi\rangle$ with a coherent state(CHS) $|q_0, p_0\rangle$ of the system [11]:

$$W_{\psi}^{Hus}(q_0, p_0) = |\langle\psi \mid q_0, p_0\rangle|^2_. \tag{16}$$

A detailed account is given in [17].

2.3. Periodic orbit theory

2.3.1. Periodic orbit theory and the density of state

Let us recapitulate Gutzwiller's periodic orbit theory [4,20]. The starting point is Feynman's path integral formula for the propagator of a particle from q' to q'' during the time interval 0 to T;

$$K(q'', q', T) \equiv \langle q''| \exp\left(-i\frac{H}{\hbar}T\right)|q'\rangle = \int_{q'}^{q''} D[q]\, e^{\frac{i}{\hbar}\int_0^T L(q,\dot{q},t)dt} \tag{17}$$

The Green function (response function) is given by the Fourier transformation of the propagator

$$G(q'', q', E) \equiv -\frac{i}{\hbar}\int_0^{\infty} dt e^{\frac{iEt}{\hbar}} K(q'', q', T)$$

where E has infinitesimally small imaginary part for convergence. Thus we have

$$G(q'', q', E) = \langle q''|\frac{1}{E + i\varepsilon - \hat{H}}|q'\rangle = -\frac{i}{\hbar}\int_0^{\infty} dt e^{\frac{iEt}{\hbar}}\left[\int_{q'}^{q''} D[q]\, e^{\frac{i}{\hbar}\int_0^T L(q,\dot{q},t)dt}\right] \tag{18}$$

By a stationary approximation we obtain a semiclassical formula for the Green function

$$\tilde{G}(q'', q', E) \simeq \sum_{\Gamma} A_{\Gamma} \exp\left(\frac{i}{\hbar}S_{\Gamma} - i\frac{\nu_{\Gamma}}{2}\right) \tag{19}$$

where Γ denotes a classically arrowed orbit, ν_{Γ} is the number of conjugate points on the orbit, and the amplitude A_{Γ} accounts for the Van Vleck determinant. Note that the principal function in (18) is changed into the action $S = \int_{q'}^{q''} pdq$ and the phase $i\pi/4$ from the stationary point approximation is shifted into A_{Γ}.

Now the density of states is given by

$$\rho(E) \equiv \sum_n \delta(E - E_n) = -\frac{1}{\pi}\mathrm{Im}\left(T_{r_n}\left(\frac{1}{E + i\varepsilon - \hat{H}}\right)\right) \tag{20}$$

where in the second equality an identity $1/(x + i\varepsilon) = P(1/x) - i\pi\delta(x)$ is used and trace is taken over all energy eigenstates $\{|n\rangle\}$. Trading this tracing with the tracing over the eigenstates of coordinate operator $\{|q\rangle\}$, we obtain a semiclassical approximation for the DOS

$$\rho(E) \approx -\frac{1}{\pi}\mathrm{Im}\int dq'\, \tilde{G}(q'', q', E)\big|_{q''=q'} = -\frac{1}{\pi}\mathrm{Im}\int dq' \sum_{\Gamma} A_{\Gamma}\, \exp\left(\frac{i}{\hbar}S_{\Gamma} - i\frac{\nu_{\Gamma}}{2}\right). \tag{21}$$

The integration over q' can be again approximated by a stationary phase approximation. Because

$$p'' = \frac{\partial S(q'', q')}{\partial q''}, \quad p' = -\frac{\partial S(q'', q')}{\partial q'} \tag{22}$$

the stationary phase condition gives

$$0 = \frac{\partial S(q', q')}{\partial q'} = p'' - p' \tag{23}$$

which dictates the periodic orbits. We obtain finally the periodic orbit theory formula for the DOS

$$\rho(E) \simeq \overline{\rho(E)} + \mathrm{Im}\sum_{r\in PO}\frac{T_r}{\pi\hbar}\sum_{n\neq 0}\frac{\exp\left\{in\left[\frac{S_r}{\hbar} - \frac{\pi}{2}l_r\right]\right\}}{[\det((M_r)^n - 1)]^{\frac{1}{2}}} \tag{24}$$

Here the first sum runs over all primitive POs and the n sum counts the repetitions of each peridic orbit; T_r, S_r, and l_r denote the period, action, and Maslov index of the primary PO, and the matrix M stands for the monodromy matrix of the primary PO.

In AKP $m = 0$ sector, the motion is restricted in a fixed plane which includes the heavy axis, and the problem essentially reduces to two dimensional one. (Later on the three dimensional feature is recovered only by the proper choice of the Maslov index [4]). As for AKP unstable periodic orbits, M has two eigenvalues e^u and e^{-u} (hyperbolic case) and the determinant in (24) is given by

$$\frac{1}{2}\left[\det(M^n - 1)\right]^{\frac{1}{2}} = -i\,\sinh\left(nu/2\right) \tag{25}$$

2.3.2. *Naming of a periodic orbit*

In AKP every PO can be coded by the sign of the heavy axis coordinate when the heavy axis is crossed by it. Note that number of the crossings must be even $(2n_c)$ for the orbit to close.

In this note we shall denote the PO according to Gutzwiller's identification number along with the Bernoulli sequence of POS. (See Table 1 in ref. [3] which gives a complete list[1] of POs up to $n_c = 5$ for the anisotropy $\gamma = 0.2$.) For instance, PO36$(++-++-)$ is the identification number 6 among $n_c = 3$ POs.

2.3.3. *The contribution of a periodic orbit to the density of state*

The contribution of a single periodic orbit r to the DOS is estimated by a resummation of the sum over the repetition j (after the approximation $\sinh x \approx e^x/2$),

$$\rho(E)|_r \approx T_r \sum_m \frac{u_r\hbar/2}{(S_r - 2\pi\hbar(m+l/4))^2 + (u_r\hbar/2)^2} . \tag{26}$$

This gives Lorentzian peak at

$$S_r = 2\pi\hbar(m + l/4) \tag{27}$$

similar to the Bohr-Sommerfeld formula. In AKP the action S is given as

$$S_r(E) = \frac{T_r}{\sqrt{-2E}}. \tag{28}$$

Hence the peak position of the Lorentzian form in the energy is given by

$$E_{r,m} = -\frac{1}{2}\left(\frac{T_r}{2\pi\hbar(m+1)}\right)^2 , \tag{29}$$

where Maslov index $l = 4$ for three dimensions is taken.

We are aware that it is meant by (24) that the exact DOS with sharp delta function peaks on the energy axis corresponds to the sum of all PO contributions [20] (assuming convergence).

It is the collective addition of all POs that gives the dos. But, still, it is amusing to observe that the localization of wave functions occurs around the classical periodic orbits as we will see below.

[1] In [3, 5] an amazing approximation formula that gives a good estimate of the action of each periodic orbit from its symbolic code is presented. The trace formula has a difficulty coming from the proliferation of POs of long length. This approximation gives a nice way of estimating the sum. The table is created to fix the two parameters involved in the approximation. We thank Professor Gutzwiller for informing us of this formula a few years ago.

2.4. Scars as observed in the probability distributions and Husimi functions

2.4.1. The manifestation of the fundamental FPO(+ −)

Let us start exploring the scars in AKP first by investigating the case of the fundamental periodic orbit FPO (+ −) which reduces the Kepler ellipse orbit in the limit $\gamma = 1$.

In Figure 1, we show the wave function squared in the $\mu\nu$ plane and the Husimi distribution in the $\mu\, p_\mu$ plane. The FPO is shown by red line and compared with the probability distributions in the in the $\mu\nu$ plane. At high anisotropy the orbit is largely distorted. Still at chosen energy levels (upper row) we find clear localization around the FPO for both anisotropies. In the lower row we have displayed other energy eigenstates. For these energy levels we see also characteristic probability distribution patterns but not around FPO. Now let us look at the Husimi distributions. In case of energies in the upper diagrams we see very clearly that around the Poincaré section of the FPO (the fixed points) the Husimi functions show clear scars, while in the lower we see anti-scars, the Husimi density is very low at the fixed points. It is clear that Husimi functions are superior observables. In this demonstration of scars we have scanned thousands of energy eigenstates and picked examples. Next task is to use the ability of POT predictions (24) to locate the scaring levels.

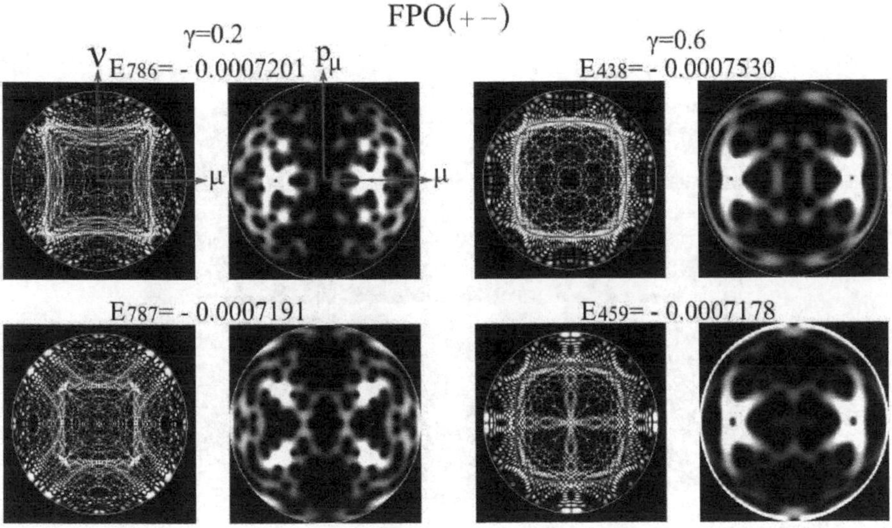

Figure 1. Scar and anti-scar phenomena with respect to the fundamental periodic orbit. The left set is for the anisotropy $\gamma = 0.2$ and right for $\gamma = 0.6$. In each set, the upper and lower row display prominent scar and anti-scar respectively, while the left and right columns exhibit the probability distribution on the $\mu\nu$ plane and Husimi distribution on the $\mu\, p_\mu$ plane ($\nu = 0$). The fundamental orbit is drawn by a red line on the $\mu\nu$ plane and its Poincaré section on the $\mu\, p_\mu$ plane by red points. The respective eigenvalues are $E_{786}, E_{787}, E_{438}, E_{459}$ in the m = 0, ℓ =even sector. Classical kinematical boundaries are shown by yellow circles.

2.4.2. *PO prediction and AKP Scars*

As for the FPO the POT works quite well. Thus for this test we have selected more complicated PO PO22 $(+++-)$ and PO36 $(++-++-)$. These orbits wind around the heavy axis forth and back and presumably correspond to the bounce orbit in the billiard.[2] The top row in Fig.2 shows the prediction from POT – the contribution of the single orbit to the DOS (26). We observe clearly the peak regions of POT prediction contains at least one energy eigenstate which shows the scar of the orbit. On the other hand we have checked that the relevant orbit pattern does not appear in the non-peak region of the POT prediction.

In this analysis the Husimi function again yields unmistakable information on the scaring.

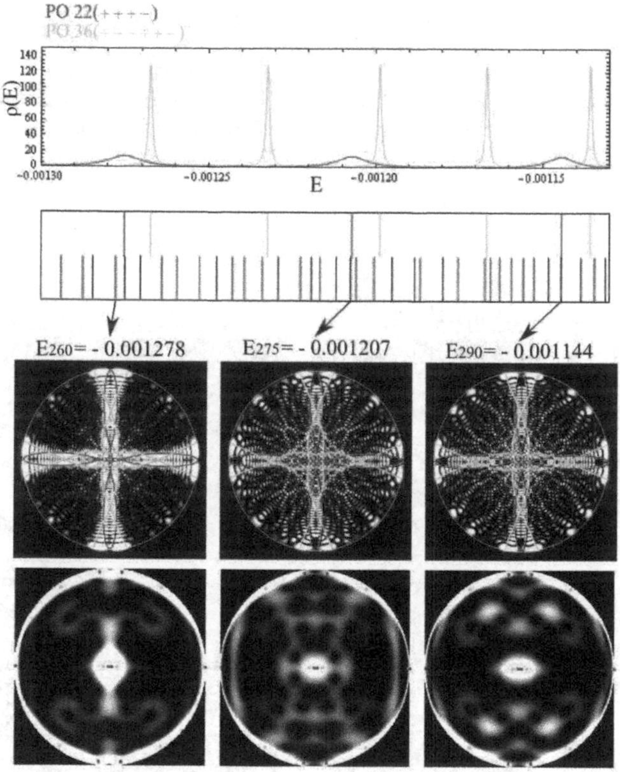

Figure 2. $\gamma = 0.6$, $l =$ even, $m = 0$. The upper diagrams: The red and green curves are contributions to the density of states $\rho(E)$ from bouncing-type periodic orbits PO22 $(+++-)$ and PO36 $(++-++-)$ respectively and the peak positions are compared with the l even $m = 0$ energy levels from matrix diagonalization (WMB with tensored harmonic oscillator basis). The lower: The quantum scars of these periodic orbits are exhibited on the probability distributions and the Husimi functions. (cf. Fig. 1).

[2] We thank Professor Toshiya Takami for explaining his articles [21,22] and pointing us this point.

2.5. Scars as analyzed by the scar strength function

2.5.1. *Scar strength functions*

In an extensive analysis of scars in the diamagnetic hydrogen, a tool called as scar strength function (SSF) is presented [11]. It is defined as

$$I_n^{PO} = \oint_{PO} d^4s\, W_{\psi_n}^{Hus}\left(\mu, \nu, p_\mu, p_\nu\right)\left(\oint_{PO} d^4s\right)^{-1}$$

where the integral is to be performed along the PO with $d^4s = \sqrt{d\mu^2 + d\nu^2 + dp_\mu^2 + dp_\nu^2}$. This quantity is exploiting to what extent a given PO is inducing localization along it in the Husimi function of a given energy eigenstate. Then spectral scar strength function is introduced as

$$I^{PO}(E) = \sum_n I_n^{PO}\delta(E - E_n)$$

This shows how the given PO affects each energy eigenstate in one function.

2.5.2. *The use of SSF $I^{PO}(E)$*

Let explore the region of high anisotropy ($\gamma = 0.2$) where the classical phase space is occupied by the unstable periodic orbits and chaoticity is rather high. We explore this region by the ability of $I^{PO}(E)$.

We start from FPO $(+-)$ in Fig. 3. The upper is the POT prediction curve, the middle is the SSF along with real eigenvalues and the bottom is as usual a direct comparison of FPO with wavefunctions squared as well as Husimi functions. The SSF is the quantum measure of scaring of a particular PO in consideration, while the FPO prediction is composed from purely classical information for the PO. When the curve of the contribution from a PO peaks, the SSF either peaks or reaches its minimum (10^{-10}). The agreement in the energy values of the peaks (or dips) is quite remarkable. But we do not know why anti-scar occurs here. This anti-scar is interesting in that it produces a bright hallow just of the same size and position of the scar but the central core is missing.

2.5.3. *$I^{PO}(E)$ for various POs and their Fourier transform*

Let us now examine the case of several POs simultaneously in Fig. 4. The profiles, the SSF, and the Fourier transforms to the action space are listed in three columns. As Wintgen et al. write as 'the scars are the rules rather than exception' [11] we find that particular energy eigenstates give salient high scar function value while the other states give very low value of order even 10^{-10}. Further more the Fourier transform $I^{PO}(S)$ of $I^{PO}(E)$shows sequential peaks at equal ΔS. We compare in Fig. 5 ΔS_r^{QM} and S_r^{Cl} (the measured spacing of the orbit and the action of the PO). They agree excellently; the POs live in quantum theory.

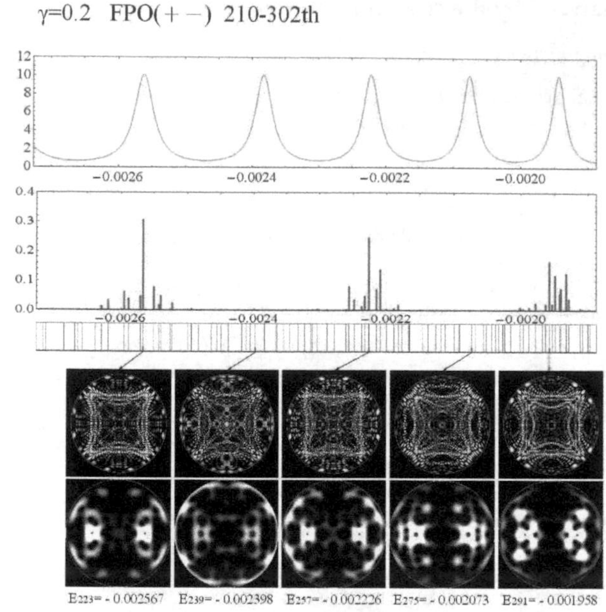

Figure 3. Contribution of FPO $(+-)$ to AKP. Upper two diagrams: The $(+-)$ contribution as a function of energy predicted by POT compared with the scar strength extraction from each of the energy eigenstates. Lower two diagrams: the scaring status of $(+-)$ at levels indicated by arrows are exhibited with respect to wave functions squared and Husimi functions. (cf. Fig. 1 and Fig. 2). Scar and anti-scar appear alternatively.

2.6. Direct phase space observation of Scaring orbit

The scar strength function is a useful tool which gives a list of numerical values which shows succinctly to which eigenstates the periodic orbit exerts its effect strongly. But we certainly want also visualized picture how the PO turns up in the 4 dimensional phase space. (Because $H = \text{const.}$, the actual independent variables are three, and we choose $\mu,\ \nu,\ p_\mu$.) The sample pictures are shown in Fig.6.

3. Robustness of Scaring under the Variation of Anisotropy Parameter

3.1. Swap of the patterns under avoiding crossings

It is well known that the patterns of wave functions (and of Husimi distributions) are swapped between the energy eigenstates via the avoiding level crossings, which is easy to demonstrate in terms a simple coupling model of two levels. Fig. 7 is a typical example of this phenomenon.

It is shown in [21,22] that with the aid of the diabatic transformation one can trace the localization on the transformed basis until very near to the minimum gap with an explicit evidence in the billiard scars. We have verified this issue in AKP. Furthermore it is conjectured that the long periodic orbits may interpolates two shorter orbits and they may be the cause of the avoiding crossings in this way. We are testing this conjecture in AKP.

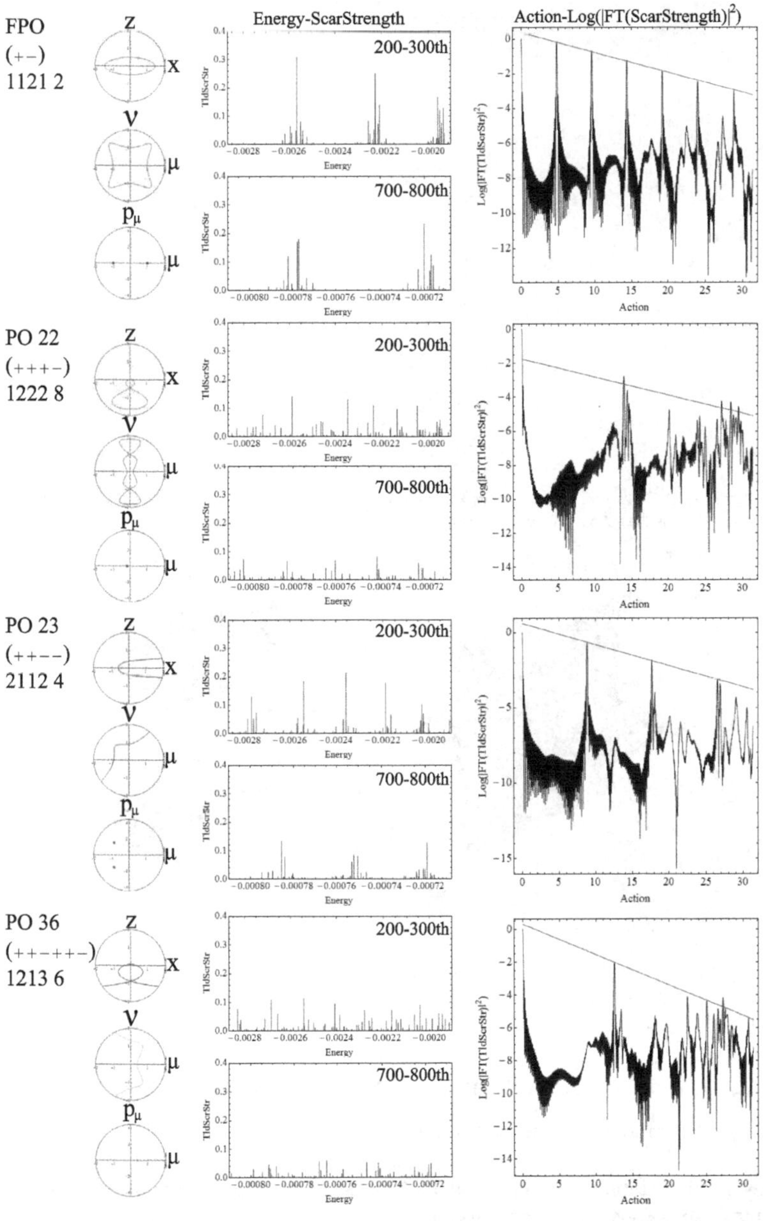

Figure 4. $\gamma = 0.2$ Profiles of PO, the scar strength function, and Fourier transformation of scar strength function to the action space.

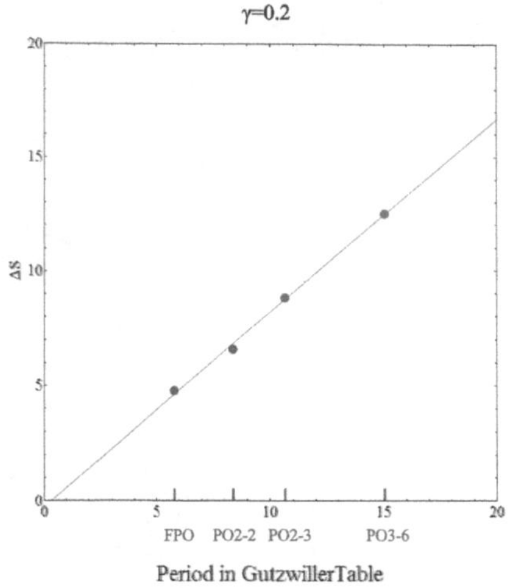

Figure 5. Plot of $(\Delta S_r^{QM}, S_r^{Cl})$ for periodic orbits $r =$ FPO, PO22, PO23, PO36, where ΔS_r^{QM} is measured from the third column of Fig.4 and S_r^{Cl} is the action value of the classical orbit.

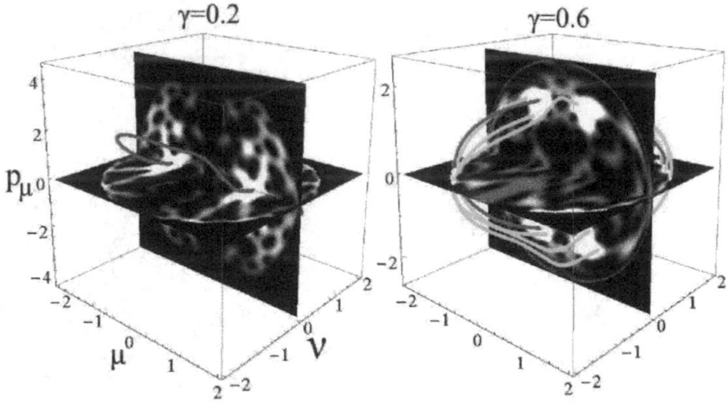

Figure 6. Two samples of density plot of Husimi functions in the 3 dimensional $\mu - \nu - p_\mu$ space. Left: γ =0.2, W_{Hus} for $E_{786} = -0.0007201$. Red orbit is the FPO $(+-)$. Right: γ=0.6, W_{Hus} for $E_{579} = -0.0005681$. Blue and green orbits are respectively PO23 $(++--)$ and PO37 $(++--+-)$.

3.2. Robust association of localization with periodic orbits

We have posed the following question in the introduction of this chapter.

1. Energy levels exhibit randomness at high anisotropy and change *their values randomly* repeating successive avoiding crossings when the anisotropy parameter is varied gradually.

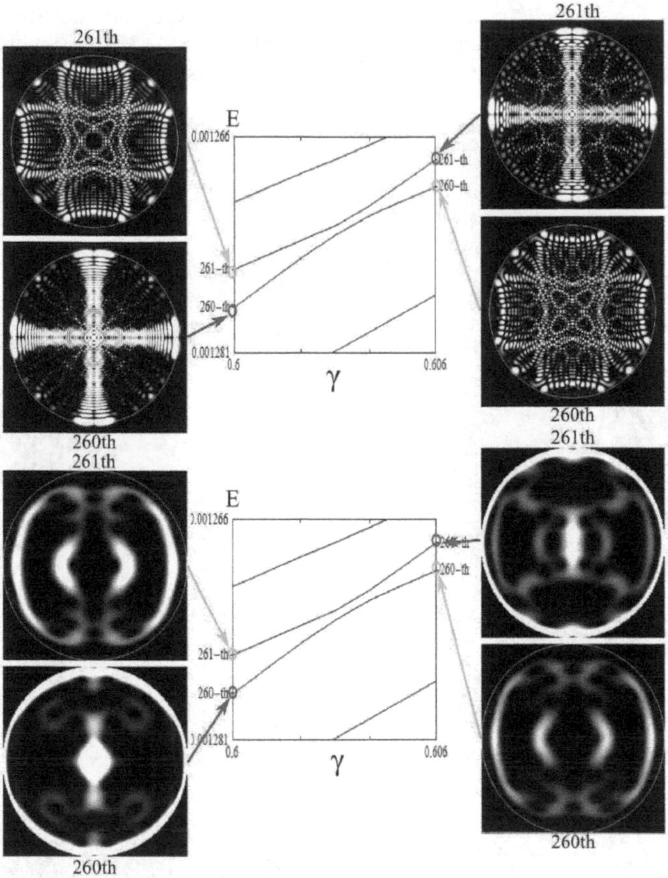

Figure 7. Avoiding level crossing between 260th and 261th energy eigen states under the variation of the anisotropy $\gamma \in [0.6, 0.606]$. Both the wave function squared and Husimi functions are swapped around the avoiding crossing.

2. On the other hand the peak locations (29) of the DOS as predicted by a single PO *change smoothly* with the anisotropy and the scar tends to be observed in the energy eigenstate around the peak position in the DOS as in Fig.2.

Aren't the two issues in contradiction? We have found that they can live together (within approximation of the fluctuation size). Most important point is that the swap of the localization patterns at avoiding crossing is in harmony with the transportation of them by the responsible PO orbits. Besides the POT prediction (29) does not imply the exact location of theappearance of the scar.

It has some allowance as recognized by the width of the *modulation* of SSF[11].

Let us explain this by Fig. 8. Here the anisotropy γ is varied from 0.6 to 0.7 with inclement 0.001. As for (1) we indeed observe both random fluctuation of energy levels as well as many avoided crossings. As for (2), we have picked the bouncing-type periodic orbit PO22 as an example. The predicted peak position (29) of its contribution to DOS varies with the change of γ as shown by a red (almost straight) curve. This PO22 produces a salient *cross-shaped scar* at E_{260} (and E_{275}) at γ=0.6. We have investigated how the cross-shaped scar travels in the spectrum space suffering many avoiding crossings. It reaches at E_{276} (and E_{291}) at γ =0.7

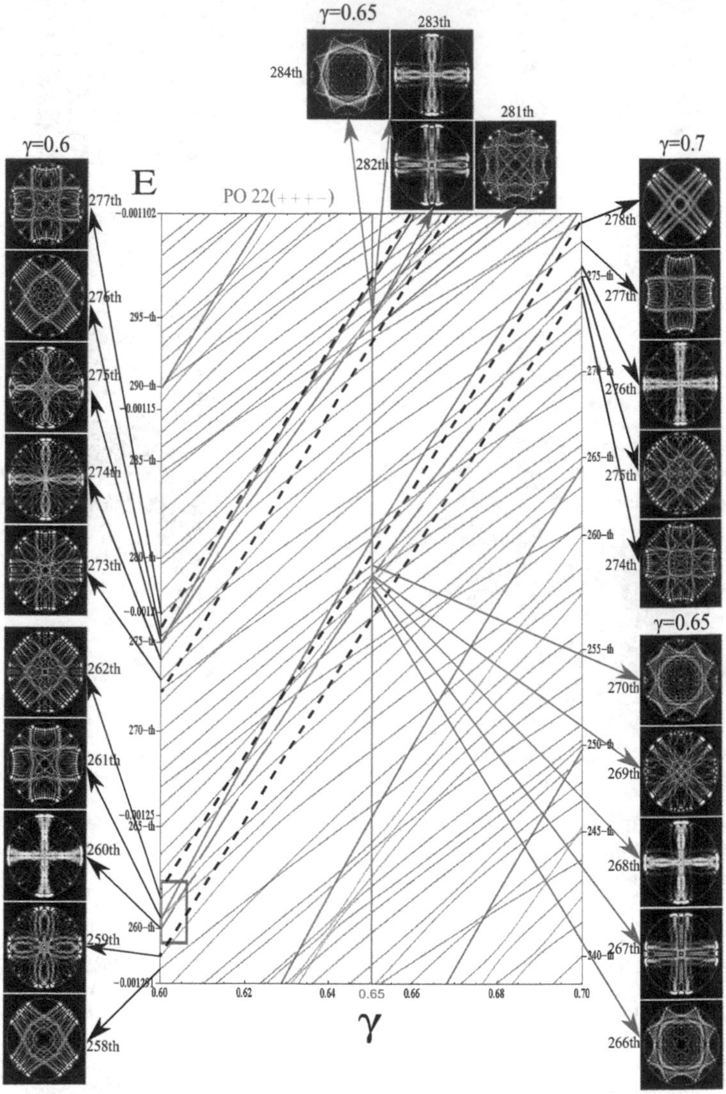

Figure 8. The spectrum lines in the wide interval $\gamma \in [0.6, 0.7]$ investigated with increment 0.001. The cross-shaped scar by PO22 travels within a belt bounded by two dashed lines. The POT prediction (24) is exhibited by a red curve.

and the track in between is enclosed by a belt shown by two dashed lines. We clearly observe that the belt is closely associated by the POT prediction curve. In this sense the association is robust.

4. Conclusion

We have presented ample examples of scaring phenomena for the first time in AKP. Especially we have found how the fluctuation of energy levels and smooth POT prediction for the scaring levels are compromised by using the advantage of AKP endowed by a chaoticity changing parameter.

Although the theme is old, the scaring phenomenon is fascinating and we hope this contribution fulfills a gap in the literature.

Acknowledgement

Both KK and TS thank Professor Toshiya Takami for sharing his wisdom with us.

Author details

Keita Sumiya, Hisakazu Uchiyama,
Kazuhiro Kubo and Tokuzo Shimada

Department of Physics, School of Science and Technology, Meiji University, Japan

References

[1] Gutzwiller, M. C. (1977). Bernoulli sequences and trajectories in the anisotropic Kepler problem, Journal of Mathematical Physics 18: 806-823.

[2] Devaney, R. L. (1979). Collision Orbits in the Anisotropic Kepler Problem, Inventions math. 45: 221-251.

[3] Gutzwiller, M. C. (1981). Periodic orbits in the anisotropic Kepler problem, in Devaney, R. L. & Nitecki, Z. H. (ed.) Classical Mechanics and Dynamical systems, Marcel Dekker, New York, pp. 69-90.

[4] Gutzwiller, M. C. (1971). Periodic orbits and classical quantization conditions, Journal of Mathematical Physics 12: 343-358.

[5] Gutzwiller, M. C. (1980). Classical quantization of a hamiltonian with ergodic behavior, Physical Review Letters 45: 150-153.

[6] Gutzwiller, M. C. (1982). The quantization of a classically ergodic system, Physica D 5: 183-207.

[7] Gutzwiller, M. C. (1990). Chaos in Classical and Quantum Mechanics, Springer.

[8] Wintgen, D., Marxer, H. & Briggs, J. S. (1987). Efficient quantisation scheme for the anisotropic Kepler problem, Journal of Physics A 20: L965-L968.

[9] Wintgen, D. & Marxer, H. (1988). Level statistics of a quantized cantori system, Physical Review Letters 60: 971-974.

[10] Heller, E. J. (1984). Bound-state eigenfunctions of classically chaotic hamiltonian systems: scars of periodic orbits, Physical Review Letters 53: 1515-1518.

[11] Müller, K. & Wintgen, D. (1994). Scars in wavefunctions of the diamagnetic Kepler problem, Journal of Physics B: Atomic, Molecular and Optical Physics 27: 2693-2718.

[12] García-García, A. M. & Verbaarschot, J. J. M. (2003). Critical statistics in quantum chaos and Calogero-Sutherland model at finite temperature, Physical Review E 67: 046104-1 -046104-13.

[13] García-García, A. M. (2007). Universality in quantum chaos and the one parameter scaling theory (a power point of a talk). http://www.tcm.phy.cam.ac.uk/ amg73/oslo2007.ppt.

[14] García-García, A. M. & Wang, J. (2008). Universality in quantum chaos and the one-parameter scaling theory, Physical Review Letters 100: 070603-1 - 070603-4.

[15] Chen, Z. et al. (2009). Realization of Anisotropic Diamagnetic Kepler Problem in a Solid State Environment , Physical Review Letters 102: 244103.

[16] Zhou, W. (2010). Magnetic Field Control of the Quantum Chaotic Dynamics of Hydrogen Analogues in an Anisotropic Crystal Field, Physical Review Letters 105: 024101.

[17] Kubo, K. & Shimada, T. (2011). Theoretical Concepts of Quantum Mechanics, InTech.

[18] Kustaanheimo, P. & Stiefel, E. (1965). Perturbation theory of Kepler motion based on spinor regularization, Journal für die reine und angewandte Mathematik 218: 204-219.

[19] Husimi, K. (1940). Some formal properties of the density matrix, Proceedings of the Physico-Mathematical Society of Japan 22: 264-314.

[20] Wintgen, D. (1988). Semiclassical Path-Integral Quantization of Nonintegrable Hamiltonian Systems, Physical Review Letters 61: 1803-1806.

[21] Takami, T. (1992). Semiclassical Interpretation of Avoided Crossings for Classically Nonintegrable Systems, Physical Review Letters 68: 3371-3374.

[22] Takami, T. (1995). Semiclassical study of avoided crossings, Physical Review E 52: 2434.

Chapter 4

The Improvement of the Heisenberg Uncertainty Principle

L. M. Arévalo Aguilar, C. P. García Quijas and
Carlos Robledo-Sanchez

Additional information is available at the end of the chapter

1. Introduction

One of the fundamental cornerstone of quantum mechanics is the Heisenberg uncertainty principle. This principle is so fundamental to quantum theory that it is believed that if a single phenomenon that could violate it is found then the whole building of quantum mechanics will fall apart. However, since the formulation of the uncertainty principle until today there is not clear and universal agreement in its formulation or interpretation. Even Heisenberg was not clear about the exact meaning of p_1 and x_1 in their first formulation of the uncertainty relations [1]:

$$p_1 q_1 \sim h, \tag{1}$$

nor in the interpretation of the uncertainty principle. According to Heisenberg, in Eq. (1) q_1 represents "*the precision with which the value of q is know (q_1 is, say, the mean error of q), therefore here the wavelength of light. Let p_1 be the precision with which the value of p is determinable; that is, here, the discontinuous change of p in the Compton effect* [1]". He also thought the uncertainty principle in terms of disturbance produced on an observable when it is measured its canonical counterpart.

The relevance of the uncertainty principle to Physics is that it introduced for the first time the indeterminacy in a physical theory, which mean the end of the era of *certainty* in Physics. That is to say, what uncertainty principle made evident was the peculiar characteristic of quantum theory of not being able to predict with certainty a property of a physical system; in words of Heisenberg: "*... canonically conjugate quantities can be determined simultaneously only with a characteristic indeterminacy. This indeterminacy is the real basis for the occurrence of statistical relations in quantum mechanics* [1]".

Since now, you can perceive two different meanings of the Uncertainty Principle in the two quoted paragraphs above. In the first one, the uncertainty comes from a statistical property (according with Heisenberg, the mean error) of quantum theory; in the second meaning the uncertainty is a restriction to simultaneously measure two physical properties.

On the other hand, to elucidate the meaning of the time-energy uncertainty relation [1] $E_1 t_1 \sim h$ is quite difficult, for, contrary to the uncertainty relation given in Eq. (1), it is not possible to deduce it from the postulates of quantum mechanics, i. e. there is not an operator for time. In Heisenberg's paper the meaning of t_1 is the "*time during which the atoms are under the influence of the deflecting field*" and E_1 refers to the accuracy in the energy measurement. Heisenberg concludes that "*a precise determination of energy can only be obtained at the cost of a corresponding uncertainty in the time* [1]".

In this Chapter of the book, we will review the evolution of the Uncertainty Principle since its inception by Heisenberg until their application to measure entanglement. We will review some problems (usually untouched by quantum mechanic's textbooks) that the usual interpretation of the Uncertainty Principle have in terms of standard deviations and its dependence of the wave function. Also, we will review the efforts made to clarify the meaning of the Uncertainty Principle using uncertainty relations.

2. The relation between the Heisenberg Uncertainty Principle and the Uncertainty Relations

The uncertainty principle is one of the fundamental issues in which quantum theory differs from the classical theories, then since its formulation has attracted considerable attention, even from areas normally outside the scientific development. This has lead to create misunderstandings about the content of the principle. Thus, it is important to mention that when we say that there is a lower limit on irreducible uncertainty in the result of a measurement, what we mean is that the uncertainty is not due to experimental errors or to inaccuracies in the laboratory. Instead, the restriction attributed to the uncertainty principle is fundamental and inherent to the theory and is based on theoretical considerations in which it is assumed that all observations are ideal and perfectly accurate.

A reading of the original Heisenber's paper shows that he writes (i. e. believes) in some pharagraps that the indeterminacies comes from the observational procedures. For, in his original paper, Heisenberg stated [1] that the concepts of classical mechanics could be used analogously in quantum mechanics to describe a mechanical system, however, the use of such concepts are affected by an indeterminacy originated *purely* by the observational procedures used to determine *simultaneously* two canonically conjugate variables. This could be contrasted with the called *Statistical Interpretation* where it is tough that the wave function represents and ensemble of identical prepared system and, therefore, the indeterminacy comes form an intrinsic indeterminacy of the physical properties.

Usually, the uncertainty principle is stated in terms of uncertainty relations. One of the first way to obtain this indeterminacy relation is due to Robertson [2]. Here, instead, we use the textbooks approach to deduce the uncertainty relations from the quantum postulates [3, 4]. This approach uses both the Schwarz inequality

$$\langle\phi|\phi\rangle \langle\varphi|\varphi\rangle \geq |\langle\phi|\varphi\rangle|^2, \tag{2}$$

and the following quantum postulates:

- The state of a quantum system is represented by a wave function $\Psi(x,t)$ ($|\Psi\rangle$, in Dirac notation).
- For every observable A there is a self-adjoint operator $\hat{A}$, its expectation value is given by $\langle\hat{A}\rangle = \int \Psi^*(x)\hat{A}\Psi(x)dx = \langle\Psi|\hat{A}|\Psi\rangle$.

Now, consider the following operators defined as [1]:

$$\begin{aligned} \Delta\hat{A} &= \hat{A} - \langle\hat{A}\rangle \\ \Delta\hat{B} &= \hat{B} - \langle\hat{B}\rangle . \end{aligned} \tag{3}$$

Let them operate on an state $|\Psi\rangle$, given:

$$\begin{aligned} \Delta\hat{A}\,|\Psi\rangle &= |\psi_a\rangle \\ \Delta\hat{B}\,|\Psi\rangle &= |\psi_b\rangle . \end{aligned} \tag{4}$$

Therefore, using the Schwarz inequality given in the Eq. (2),

$$\langle\psi_a|\psi_a\rangle\,\langle\psi_b|\psi_b\rangle \geq |\langle\psi_a|\psi_b\rangle|^2 \tag{5}$$

we arrive to:

$$\left\langle\Delta\hat{A}^2\right\rangle\left\langle\Delta\hat{B}^2\right\rangle \geq |\langle\Delta\hat{A}\Delta\hat{B}\rangle|^2, \tag{6}$$

where $\langle\Delta\hat{A}^2\rangle = \langle\hat{A}^2\rangle - \langle\hat{A}\rangle^2 = \delta A^2$ is the variance, the same for the operator $\hat{B}$. From the Eq. (6), it is not difficult to show that:

$$\delta A\delta B \geq \sqrt{|\langle[\hat{A},\hat{B}]\rangle|^2 + |\langle\{\hat{A},\hat{B}\}\rangle|^2}, \tag{7}$$

where $\{\hat{A},\hat{B}\} = \hat{A}\hat{B} + \hat{B}\hat{A}$, and δA and δB are the standard deviation. It is worth to notice that the association of the standard deviation whit the uncertainty relations was not proposed by Heisenberg, it was Kennard and Robertson [2] who made this association. Although Heisenberg endorse it later. As it was stated above, Heisenberg associates p_1 and q_1 with the mean error, also in the same paper he associates these quantities with the widths of Gaussian functions representing the quantum states of the system.

Some problems arises with the textbooks uncertainty relations: *i)* They are given in terms of the standard deviation, *ii)* They depend on the state of the system. Additionally, *iii)* They

[1] There are others forms to obtain the uncertainty relations, this begin by defining an operator as $\hat{D} = \Delta\hat{A} + \lambda\Delta\hat{B}$ and, then, requiring that $\langle\hat{D}^{\dagger}\hat{D}\rangle \geq 0$.

does not represent the meaning of the impossibilities of simultaneous measurement of two observables, *iv)* They does not quantify the role of the disturbance in the state after the measurement process. Finally, *v)* They does not address the concept of complementarity. There have been proposed some criteria to solve this problems, we are going to review this proposals in the next sections.

3. Reformulations to the uncertainty principle

In this section we will review some proposed solutions to the problems stated in the last paragraph of the previous section.

3.1. The dependence on the standard deviation

The principal criticism to the dependence of the uncertainty relation on the standard deviation comes from J. Hilgevoord and J. M. B. Uffink [5, 6]. Their argument is based on two reason, first, they argue that the standard deviation is an appropriate measure of the error of a measurement because errors usually follow a Gaussian distribution, and the standard deviation is an appropriate measurement of the spread of a Gaussian; however, this is not true for a general distribution. Secondly, they gave as a principal counter argument the fact that even for simple phenomenon as the single slit the standard deviation of momentum diverges. Their approach is inside the thinking that the uncertainty relations are the measure of the spread of the probability distribution, i. e. it is believed that Δx and Δp represents the probability distribution of the possibles properties of the system. In short, it represents the spread of values (of $\hat{x}$ or $\hat{p}$) that are intrinsic in the physical system that are available to appear after a measurement.

The principal counter argument with regard to the standard deviations comes from the single-slit experiment. In this case, it is supposed that the state of an income beam of particles is represented by plane waves. This plane wave represents a particle of precise momentum p_0. Then, the plane wave arrives at the single-slit and is diffracted by it. Therefore, the wave function at the screen, according to Hilgevoord and Uffink, is:

$$\psi(x) = \begin{cases} (2a)^{-1/2}, & \text{if } |x| \leq a; \\ 0, & \text{if } |x| < a. \end{cases}$$

and

$$\phi(p) = (a/\pi)^{1/2} \frac{\sin ap}{ap}. \tag{8}$$

Now, the problem with the standard deviation, as defined in quantum mechanics, in this case is that it diverges:

$$\Delta p = \left\langle \hat{p}^2 \right\rangle - \langle \hat{p} \rangle^2 \rightarrow \infty. \tag{9}$$

Therefore, these authors defined, instead of the standard deviation, the overall width (W_ψ) and the mean peak width of ψ as the smaller W and w that satisfies the following equations [7]:

$$\int_{x_0-W/2}^{x_0+W/2} |\psi(x)|^2\, dx = N$$
$$\left| \int \psi^*(x')\psi(x'-w)dx' \right|^2 = M^2 \tag{10}$$

These quantities, i. e. W and w, provides a better characterization of the spread of the possible values of $\hat{x}$ and $\hat{p}$, in particular there is not any divergence in these numbers. Based in these definitions Hilgevoord and Uffink give the following uncertainty relations, that they propose as a substitute to the uncertainty relation given by Kennard ($\Delta x \Delta p \geq 1/2$), [7]:

$$w_\phi W_\psi \geq \arccos\left(\frac{M+1-N}{N}\right)$$
$$w_\psi W_\phi \geq \arccos\left(\frac{M+1-N}{N}\right) \tag{11}$$

these uncertainty relations works well for the single-slit and double-slit experiments.

3.2. Entropic Uncertainty Relations

In the quantum literature, there are many defined Entropic Uncertainty Relations. Mostly, they are based in terms of Shannon entropy [8, 9], although in last ten years there has been extension to other forms of entropy, like Renyi entropy [10]. In reference [11] there is a recent review of this research area.

One of the important result in this area was the one found by Deutsch [8]. What Deutsch pursuit was a quantitative expression of the Heisenberg uncertainty principle, he notice that the customary generalization has the drawback that the lower limit depends on the quantum state, that is:

$$\Delta A \Delta B \geq \frac{1}{4} \left|\langle[\hat{A}, \hat{B}]\rangle\right|^2. \tag{12}$$

Deutsch stress that the right hand side of the Equ. (12) does not has a lower bound but is a function of the state $|\psi\rangle$, even it vanishes for some choices of $|\psi\rangle$. So, in search of a quantity that could represent the uncertainty principle Deutsch propose some elementary properties, like for example that the lower limit must vanishes if the observables have an eigenstate in common. Based in this considerations he proposed the following entropic uncertainty relation:

$$S_{\hat{A}} + S_{\hat{B}} \geq 2Ln\left(\frac{2}{1+sup\{|\langle a|b\rangle|\}}\right), \tag{13}$$

where $S_{\hat{A}} = -\sum_a |\langle a|\psi\rangle|^2 Ln\,|\langle a|\psi\rangle|^2$ and $S_{\hat{B}} = -\sum_b |\langle b|\psi\rangle|^2 Ln\,|\langle b|\psi\rangle|^2$ are the Shanon entropy, and $|a\rangle$ and $|b\rangle$ are, respectively, the eigenstates of $\hat{A}$ and $\hat{B}$.

The next step in this line of research, was quite soon given by Hossein Partovi [12], he points out that the above uncertainty relation does not take into account the measurement process. Then, considering that the measuring device realizes a partitioning of the spectrum of the observable and the assignation of their corresponding probabilities, he proposes the following definition of entropy [12]:

$$S_A = -\sum_i p_i ln\left\{p_i\right\}. \tag{14}$$

where $p_i = \langle\psi\left|\hat{\pi}_i^A\right|\psi\rangle / \langle\psi|\psi\rangle$ and $\hat{\pi}_i^A$ is the projection onto the subspaces spanned by the states corresponding to the partition induced by the measuring apparatus [12]. In this case, p_i gives the probability of obtaining the outcome of a measurement in a subset of the partition realized by the measuring apparatus. In this approach, the whole spectrum correspond to the observable $\hat{A}$ but its partitioning correspond to the measuring device. Using these considerations Hossein Patrovi proses the following lower bound for the uncertainty relation:

$$S_{\hat{A}} + S_{\hat{B}} \geq 2Ln\left(\frac{2}{1+sup_{ij}\{||\hat{\pi}_i^A + \hat{\pi}_j^B||\}}\right). \tag{15}$$

In the special case where the partition realized by the measuring device includes only one point of the spectrum of $\hat{A}$, i. e. $\hat{\pi}_i^A = |a_i\rangle\langle a_i|$ and $\hat{B}$, i. e. $\hat{\pi}_j^B = \left|b_j\right\rangle\left\langle b_j\right|$, then Equ. (15) reduces to Equ. (13). Finally, it is worth to mention that the Patrovi's formulation requires a formulation of the details of the measuring devices, specifically, the kind of partition that induces (or could be used) in the spectrum of the observable.

There were two additional improvement on the lower bound of the entropic uncertainty relations defined above. The first one was due to Bialynicki-Birula who presented, based in his earlier wok [9], a lower bound for the angle-angular momentum pair [13] $S^{\phi} + S^{L_z} \geq -ln(\Delta\phi/2\pi)$ and an improved lower bound for the position-momentum pair $S^x + S^p \geq 1 - ln(2) - ln(\gamma)$, where $\gamma = \Delta x \Delta p/h$. The second one was proposed by Maasen and Uffink [14] who demonstrated, based on a previous work of Kraus [15], that

$$S^A + S^B \geq -2ln(c), \tag{16}$$

where $c = max_{jk}\left|\left\langle a_j|b_k\right\rangle\right|$.

3.3. Simultaneous measurement

Whereas in the previous two subsection we treated the face of the Uncertainty Principle that is related with the probability distribution of observables of a given wave function, in this subsection we talk a bout a second version of The Uncertainty Principle. This version is related with the fact that it is not possible to determine simultaneously, with precision, two canonically conjugate observable and usually called *joint measurement*. This is stated, generally, as: "*It is impossible to measure simultaneously two observables like, for example, position and momentum.*" So, this sub-research area is concerned with the simultaneous measurement of two observables.

One of the first work in this approach was that of Arthurs and Kelly [16], they analyze this problems as follows: First, they realize that as the problem is the measurement of two observables, then it is required two devices to perform the measurement. That is, the system is coupled to two devises. Then, they consider that as the two meter position commutes then it is possible to perform two simultaneous measurements of them. Therefore, the simultaneous measurement of the two meters constitutes a simultaneous measurement of two non-commuting observables of the system. As the two meters interacts with the quantum system, they consider the following Hamiltonian:

$$\hat{H}_{int} = K\left(\hat{q}\hat{P}_x + \hat{p}\hat{P}_y\right) \tag{17}$$

where $\hat{q}$ and $\hat{p}$ correspond to the position and momentum of the quantum system, respectively, and $\hat{P}_x$ and $\hat{P}_y$ are the momentum of the two independent meters. Using two Gaussian function as the initial wave function of the meters they arrive at the following uncertainty relation for the simultaneous measurement of two observables:

$$\sigma_x \sigma_p \geq 1. \tag{18}$$

Therefore, the uncertainty relation of the simultaneous measurement of $\hat{q}$ and $\hat{p}$ is greater (by a factor of two) than the uncertainty relations based on the probability distribution of the two observables, the topic of the previous two sub-sections.

The next step in this approach was given by Arthurs and Goodman [17]. In this case, the approach is as follow: To perform a measurement, the system observables, $\hat{C} = \hat{C}_1 \otimes \hat{I}_2$ and $\hat{D} = \hat{D}_1 \otimes \hat{I}_2$, must be coupled to a measuring apparatus which is represented by the operators $\hat{R} = \hat{I}_1 \otimes \hat{R}_2$ and $\hat{C} = \hat{C}_1 \otimes \hat{I}_2$. Then, if we consider that there is access only to the meter operators then there must exist an uncertainty relations for these operators that puts a limit to the available information. Based in this consideration, they prove what they call a generalized uncertainty relation. To prove it they defined a a noise operator by

$$\begin{aligned} \hat{N}_R &= \hat{R} - G_R \hat{C}(0), \\ \hat{N}_S &= \hat{S} - G_S \hat{D}(0) \end{aligned} \tag{19}$$

where $\hat{C}(0)$ and $\hat{D}(0)$ are the system observables and $\hat{R}$ and $\hat{S}$ are the tracking apparatus observables, the latter obey the commutation rule $[\hat{R}, \hat{S}] = 0$. Also, it is required that the correlation between the system observables and the meter has, on average, a perfect match, that is:

$$Tr\left(\hat{\rho}\hat{N}_{R,S}\right) = \langle \hat{R} \rangle - G_R \left\langle \hat{C}(0) \right\rangle = 0. \tag{20}$$

Using the previous condition, i. e. Equ (20), it is possible to show that the noise operator is uncorrelated with all system operators like $\hat{C}$ and $\hat{D}$. Using all the previous properties of the system, meter and noise operators they arrive to the following generalized Heisenberg uncertainty relation:

$$\sigma_\xi \sigma_\eta \geq \left| Tr\left(\hat{\rho}\left[\hat{C},\hat{D}\right]\right)\right|, \tag{21}$$

where $\hat{\rho}$ is the state of the system, σ_ξ and σ_η are, respectively the standard deviation of the normalized operators $\xi = \hat{R}/G_R$ and $\eta = \hat{R}/G_R$. This uncertainty relation is four times the corresponding uncertainty relation for $\hat{C}$ and $\hat{D}$. Notice that in the left hand side of the Eq. (21) there is information of the meter operator whereas in the right hand side there is information of the system operators and that we have access only to the meter system. In reference [18] there was published an experimental verification of this uncertainty relation.

3.4. Disturbance due to measurement

The disturbance produced on an observable due to the measurement of another observable is, perhaps, the face of the uncertainty principled most talked about but the least studied. This comes from the fact that in quantum mechanics any measurement introduces an unforeseeable disturbance in the measured quantum system. It was only recently that there have been some research and understanding of this effect.

Originally, the idea that the measuring process disturb observables comes from Heisenberg's analysis of the observation of an electron by means of a microscope. This kind of uncertainty principle is written down, to use recent terminology, as [19]:

$$\epsilon(x)\eta(p) \geq \frac{1}{2}\left|\langle\psi\left|[\hat{x},\hat{p}]\right|\psi\rangle\right|, \tag{22}$$

where $\epsilon(x)$ is the noise in the measurement in position and $\eta(p)$ is the disturbance caused by the apparatus [19]. Using a general description of measurement Ozawa demonstrated that the uncertainty relation for disturbance and noise given by the Eq. (22) does not accurately represent the disturbance process. He has show that this kind of uncertainty relation includes additional terms not present in Eq. (22). In the measurement process, the quantum system interacts with a measuring device. He considers that this devices measures observable A *precisely* if its experimental probabilty distribution coincides with the theoretical probability distribution of the observable. In the measurement process, when the interaction have been turned off, the device is subject to a measurement of an observable M. Then, $\hat{A}^{in} = \hat{A} \otimes \hat{I}$ is the input observable, $\hat{A}^{out} = \hat{U}^\dagger\left(\hat{A} \otimes \hat{I}\right)\hat{U}$ is the observable after the mesaurement, $\hat{M}^{in} = \hat{I} \otimes \hat{M}$ is the device observable when the interaction begin, $\hat{M}^{out} = \hat{U}^\dagger(\hat{I} \otimes \hat{M})\hat{U}$ and $\hat{U}$ is the unitary time evolution operator

To show that the original uncertainty relation need additional terms, he introduces the following noise $N(\hat{A})$ and disturbance $D(\hat{B})$ operators:

$$\begin{aligned} N(\hat{A}) &= \hat{M}^{out} - \hat{A}^{in}, \\ D(\hat{B}) &= \hat{B}^{out} - \hat{B}^{in}. \end{aligned} \tag{23}$$

Using this operators, and considering that $\left[\hat{M}^{out}, \hat{B}^{out}\right] = 0$, Ozawa was able to show the following uncertainty relation [19]:

$$\epsilon(A)\eta(B) + \frac{1}{2}\left|\left\langle\left[N(\hat{A}), \hat{B}^{in}\right]\right\rangle + \left\langle\left[\hat{A}^{in}, D(\hat{B})\right]\right\rangle\right| \geq \frac{1}{2}\left|\langle\psi\,|\,[\hat{A}, \hat{B}]\,|\,\psi\rangle\right|. \tag{24}$$

where the noise $\epsilon(A)$ was defined by Ozawa as the root-mean-square deviation of the experimental variable $\hat{M}^{out}$ from the theoretical variable $\hat{A}^{in}$:

$$\epsilon(A) = \left\langle\left(\hat{M}^{out} - \hat{A}^{in}\right)^{1/2}\right\rangle \tag{25}$$

and the disturbance $\eta(B)$ on observable $\hat{B}$ is the change in the observable caused by the measurement process:

$$\eta(B) = \left\langle\left(\hat{B}^{out} - \hat{B}^{in}\right)^{1/2}\right\rangle. \tag{26}$$

This uncertainty relation has been recently experimentally tested, see reference [20]

4. Entanglement determination using entropic uncertainty relations

Nowadays entanglement is considered as an useful resource to make non-clasical task. As a resource it is convenient to have adequate measures to quantify how much entanglement are in a given entangled state. However, until recently the most known proposed measures have the unwanted fact of being difficult to apply in experimental settings. Therefore, it was necessary to find out new ways of entanglement determination that enable that the amount of entanglement in a quantum state could be experimentally tested.

Recently there has been much research to proposed new entanglement determination based, mostly, in uncertainty relations. In this case, the entropic uncertainty relations helps to realize this task. Recently, Berta et. al. [21] have proposed a new uncertainty relation (based on that proposed in references [8, 14]) to test the entanglement:

$$S(R|B) + S(S|B) log2c + S(A|B) \tag{27}$$

to propose this equation Berta et. al. consider that the system, with observables S and R, is entangled with a memori, with observable B, so in equation $S(R, B)$ is the von Neumann entropy and gives the uncertainty about the measurement of R given information stored in a quantum memory, B. The term S(A | B) quantifies the amount of entanglement between the particle and the memory. This relation was experimentally tested in reference [22].

5. Conclusions

In this chapter we review some of the most important improvements of the Heisenberg uncertainty relation. Although there are advances in their understanding and formulation, it remains yet as an open research area, specially in the quantification of entanglement.

Acknowledgements

We thanks Consejo Nacional de Ciencia y Tecnologia (CONACYT). L. M. Arévalo Aguilar acknowledge the support from Vicerrectoria de Investigación y Posgrado VIEP-BUAP under grand ARAL-2012-I. P. C. Garcia Quijas acknowledges CONACYT for a posdoctoral scholarship at the Universidad Autonoma de Guadalajara.

Author details

L. M. Arévalo Aguilar[1,*],
C. P. García Quijas[2] and Carlos Robledo-Sanchez[3]

* Address all correspondence to: olareva@yahoo.com.mx

1 Facultad de Ciencias Físico Matemáticas, Benemérita Universidad Autónoma de Puebla, Puebla, México
2 Departamento de Fśica, Universidad de Guadalajara, Guadalajara, Jalisco, México
3 Facultad de Ciencias Fisico Matemáticas, Benemérita Universidad Autónoma de Puebla, Puebla, México

References

[1] W. Heisenberg, Zeitschrift fur Physik 43, 172 (1927). Translated in the book, Quantum Measurement, Weeler and Zurek editors.

[2] H. P. Robertson, Phys. Rev. 46, 794 (1934).

[3] D. Griffiths, *Introduction to Quantum Mechanics*, Addison-Wesley; 2nd edition (April 10, 2004).

[4] N. Zettili, *Quantum Mechanics: Concepts and Applications*, Wiley; 2 edition (March 4, 2009).

[5] J. Hilgevoord and J. M. M. Uffink, Eur. J. Phys. 6, 165 (1985).

[6] J. Hilgevoord and J. M. M. Uffink, The mathematical expression of the uncertainty principle, in the book *Microphysical Reality and Quantum Formalism*, 91-114, A. van der Merve et. al. Eds.

[7] J. M. M. Uffink and J. Hilgevoord, Found. Phys. 15, 925 (1985).

[8] D. Deutsch, Phys. Rev. Lett. 50, 631 (1983).

[9] I. Bialynicki-Birula and J. Mycielski, Commun. math. Phys. 44, 129 (1975).

[10] I. Bialynicki-Birula, Phys. Rev. A 74, 052102 (2006).

[11] S. Wehner, New Jour. of Phys. 12, 025009 (2010).

[12] M. Hossein Partovi, Phys. Rev. Lett. 50, 1883 (1983).

[13] I. Bialynicki-Birula, Phys. Lett. A 103, 253 (1984).

[14] H. Maasen and J. B. Uffink, Phys. Rev. Lett. 60, 1103 (1988).

[15] Krauss, Phys. Rev. D 35, 3070 (1987).

[16] E. Arthurs and J. L. Kelly JR., Bell. Syst. Tech. J. 44, 725 (1965).

[17] E. Arthurs, M. S. Goodman, Phys. Rev. Lett. 60, 2447 (1988).

[18] A. Trifonov, G. Björk and J. Söderholm, Phys. Rev. Lett. 86, 4423 (2001).

[19] M. Ozawa, Phys. Rev. A67, 042105 (2003).

[20] J. Erhart, S. Sponar, G. Sulyok, G. Badurek, M. Ozawa and Y. Hasegawa, Nature Physics 8, 185Ð189 (2012).

[21] M. Berta, M. Christandl, R. Colbeck, J. M. Renes and R. Renner, Phys. Nature Physics 6, 659 (2010).

[22] Chuan-Feng Li, Jin-Shi Xu, Xiao-Ye Xu, Ke Li & Guang-Can Guo, Nature Physics 7, 752 (2011).

Section 2

The Schrödinger Equation

Chapter 5

Schrödinger Equation as a Hamiltonian System, Essential Nonlinearity, Dynamical Scalar Product and some Ideas of Decoherence

Jan J. Sławianowski and Vasyl Kovalchuk

Additional information is available at the end of the chapter

1. Introduction

Our main idea is to suggest some new model of nonlinearity in quantum mechanics. The nonlinearity we discuss is non-perturbative and geometrically motivated, in any case it is not an auxiliary correction to the linear background. It has a group-theoretic motivation based on the assumption of the "large" symmetry group. In a sense, it develops further our ideas suggested earlier in [1–4].

It is well known that quantum mechanics is still plagued by some paradoxes concerning decoherence, measurement process and the reduction of the state vector. In spite of certain optimistic opinions, the problem is still unsolved, although many interesting ideas have been formulated, like that about subsystems of a large (infinite) quantum system or stochastic quantum Markov processes with the spontaneous reduction of state vectors. There is still an opinion that the main problem is the linearity of the Schrödinger equation, which seems to be drastically incompatible with the mentioned problems [5, 6]. But at the same time, that linearity works beautifully when describing the unobserved unitary quantum evolution, finding the energy levels and in all statistical predictions. It seems that either we are faced here with some completely new type of science, roughly speaking, based on some kind of solipsism with the irreducible role of human being in phenomena, or perhaps we deal with a very sophisticated and delicate nonlinearity which becomes active and remarkable just in the process of interaction between quantum systems and "large" classical objects.

The main idea is to analyse the Schrödinger equation and corresponding relativistic linear wave equations as usual self-adjoint equations of mathematical physics, thus ones derivable from variational principles. It is easy to construct their Lagrangians. Some problems appear when trying to formulate Hamiltonian formalism, because Lagrangians for the Schrödinger or Dirac equations are highly degenerate and the corresponding Legendre transformation is uninvertible and leads to constraints in the phase space. Nevertheless, using the Dirac formalism for such Lagrangians, one can find the corresponding Hamiltonian formalism. Incidentally, it turns out that introducing the second-order time derivatives to dynamical equations, even as small corrections, one can obtain the regular Legendre transformation. In non-relativistic quantum mechanics there are certain hints suggesting just such a modification in the nano-scale physics [1, 7, 8]. One can also show that in $SU(2,2)$-invariant gauge models, i.e., roughly-speaking, in conformal theory, it is more natural to begin with the four-component Klein-Gordon amplitude and then to derive the Dirac behavior as an unexpected aspect of the Klein-Gordon theory [4, 9]. This leads us to certain interesting statements concerning the pairing of fundamental quarks and leptons in electroweak interactions.

We begin just like in [1] from the first- and second-order (in time) Schrödinger equations for a finite-level system, i.e., for the finite "configuration space". We construct the "direct nonlinearity" as a non-quadratic term in Lagrangian, but further on we concentrate on our main idea. It consists in that we follow the conceptual transition from the special to general relativity. Namely, just like in the passing to the theory of general relativity, the metric tensor loses its status of the absolute geometric object and becomes included into degrees of freedom (gravitational field), so in our treatment the Hilbert-space scalar product becomes a dynamical quantity which satisfies together with the state vector the system of differential equations. The main idea is that there is no fixed scalar product metric and the dynamical term of Lagrangian, describing the self-interaction of the metric, is invariant under the total group $GL(n, \mathbb{C})$. But this invariance is possible only for models non-quadratic in the metric, just like in certain problems of the dynamics of "affinely-rigid" body [2, 3]. There is a natural metric of this kind and it introduces to the theory a very strong nonlinearity which induces also the effective nonlinearity of the wave equation, even if there is no "direct nonlinearity" in it. The structure of Lagrangian and equations of motion is very beautiful, as usual in high-symmetry problems. Nevertheless, the very strong nonlinearity prevents us to find a rigorous solution. Nevertheless, there are some partial results, namely, if we fix the behaviour of wave function to some simple form and provide an academic discussion of the resulting behaviour of the scalar product, then it turns out that there are rigorous exponential solutions, including ones infinitely growing and ones exponentially decaying in future. This makes some hope for describing, e.g., some decay/reduction phenomena. Obviously, the full answer will be possible only when we will be able to find a rigorous solution for the total system. We are going to repeat the same discussion for the more realistic infinite-level system, when the wave function is defined somehow on the total configuration space like, e.g., the arithmetic space. As usual when passing from the finite to infinite dimensions, some essentially new features appear then, nevertheless, one can hope that the finite-dimensional results may be to some extent applicable. This will be done both in the usual non-relativistic Schrödinger wave equation and for the relativistic Klein-Gordon and Dirac equations. In any of those cases we are dealing with two kinds of degrees of freedom, i.e., dynamical variables: wave function and scalar product. They are mutually interacting.

All said above concerns the self-adjoint model of the Schrödinger equation, derivable from variational principle. However, one can also ask what would result if we admitted "dissipative" models, where the Schrödinger equation does possess some "friction-like" term. As yet we have not a ready answer, nevertheless, the question is well formulated and we will try to check what might appear in a consequence of such a generalization. Maybe some quantum model of dissipation, i.e., of the open system, but at a moment we are unable to answer the question.

2. Nonlinear Schrödinger equation as a self-adjoint equation of mathematical physics

It is known, although not often noticed and declared, that the Schrödinger equation and other equations of quantum mechanics, including relativistic ones, are self-adjoint, i.e., derivable from variational principles. Therefore, they may be expressed in Hamiltonian terms, i.e., quantum mechanics becomes a kind of analytical mechanics, usually with an infinite number of degrees of freedom (excepting finite-level systems). So, as far as one deals with the unobserved quantum system, its evolution may be described within the classical mathematical framework of Hamiltonian mechanics and canonical transformations. Of course, this breaks down when quantum springs, jumps, occur, i.e., when one is faced with phenomena like macroscopic observation, measurements and decoherence. This happens when a small/quantum system interacts with a large/classical object showing some characteristic instability. There were various ways of explaining those catastrophic phenomena and their statistical rules. It is very interesting that those rules are based on the Hilbert space geometry, but in addition some unpredictable, statistical phenomena appear. There were many attempts of explanation, based either on the extension to larger, infinite systems or on the idea of spontaneous stochastic reduction. But one of the permanent motives is the hypothesis of nonlinearity, especially one which is "silent" in the evolution of the unobserved quantum system, but becomes essential in the process of interaction with the large and unstable classical system. This is also our line in this paper. Mathematical methods of nonlinear analytical mechanics just seem to suggest some attempts of solution.

To explain the main ideas we start from the simple finite-level system, i.e., one with a finite-dimensional unitary space of states. Let us denote this complex linear space by W and put $\dim_{\mathbb{C}} W = n$. The dual, antidual and complex-conjugate spaces will be denoted respectively by W^*, $\overline{W}^* = \overline{W^*}$, and $\overline{W}$. As usual, W^* is the space of $\mathbb{C}$-linear functions on W. Having the same finite dimension, the spaces W, W^* are isomorphic, however in a non-canonical way until we introduce some unitary structure to W. But some comments are necessary concerning the complex conjugate spaces $\overline{W}^* = \overline{W^*}$, $\overline{W}$. It must be stressed that in general nothing like the complex structure is defined in W. It is a structure-less space and the half-linear (semi-linear) bar-operations are defined pointwisely. Therefore, for any $f \in W^*$ and for any $u \in W$ the corresponding $\overline{f} \in \overline{W}^*$, $\overline{u} \in \overline{W}$ are given by

$$\left(\overline{f}\right)(w) = \overline{f(w)}, \qquad \overline{u}(g) = \overline{u(g)}, \tag{1}$$

where w, g are arbitrary elements of W and W^*. Therefore, under the bar-operation $\overline{W}$ is canonically anti-isomorphic with W and $\overline{W}^*$ with W^*. Nevertheless, the bar-operation acts between different linear spaces and this is often essential. The spaces W and $\overline{W}$, and similarly W^* and $\overline{W}^*$ may be mutually identified only in important, nevertheless mathematically exceptional, situations when by the very definition W is a linear subspace of the space of $\mathbb{C}$-valued functions on a given "configuration space" Q. Then we simply define pointwisely

$$\overline{\psi}(q) := \overline{\psi(q)} \tag{2}$$

and so W becomes identical with $\overline{W}$. In general this is impossible. Let us mention of course that for the n-level system, Q is an n-element set.

Let us quote a few analytical formulas. We choose a pair of mutually dual bases $(e_1, \ldots, e_n)$, $(e^1, \ldots, e^n)$ in W, W^* and induced pair of dual bases $(\overline{e}_{\overline{1}}, \ldots, \overline{e}_{\overline{1n}})$, $\left(\overline{e}^{\overline{1}}, \ldots, \overline{e}^{\overline{1n}}\right)$ in the complex-conjugate spaces $\overline{W}$ and $\overline{W}^* \simeq \overline{W^*}$. Then the complex conjugates of vectors

$$u = u^a e_a \in W, \qquad f = f_a e^a \in W^* \tag{3}$$

are analytically expressed as

$$\overline{u} = \overline{u}^{\overline{a}} \overline{e}_{\overline{a}}, \qquad \overline{f} = \overline{f}_{\overline{a}} \overline{e}^{\overline{a}}, \tag{4}$$

where, obviously, $\overline{u}^{\overline{a}}, \overline{f}_{\overline{a}}$ are the usual complex conjugates of numbers u^a, f_a.

In quantum mechanics one uses often sesquilinear forms, usually Hermitian ones. Usually our sesquilinear forms are antilinear (half-linear) in the first argument and linear in the second, therefore,

$$F(au + bw, v) = \overline{a}F(u, v) + \overline{b}F(w, v), \tag{5}$$

i.e., analytically

$$F(u, w) = F_{\overline{a}b}\overline{u}^{\overline{a}} w^b. \tag{6}$$

So, they are elements of $\overline{W}^* \otimes W^*$. For Hermitian forms we have

$$F(u, w) = \overline{F(w, u)}, \qquad F_{\overline{a}b} = \overline{F}_{b\overline{a}}. \tag{7}$$

If F is non-degenerate,

$$\det\left[F_{\overline{a}b}\right] \neq 0, \tag{8}$$

then the inverse form $F^{-1} \in W \otimes \overline{W}$ does exist with coefficients $F^{a\overline{b}}$ such that

$$F^{a\overline{c}} F_{\overline{c}b} = \delta^a{}_b, \qquad F_{\overline{a}c} F^{c\overline{b}} = \delta_{\overline{a}}{}^{\overline{b}}. \tag{9}$$

For any quantum system there are two Hermitian forms: a) the scalar product $\Gamma \in \overline{W}^* \otimes W^*$ and b) the Hamiltonian form ${}_{\Gamma}H$ obtained by the Γ-lowering of the first index of the Hamilton operator $H \in L(W) \simeq W \otimes W^*$. The Hamilton operator H is Γ-Hermitian, i.e.,

$$\Gamma(H\psi, \varphi) = \Gamma(\psi, H\varphi). \tag{10}$$

Analytically the sesquilinear form ${}_{\Gamma}H$,

$${}_{\Gamma}H_{\overline{a}b} = \Gamma_{\overline{a}c} H^c{}_b, \tag{11}$$

is simply Hermitian without any relationship to Γ, and from the Langrangian point of view it is more fundamental than H itself. The finite-level Schrödinger equation

$$i\hbar \frac{d\psi^a}{dt} = H^a{}_b \psi^b \tag{12}$$

is derivable from the Lagrangian

$$L(1) = \frac{i\hbar}{2} \Gamma_{\bar{a}b} \left(\overline{\psi}^{\bar{a}} \dot{\psi}^b - \dot{\overline{\psi}}^{\bar{a}} \psi^b \right) - {}_{\Gamma}H_{\bar{a}b} \overline{\psi}^{\bar{a}} \psi^b. \tag{13}$$

Having in view some kind of "generality" it may be convenient to admit some general constant coefficients α, γ:

$$L(1) = i\alpha \Gamma_{\bar{a}b} \left(\overline{\psi}^{\bar{a}} \dot{\psi}^b - \dot{\overline{\psi}}^{\bar{a}} \psi^b \right) - \gamma {}_{\Gamma}H_{\bar{a}b} \overline{\psi}^{\bar{a}} \psi^b. \tag{14}$$

It is seen that unlike in the Schrödinger equation, from the variational point of view ${}_{\Gamma}H$ is more fundamental. It should be denoted rather as $\chi_{\bar{a}b}$, and $H^a{}_b$ with the convention of the Γ-raised first index of χ, as

$$H^a{}_b = \left({}^{\Gamma}\chi \right)^a{}_b = \Gamma^{a\bar{c}} \chi_{\bar{c}b}. \tag{15}$$

One does not do so because of the prevailing role of Schrödinger equation over its variational interpretation. The Hermitian structure of Γ and ${}_{\Gamma}H$ imply that $L(1)$ is real. The descriptor (1) refers to the first-order polynomial dependence of $L(1)$ on the time derivatives of ψ. Obviously, the corresponding Legendre transformation leads to phase-space constraints and to the Dirac procedure in canonical formalism. It is interesting to admit some regularization by allowing L to contain the terms quadratic in generalized velocities, just in the spirit of analytical mechanics. The corresponding Lagrangian will have the following form:

$$L(1,2) = i\alpha \Gamma_{\bar{a}b} \left(\overline{\psi}^{\bar{a}} \dot{\psi}^b - \dot{\overline{\psi}}^{\bar{a}} \psi^b \right) + \beta \Gamma_{\bar{a}b} \dot{\overline{\psi}}^{\bar{a}} \dot{\psi}^b - \gamma {}_{\Gamma}H_{\bar{a}b} \overline{\psi}^{\bar{a}} \psi^b. \tag{16}$$

To be more precise, in the term quadratic in velocities one can admit some more general Hermitian form, not necessarily the one proportional to $\Gamma_{\bar{a}b}$. However, we do not do things like those in this paper. Let us stress that α, β, γ are real constants.

One circumstance must be stressed: we use as "independent" components ψ^a, $\overline{\psi}^{\bar{a}}$. The procedure is not new. The same is done in variational principles of field theory [10]. Lagrangians are real, based on Hermitian forms, therefore in variational procedure it is sufficient to subject, e.g., only $\overline{\psi}^{\bar{a}}$ to the modification $\overline{\psi}^{\bar{a}} \mapsto \overline{\psi}^{\bar{a}} + \delta\overline{\psi}^{\bar{a}}$. Then, e.g., for the action functional

$$I(1,2) = \int L(1,2) dt \tag{17}$$

one obtains

$$\frac{\delta I(1,2)}{\delta \overline{\psi}^{\bar{a}}(t)} = 2i\alpha \Gamma_{\bar{a}b} \frac{d\psi^b}{dt} - \beta \Gamma_{\bar{a}b} \frac{d^2\psi^b}{dt^2} - \gamma {}_{\Gamma}H_{\bar{a}b} \psi^b \tag{18}$$

and the resulting Schrödinger equation:

$$2i\alpha\frac{d\psi^a}{dt} - \beta\frac{d^2\psi^a}{dt^2} = \gamma H^a{}_b\psi^b. \tag{19}$$

And this is all, because the variation with respect to ψ^a leads to the complex-conjugate equation. This is a convenient and commonly used procedure.

The language of analytical mechanics, in this case finite-dimensional one, opens some possibility of introducing nonlinearity to quantum-mechanical equations. The simplest way is to believe in Schrödinger equation but reinterpreting it in terms of Hamiltonian mechanics, to introduce some naturally looking nonlinear perturbations to it. The simplest way is to introduce to L some non-quadratic potential term $\mathcal{V}\left(\psi,\overline{\psi}\right)$ and the corresponding action term to I:

$$I(\mathcal{V}) = \int \mathcal{V}dt. \tag{20}$$

For example, the simplest possibility is to use the term like

$$\mathcal{V}\left(\psi,\overline{\psi}\right) = f\left(\Gamma_{\overline{a}b}\overline{\psi}^{\overline{a}}\psi^b\right), \tag{21}$$

with some model function $f : \mathbb{R} \rightarrow \mathbb{R}$. In various physical applications one uses often the quartic term:

$$f(y) = \frac{\varkappa}{2}(y-b)^2. \tag{22}$$

When using the $\mathcal{V}$-term, one obtains after the variational procedure the following nonlinear Schrödinger equation:

$$2i\alpha\frac{d\psi^a}{dt} - \beta\frac{d^2\psi^a}{dt^2} = \gamma H^a{}_b\psi^b + f'\psi^a, \tag{23}$$

where f' denotes the usual first-order derivative of f. This is the simplest model containing the superposition of first- and second-order time derivatives of ψ. It is very simple because of being a finite-level system and because of the direct introduction of nonlinearity as a perturbation of the primarily linear model. Nevertheless, it demonstrates some interesting features of nonlinearity and of the mixing of derivatives order.

The problem of the order of derivatives is strongly related to the structure of Hamiltonian mechanics of our systems. It occurs also in corresponding problems of field theory. Let us mention some elementary facts. As usual, it is convenient to use the doubled number of degrees of freedom ψ^a, $\overline{\psi}^{\overline{a}}$ and the corresponding canonical momenta π_a, $\overline{\pi}_{\overline{a}}$. The symplectic form is given by

$$\omega = d\pi_a \wedge d\psi^a + d\overline{\pi}_{\overline{a}} \wedge d\overline{\psi}^{\overline{a}}, \tag{24}$$

and the resulting Poisson bracket is expressed as follows:

$$\{F,G\} = \frac{\partial F}{\partial \psi^a}\frac{\partial G}{\partial \pi_a} + \frac{\partial F}{\partial \overline{\psi}^{\bar{a}}}\frac{\partial G}{\partial \overline{\pi}_{\bar{a}}} - \frac{\partial F}{\partial \pi_a}\frac{\partial G}{\partial \psi^a} - \frac{\partial F}{\partial \overline{\pi}_{\bar{a}}}\frac{\partial G}{\partial \overline{\psi}^{\bar{a}}}. \tag{25}$$

The Hamiltonian vector field is given by

$$X_F = \frac{\partial F}{\partial \pi_a}\frac{\partial}{\partial \psi^a} + \frac{\partial F}{\partial \overline{\pi}_{\bar{a}}}\frac{\partial}{\partial \overline{\psi}^{\bar{a}}} - \frac{\partial F}{\partial \psi^a}\frac{\partial}{\partial \pi_a} - \frac{\partial F}{\partial \overline{\psi}^{\bar{a}}}\frac{\partial}{\partial \overline{\pi}_{\bar{a}}}. \tag{26}$$

It must be stressed that all dynamical quantities in this formalism are considered as independent on their complex conjugates:

$$\left\langle d\psi^a, \frac{\partial}{\partial \psi^b}\right\rangle = \delta^a{}_b, \qquad \left\langle d\overline{\psi}^{\bar{a}}, \frac{\partial}{\partial \overline{\psi}^{\bar{b}}}\right\rangle = \delta^{\bar{a}}{}_{\bar{b}}, \tag{27}$$

but

$$\left\langle d\psi^a, \frac{\partial}{\partial \overline{\psi}^{\bar{b}}}\right\rangle = 0, \qquad \left\langle d\overline{\psi}^{\bar{a}}, \frac{\partial}{\partial \psi^b}\right\rangle = 0, \tag{28}$$

and similarly,

$$\left\langle d\pi_a, \frac{\partial}{\partial \pi_b}\right\rangle = \delta_a{}^b, \qquad \left\langle d\overline{\pi}_{\bar{a}}, \frac{\partial}{\partial \overline{\pi}_{\bar{b}}}\right\rangle = \delta_{\bar{a}}{}^{\bar{b}}, \tag{29}$$

$$\left\langle d\pi_a, \frac{\partial}{\partial \overline{\pi}_{\bar{b}}}\right\rangle = 0, \qquad \left\langle d\overline{\pi}_{\bar{a}}, \frac{\partial}{\partial \pi_b}\right\rangle = 0. \tag{30}$$

All the remaining basic evaluations are vanishing, in particular those for $d\psi^a$, $d\overline{\psi}^{\bar{a}}$ with $\partial/\partial\pi_b$, $\partial/\partial\overline{\pi}_{\bar{b}}$, and similarly, for $d\pi_a$, $d\overline{\pi}_{\bar{a}}$ with $\partial/\partial\psi^b$, $\partial/\partial\overline{\psi}^{\bar{b}}$.

Let us write down the Hamilton equations of motion. Their form depends strongly on the occurrence of second time derivatives in the "Schrödinger equation". For simplicity let us begin with the assumption that $\beta \neq 0$ and our equation is second-order in time derivatives. Then the Legendre transformation is given by the formulas:

$$\pi_a = i\alpha\overline{\psi}^{\bar{b}}\Gamma_{\bar{b}a} + \beta\dot{\overline{\psi}}^{\bar{b}}\Gamma_{\bar{b}a}, \qquad \overline{\pi}_{\bar{a}} = -i\alpha\Gamma_{\bar{a}b}\psi^b + \beta\Gamma_{\bar{a}b}\dot{\psi}^b. \tag{31}$$

They are invertible and

$$\dot{\psi}^a = \frac{1}{\beta}\Gamma^{a\bar{b}}\overline{\pi}_{\bar{b}} + \frac{i\alpha}{\beta}\psi^a, \qquad \dot{\overline{\psi}}^{\bar{a}} = \frac{1}{\beta}\pi_b\Gamma^{b\bar{a}} - \frac{i\alpha}{\beta}\overline{\psi}^{\bar{a}}. \tag{32}$$

The Lagrangian "energy" function is given by

$$\mathcal{E} = \beta\Gamma_{\overline{a}b}\dot{\overline{\psi}}^{\overline{a}}\dot{\psi}^{b} + \gamma_{\Gamma}H_{\overline{a}b}\overline{\psi}^{\overline{a}}\psi^{b} + \mathcal{V}\left(\psi,\overline{\psi}\right), \tag{33}$$

and substituting here the above inverse formula we obtain the "Hamilton function" in the sense of analytical mechanics:

$$\mathcal{H} = \frac{1}{\beta}\left(\Gamma^{a\overline{b}}\pi_{a}\overline{\pi}_{\overline{b}} + i\alpha\left[\pi_{a}\psi^{a} - \overline{\pi}_{\overline{a}}\overline{\psi}^{\overline{a}}\right]\right) + \left(\frac{\alpha^{2}}{\beta}\Gamma_{\overline{a}b} + \gamma_{\Gamma}H_{\overline{a}b}\right)\overline{\psi}^{\overline{a}}\psi^{b} + \mathcal{V}\left(\psi,\overline{\psi}\right). \tag{34}$$

It is clear that the "energy" function is always globally defined in the tangent bundle, but the "Hamiltonian" $\mathcal{H}$ does exist as a function on the cotangent bundle only if β does not vanish. The special case $\beta = 0$ is essentially singular. Let us mention that it is a general rule that differential equations are catastrophically sensitive to the vanishing of coefficients at highest-order derivatives. In any case the Schrödinger equation modified by terms with second derivatives is essentially different than the usual, first-order equation. The problem has to do both with some doubts concerning the occurrence of second derivatives but also with certain hopes and new physical ideas. Obviously, if $\beta \neq 0$, the second-order Schrödinger equation is equivalent to the following canonical Hamilton equations:

$$\frac{d\psi^{a}}{dt} = \{\psi^{a}, \mathcal{H}\} = \frac{\partial\mathcal{H}}{\partial\pi_{a}}, \qquad \frac{d\pi_{a}}{dt} = \{\pi_{a}, \mathcal{H}\} = -\frac{\partial\mathcal{H}}{\partial\psi^{a}}. \tag{35}$$

Let us mention that there are various arguments for the second-order differential equations as fundamental ones for quantum theory. In a sense, in conformal $\mathrm{SU}(2,2)$-ruled geometrodynamics, some kind of Dirac behaviour is a byproduct of the quadruplet of the gauge Klein-Gordon equation [4, 9]. Besides, in nano-physics there are also some other arguments for the mixing of first- and second-order Schrödinger equations [1, 8]. In the $\mathrm{SU}(2,2)$-gauge theory there are also some interesting consequences of this mixing within the framework of the standard model.

Nevertheless, it is also convenient to discuss separately the degenerate Schrödinger (Schrödinger-Dirac?) model based on the first derivatives. Our Legendre transformation becomes then

$$\pi_{a} = i\alpha\overline{\psi}^{\overline{b}}\Gamma_{\overline{b}a}, \qquad \overline{\pi}_{\overline{a}} = -i\alpha\Gamma_{\overline{a}b}\psi^{b}. \tag{36}$$

It does not depend on velocities at all. The same concerns the energy function:

$$\mathcal{E} = \gamma_{\Gamma}H_{\overline{a}b}\overline{\psi}^{\overline{a}}\psi^{b} = \gamma\Gamma\left(\psi, \widehat{H}\psi\right). \tag{37}$$

Strictly speaking, Hamiltonian is defined only on the manifold of "primary constraints" in the sense of Dirac, $M = \mathcal{L}\left(W \times \overline{W} \times W \times \overline{W}\right) \subset W \times \overline{W} \times W^{*} \times \overline{W}^{*}$, where $\mathcal{L}$ is just the above Legendre transformation. Some authors, including Dirac himself, define Hamiltonian $\mathcal{H}$ all over the phase space, however, it is then non-unique, and namely

$$\mathcal{H} = \mathcal{H}_0 + \lambda^a \left(\pi_a - i\alpha \overline{\psi}^{\overline{b}} \Gamma_{\overline{b}a} \right) + \overline{\lambda}^{\overline{a}} \left(\overline{\pi}_{\overline{a}} + i\alpha \Gamma_{\overline{a}b} \psi^b \right), \tag{38}$$

where $\lambda^a, \overline{\lambda}^{\overline{a}}$ are Lagrange multipliers and

$$\mathcal{H}_0 = \gamma_\Gamma H_{\overline{a}b} \overline{\psi}^{\overline{a}} \psi^b + \mathcal{V}(\psi, \overline{\psi}). \tag{39}$$

One can easily show that the Dirac secondary constraints coincide with the primary ones, $M_S = M$, and the multipliers are given by

$$\lambda^a = -\frac{i\gamma}{2\alpha} H^a{}_b \psi^b - \frac{i}{2\alpha} \Gamma^{a\overline{c}} \frac{\partial \mathcal{V}}{\partial \overline{\psi}^{\overline{c}}}, \tag{40}$$

$$\overline{\lambda}^{\overline{a}} = \frac{i\gamma}{2\alpha} \overline{\psi}^{\overline{b}} H_{\overline{b}}{}^{\overline{a}} + \frac{i}{2\alpha} \frac{\partial \mathcal{V}}{\partial \psi^c} \Gamma^{c\overline{a}}, \tag{41}$$

where operations on indices of H are meant in the sense of the metric tensor Γ.

It is clear that $M = M_S$ has the complex dimension n, but its real dimension is $2n$, always even, as it should be with symplectic manifolds. The following quantities, π-s doubled in a consequence of this "complex-real",

$$\Pi_a = 2i\alpha \overline{\psi}^{\overline{b}} \Gamma_{\overline{b}a}, \qquad \overline{\Pi}_{\overline{a}} = -2i\alpha \Gamma_{\overline{a}b} \psi^b, \tag{42}$$

may be used to represent the canonical conjugate momenta. It follows in particular, that on the constraints submanifold M we have the following Poisson brackets:

$$\{\psi^a, \psi^b\}_M = 0, \qquad \{\overline{\psi}^{\overline{a}}, \overline{\psi}^{\overline{b}}\}_M = 0, \qquad \{\psi^a, \overline{\psi}^{\overline{b}}\}_M = \frac{1}{2i\alpha} \Gamma^{a\overline{b}}. \tag{43}$$

Therefore, it is seen that up to normalization the complex conjugates $\overline{\psi}^{\overline{a}}$ coincide with canonical momenta conjugate to ψ^a. Using the standard properties of Poisson brackets we can write the resulting canonical equations in the bracket form:

$$\frac{d\psi^a}{dt} = \{\psi^a, \mathcal{H}\}_M, \qquad \frac{d\overline{\psi}^{\overline{a}}}{dt} = \{\overline{\psi}^{\overline{a}}, \mathcal{H}\}_M. \tag{44}$$

This implies, of course, the following well-known equation:

$$i\hbar \frac{d\psi^a}{dt} = H^a{}_b \psi^b + \frac{1}{2} \Gamma^{a\overline{b}} \frac{\partial \mathcal{V}}{\partial \overline{\psi}^{\overline{b}}} \tag{45}$$

and, equivalently, its complex conjugate.

Let us stress that due to the $\mathcal{V}$-term, this is a nonlinear Schrödinger equation. The nonlinearity and its possible consequences for the decoherence and measurement problems depend on our invention in constructing the $\mathcal{V}$-model. Of course, the procedure is more promising for large, in particular infinite, systems, and the above finite-level framework is rather a toy model. This concerns both the first- and second-order Schrödinger equations. Nevertheless, in the above models nonlinearity was more or less introduced "by hand", as an additional perturbation term. Our main idea, we are going to describe it now, consists in introducing of nonlinearity in analogy to the passing from special to general relativity.

3. Non-direct nonlinearity and the dynamical scalar product

Let us remind some other, well-established nonlinearities of intrinsically geometric origin, appearing in physics. One of them is Einstein-Hilbert general relativity. It is well known that majority of well-established field theories is originally linear, and the nonlinearity appears in a consequence of their mutual interactions and symmetry principles. But there is one exceptional nonlinearity, namely that of gravitation theory. In special-relativistic physics the space-time arena is given by the flat Minkowski space. Its geometry is an absolute factor which restricts the symmetry to the Poincare group. But it is a strange and originally surprizing fact that physics does not like absolute objects. In general relativity the metric tensor becomes a dynamical quantity with the dynamics ruled by the Hilbert Lagrangian. It is so-to-speak an essentially nonlinear centre of physical reality. Its dynamics is essentially, non-perturbatively nonlinear and invariant under the infinite-dimensional group of the space-time diffeomorphisms (general covariance group). And automatically it becomes the group of symmetry of the whole physics. The dynamics is quasilinear, nevertheless by necessity nonlinear. The relationship between essential nonlinearity and large symmetry groups seems to be a general rule. Let us mention now two another, simpler examples from different branches of physics.

The first example belongs to mechanics of continua, first of all to plasticity theory, although elastic applications are also possible [11]. Let us consider a real linear space V and the set $\mathrm{Sym}\,(V^* \otimes V^*)$ of symmetric metric tensors on V. It is obviously non-connected and consists of components characterized by the signature. Let us consider the manifold $\mathrm{Sym}^+\,(V^* \otimes V^*)$ of positively definite metrics. And now, assuming that the metrics elements of $\mathrm{Sym}^+\,(V^* \otimes V^*)$ describe some physical reality, let us ask for the metric structures, i.e., kinetic energy forms on $\mathrm{Sym}^+\,(V^* \otimes V^*)$. Of course, the simplest possibility is

$$ds^2 = G^{ijkl} dg_{ij} dg_{kl}, \qquad \text{i.e.,} \qquad G = G^{ijkl} dg_{ij} \otimes dg_{kl}, \tag{46}$$

where G^{ijkl} is constant and satisfies the natural nonsingularity and symmetry conditions:

$$G^{ijkl} = G^{klij}, \qquad G^{ijkl} = G^{jikl} = G^{ijlk}. \tag{47}$$

This metric on the manifold of metrics is flat. But this is rather strange and non-aesthetic. The natural question appears why not to use the following intrinsic metric:

$$ds^2 = \lambda g^{jk} g^{li} dg_{ij} dg_{kl} + \mu g^{ji} g^{lk} dg_{ij} dg_{kl}, \tag{48}$$

or, in more sophisticated terms:

$$G = \lambda g^{jk} g^{li} dg_{ij} \otimes dg_{kl} + \mu g^{ji} g^{lk} dg_{ij} \otimes dg_{kl}, \tag{49}$$

where λ, μ are constants and $\left[g^{ij}\right]$ is the contravariant inverse of $\left[g_{ij}\right]$, i.e., $g^{ik} g_{kj} = \delta^i{}_j$. This metric structure is evidently non-Euclidean, Riemannian in $\mathrm{Sym}^+ (V^* \otimes V^*)$, but it does not contain anything a priori fixed, but λ, μ. To be more precise, it is only λ that is essential up to normalization, because the μ-term, being degenerate, is only an auxiliary correction. The corresponding kinetic energy of the g-process will be

$$T = \frac{\lambda}{2} g^{jk} g^{li} \frac{dg_{ij}}{dt} \frac{dg_{kl}}{dt} + \frac{\mu}{2} g^{ji} g^{lk} \frac{dg_{ij}}{dt} \frac{dg_{kl}}{dt}. \tag{50}$$

Expressions of this type are used, e.g., in incremental approaches to plasticity. They are also interesting in certain elastic problems and in defect theory.

Let us also mention about some other application. Consider the motion of material point with the mass m and internal g-degrees of freedom. The corresponding kinetic energy will be given by

$$T = \frac{m}{2} g_{ij} \frac{dx^i}{dt} \frac{dx^j}{dt} + \frac{\lambda}{2} g^{jk} g^{li} \frac{dg_{ij}}{dt} \frac{dg_{kl}}{dt} + \frac{\mu}{2} g^{ji} g^{lk} \frac{dg_{ij}}{dt} \frac{dg_{kl}}{dt}, \tag{51}$$

obviously x^i are here coordinates of the centre of mass. One can also introduce some potential term built of x^a, g_{ij}. In a sense, the structure of (50), (51) resembles that of generally-relativistic Lagrangians, obviously with the proviso that only the time derivatives occur, as we are dealing here with a system which does not possess any other continuous independent variables. Indeed, the main term of Hilbert Lagrangian begins from the expression proportional to

$$g^{\nu\varkappa} g^{\mu\lambda} g^{\alpha\beta} g_{\mu\nu,\alpha} g_{\varkappa\lambda,\beta}, \tag{52}$$

where $g_{\mu\nu}$ is the space-time metric and $g^{\alpha\beta}$ is its contravariant inverse. Differentiation with respect to the space-time coordinates is meant here. The structural similarity to the prescription (50), (51) is obvious.

It is important that the both last expressions for T are invariant under the total group $\mathrm{GL}(V)$, or rather under the semi-direct product $\mathrm{GL}(V) \times_s V$. Again the "large" symmetry group is responsible for the essential nonlinearity even of the geodetic models described by the expressions for T.

It is interesting to ask what changes appear when we assume V to be a complex linear space and g a sesquilinear Hermitian form. Obviously, instead of (50), (51) we will have then

$$T = \frac{\lambda}{2} g^{j\bar{k}} g^{l\bar{i}} \frac{dg_{\bar{i}j}}{dt} \frac{dg_{\bar{k}l}}{dt} + \frac{\mu}{2} g^{j\bar{i}} g^{l\bar{k}} \frac{dg_{\bar{i}j}}{dt} \frac{dg_{\bar{k}l}}{dt}, \tag{53}$$

$$T = \frac{m}{2} g_{\bar{i}j} \frac{dx^{\bar{i}}}{dt} \frac{dx^j}{dt} + \frac{\lambda}{2} g^{j\bar{k}} g^{l\bar{i}} \frac{dg_{\bar{i}j}}{dt} \frac{dg_{\bar{k}l}}{dt} + \frac{\mu}{2} g^{j\bar{i}} g^{l\bar{k}} \frac{dg_{\bar{i}j}}{dt} \frac{dg_{\bar{k}l}}{dt}. \tag{54}$$

As usual, the matrix $\left[g^{i\bar{j}}\right]$ is reciprocal to $\left[g_{\bar{k}l}\right]$.

Let us observe that the models (50), (51), (53), (54) are structurally similar to our earlier affinely-invariant models of the affinely-rigid body [2, 3], i.e., roughly speaking to affinely-invariant geodetic models on the affine group. The idea there was that the material point was endowed with additional internal or collective degrees of freedom described by the attached linear basis $(\ldots, e_A, \ldots)$, or equivalently its dual $(\ldots, e^A, \ldots)$. The affinely-invariant kinetic energy was given by

$$T = \frac{m}{2} C_{ij} \frac{dx^i}{dt} \frac{dx^j}{dt} + \frac{A}{2} \Omega^i{}_j \Omega^j{}_i + \frac{B}{2} \Omega^i{}_i \Omega^j{}_j, \tag{55}$$

where $C_{ij} = \eta_{AB} e^A{}_i e^B{}_j$ is the Cauchy deformation tensor, η_{AB} is the fixed reference (material) metric, and $\Omega^i{}_j$ is so-called affine velocity (affine generalization of angular velocity),

$$\Omega^i{}_j = \frac{de^i{}_A}{dt} e^A{}_j. \tag{56}$$

This expression for T is affinely invariant and in spite of its apparently strange structure it is dynamically applicable, due to its strongly non-quadratic prescription (nonlinearity of equations of motion).

Let us now go back to our quantum problem. First of all, let us notice that even in our finite-level system with the nonlinearity directly introduced to the Schrödinger equation, the procedure is in general non-trivial. There are two reasons for that: the possible time-dependence of the Hamiltonian $H^a{}_b$, and the non-quadratic term $\mathcal{V}\left(\psi, \overline{\psi}\right)$. Nevertheless, it is still a provisional solution.

Much more geometric is the following reasoning. To give up the fixed scalar product and to introduce instead the dynamical one, in analogy to general relativity and continuum mechanics. And then to define the kinetic energy for Γ in analogy to (53):

$$T = L[\Gamma] = \frac{A}{2} \Gamma^{b\bar{c}} \Gamma^{d\bar{a}} \dot{\Gamma}_{\bar{a}b} \dot{\Gamma}_{\bar{c}d} + \frac{B}{2} \Gamma^{b\bar{a}} \Gamma^{d\bar{c}} \dot{\Gamma}_{\bar{a}b} \dot{\Gamma}_{\bar{c}d}. \tag{57}$$

Therefore, the configuration space of our system consists of pairs $\left(\psi^a, \Gamma_{\bar{a}b}\right)$. The Lagrangian may be given by

$$L[\psi, \Gamma] = L(1,2)[\psi, \Gamma] + \mathcal{V}[\psi, \Gamma] + L[\Gamma], \tag{58}$$

where $L(1,2)\,[\psi,\Gamma]$, $\mathcal{V}\,[\psi,\Gamma]$ are just the previously introduced models (16), (21), however with the dynamical, non-fixed Γ subject to the variational procedure. Obviously, this complicates the Euler-Lagrange equations even in their parts following only from $L(1,2)+\mathcal{V}$.

The resulting theory is essentially nonlinear and invariant under the group $\mathrm{GL}(W)$, instead of the unitary group $\mathrm{U}(W,\Gamma)$ which preserves the traditional quantum mechanics. In a sense, the gap between quantum and classical mechanics diffuses. Not everything is quantum, not everything is classical. And the effective nonlinearity creates some hope for explaining the quantum paradoxes. Let us observe that the variation of $I[\Gamma]=\int L[\Gamma]dt$ leads to the following equations:

$$-A\Gamma^{b\overline{n}}\left(\ddot{\Gamma}_{\overline{n}k}-\dot{\Gamma}_{\overline{n}l}\Gamma^{l\overline{c}}\dot{\Gamma}_{\overline{c}k}\right)\Gamma^{k\overline{a}}-B\Gamma^{l\overline{n}}\left(\ddot{\Gamma}_{\overline{n}l}-\dot{\Gamma}_{\overline{n}k}\Gamma^{k\overline{c}}\dot{\Gamma}_{\overline{c}l}\right)\Gamma^{b\overline{a}}=0. \tag{59}$$

As mentioned, the second term is merely a correction; the first term is essential. Let us notice that we did not perform variation in any other term of Lagrangian (58). It is so as if the ψ^{a}-degrees of freedom were non-excited. Of course, this is more than academic assumption, nevertheless convenient as a toy model. It is clear that the above equation (59) possesses solutions of the form:

$$\Gamma_{\overline{r}s}=G_{\overline{r}z}\exp(Et)^{z}{}_{s}=\exp(Ft)_{\overline{r}}{}^{\overline{z}}G_{\overline{z}s}, \tag{60}$$

where the initial condition for the scalar product $G=\Gamma(0)$ is a Hermitian sesquilinear form, $G\in\mathrm{Herm}\left(\overline{W}^{*}\otimes W^{*}\right)$. But this Hermitian property is to be preserved during the whole evolution. This will be the case when the linear mappings $E\in L(W)$, $F\in L\left(\overline{W}^{*}\right)$ will be G-Hermitian, i.e., when the sesquilinear forms analytically given by

$${}_{G}E_{\overline{r}s}=G_{\overline{r}z}E^{z}{}_{s},\qquad (F_{G})_{\overline{r}s}=F_{\overline{r}}{}^{\overline{z}}G_{\overline{z}s} \tag{61}$$

are Hermitian. Such solutions are analogous to our geodetic solutions in affinely-invariant models of the homogeneously deformable body [2, 3]. Obviously, there is a deep geometric difference, because in mechanics of homogeneously deformable bodies one deals with real mixed tensors describing configurations, while here we are doing with complex sesquilinear forms. Nevertheless, the general philosophy is the same. Let us observe some interesting facts. Namely, the above evolution of Γ may show all possible modes: it may be exponentially increasing, exponentially decaying, and even oscillatory. The point is how the initial data for $\Gamma(0)=G$, E, F are fixed. Obviously, the academic model of the evolution of Γ when ψ is fixed is rather non-physical, nevertheless, there is a hope that the mentioned ways of behaviour may have something to do with decoherence and measurement paradoxes. Obviously, this hypothesis may be confirmed only a posteriori, by solving, at least approximately, the total system of equations derived from (58) and (63) below. In any case, it is almost sure that the supposed dynamics of Γ should be based on (57) in (58). This follows from our demand of $\mathrm{GL}(W)$-invariance and from the analogy with general relativity and affine body dynamics. For example, we could try to use some fixed background metric G and assume:

$$L[G,\Gamma]=\frac{I}{2}G^{b\overline{c}}G^{d\overline{a}}\dot{\Gamma}_{\overline{a}b}\dot{\Gamma}_{\overline{c}d}+\frac{K}{2}G^{b\overline{a}}G^{d\overline{c}}\dot{\Gamma}_{\overline{a}b}\dot{\Gamma}_{\overline{c}d}. \tag{62}$$

When having G at disposal, we can also define potential-like terms $\mathrm{Tr}\left({}^{G}\Gamma^{p}\right)$, where ${}^{G}\Gamma^{r}{}_{s} := G^{r\bar{z}}\Gamma_{\bar{z}s}$. But of course, it seems aesthetically superfluous to fix some scalar products G taken from nowhere, when the dynamical one is used. And it is only $L[\Gamma]$ (57) that seems to have a chance for solving the decoherence problem due to its strong, geometrically motivated nonlinearity.

In any case, the simplest $\mathrm{GL}(W)$-invariant Lagrangian seems to have the form:

$$L = i\alpha_1\Gamma\left(\overline{\psi}^{\bar{a}}\dot{\psi}^{b} - \dot{\overline{\psi}}^{\bar{a}}\psi^{b}\right) + \alpha_2\Gamma_{\bar{a}b}\dot{\overline{\psi}}^{\bar{a}}\dot{\psi}^{b} + \left(\alpha_3\Gamma_{\bar{a}b} + \alpha_4 H_{\bar{a}b}\right)\overline{\psi}^{\bar{a}}\psi^{b}$$
$$+ \alpha_5\Gamma^{d\bar{a}}\Gamma^{b\bar{c}}\dot{\Gamma}_{\bar{a}b}\dot{\Gamma}_{\bar{c}d} + \alpha_6\Gamma^{b\bar{a}}\Gamma^{d\bar{c}}\dot{\Gamma}_{\bar{a}b}\dot{\Gamma}_{\bar{c}d} - \mathcal{V}(\psi,\Gamma), \tag{63}$$

where, e.g.,

$$\mathcal{V}(\psi,\Gamma) = \frac{\varkappa}{2}\left(\Gamma_{\bar{a}b}\overline{\psi}^{\bar{a}}\psi^{b} - b\right)^{2}. \tag{64}$$

This expression (63) contains all the structural terms mentioned above and is $\mathrm{GL}(W)$-invariant. All quantities in it (except real constants) are dynamical variables and are subject to the variational procedure. We do not investigate in detail the resulting equations of motion. They are very complicated and describe the mutual interaction between ψ^{a}, $\Gamma\overline{ab}$. Nevertheless, their structure is interesting and instructive. Let us quote them for the above Lagrangian (63):

$$\frac{\partial L}{\partial\overline{\psi}^{\bar{a}}} = \left(2i\alpha_1\Gamma_{\bar{a}b} - \alpha_2\dot{\Gamma}_{\bar{a}b}\right)\dot{\psi}^{b} - \alpha_2\Gamma_{\bar{a}b}\ddot{\psi}^{b}$$
$$+ \left(i\alpha_1\dot{\Gamma}_{\bar{a}b} + \alpha_3\Gamma_{\bar{a}b} + \alpha_4\Gamma H_{\bar{a}b} - \mathcal{V}'\Gamma_{\bar{a}b}\right)\psi^{b} = 0, \tag{65}$$
$$\frac{\partial L}{\partial\Gamma_{\bar{a}b}} = -A\Gamma^{b\bar{n}}\left(\ddot{\Gamma}_{\bar{n}k} - \dot{\Gamma}_{\bar{n}l}\Gamma^{l\bar{c}}\dot{\Gamma}_{\bar{c}k}\right)\Gamma^{k\bar{a}} - B\Gamma^{l\bar{n}}\left(\ddot{\Gamma}_{\bar{n}l} - \dot{\Gamma}_{\bar{n}k}\Gamma^{k\bar{c}}\dot{\Gamma}_{\bar{c}l}\right)\Gamma^{b\bar{a}}$$
$$+ i\alpha_1\left(\overline{\psi}^{\bar{a}}\dot{\psi}^{b} - \dot{\overline{\psi}}^{\bar{a}}\psi^{b}\right) + \alpha_2\dot{\overline{\psi}}^{\bar{a}}\dot{\psi}^{b} + \left(\alpha_3 - \mathcal{V}'\right)\overline{\psi}^{\bar{a}}\psi^{b} = 0. \tag{66}$$

In spite of their relatively complicated structure, these nonlinear equations are readable. For any case we have retained the direct nonlinearity term derived from $\mathcal{V}$. But the main idea of nonlinearity and large $\mathrm{GL}(W) \simeq \mathrm{GL}(n,\mathbb{C})$-symmetry is just the interaction between ψ and Γ. And it is just the interaction of the generally-relativistic type. As seen, at the same time it is structurally similar to affinely-invariant geodetic models of elastic vibrations of the homogeneously deformable body [2, 3]. Even independently on our quantum programme, this model is interesting in itself as an example of highly-symmetric dynamical system on a homogeneous space. Nonlinearity of the system is rational because the inverse matrix $\left[\Gamma^{a\bar{b}}\right]$ is a rational function of $\left[\Gamma_{\bar{c}d}\right]$. Therefore, there are some hopes for a solvability, perhaps at least qualitative or approximate, of the system (65), (66).

Let us notice that the α_3-controlled term may be included into the α_4-expression. We simply decided to write it separately to stress the special role of the Hamiltonian terms proportional to the identity operator. Let us stress that the Lagrangian (63) is not the only expression with the above enumerated properties. Rather, it is the simplest one. For example, one might replace $\Gamma^{a\bar{b}}$ by $\Gamma^{a\bar{b}} + \alpha_7 \psi^a \overline{\psi}^{\bar{b}}$, etc. Perhaps this will modify somehow the resulting equations, but in a rather non-essential way. All important features are already predicted by equations following from (63). Let us remind that in (63) we have according to quantum mechanics

$$\alpha_1 = \frac{\hbar}{2}, \qquad \alpha_4 = -1. \tag{67}$$

The nonlinearity contained in (63) resembles in a sense the Thomas-Fermi approximation in quantum mechanics [12, 13]. It may be considered as an alternative way of describing open systems.

When concentrating on the dynamical system model of Schrödinger equation, one is faced with the interesting question as to what might be described by admitting non-Hamiltonian forces. Perhaps nothing physical. There is nevertheless some possibility that such forces might be useful for describing quantum dissipative phenomena. However, at this stage we have no idea concerning this problem.

The $\mathrm{GL}(W) \simeq \mathrm{GL}(n, \mathbb{C})$-invariance of the Hamiltonian system implies the existence of n^2 complex constants of motion. We do not quote here their explicit form to avoid writing unnecessary and complicated formulas. Nevertheless, their existence is a remarkable property of the theory.

Let us stress that the strong and geometrically implied nonlinearity of equations following from (63) gives a chance for the macroscopic reinforcement and enhancement of quantum events.

It is very important to remember that in the model (63), (65), (66), based on the mutual interaction between ψ and Γ, the "scalar product" Γ is not a constant of motion. And there is an exchange of energy between ψ- and Γ-degrees of freedom, especially in situations when the Γ-motion is remarkably excited. Therefore, if $\alpha_5 \neq 0$, $\alpha_6 \neq 0$, the quantity Γ, although fundamental for physical interpretation, in a sense loses its physical meaning of the scalar product. If $\alpha_2 = 0$, $\alpha_5 = 0$, $\alpha_6 = 0$, then the invariance of $\mathcal{L}$ under the $\mathrm{U}(1)$-group

$$\psi \mapsto \exp\left(-\frac{i}{\hbar} e\chi\right) \psi \tag{68}$$

implies, via Noether theorem, that

$$e\frac{2\alpha_1}{\hbar}\langle\psi|\psi\rangle = e\frac{2\alpha_1}{\hbar}\Gamma_{\bar{a}b}\overline{\psi}^{\bar{a}}\psi^b \tag{69}$$

is a constant of motion. And then the usual polarization formula for quadratic forms implies that $\langle\psi|\varphi\rangle = \Gamma_{\bar{a}b}\overline{\psi}^{\bar{a}}\varphi^b$ is preserved. If $\alpha_2 \neq 0$, this procedure leads to the scalar product

$$\Gamma_{\bar{a}b}\overline{\psi}^{\bar{a}}\varphi^b + \frac{i\alpha_2}{2\alpha_1}\Gamma_{\bar{a}b}\left(\overline{\psi}^{\bar{a}}\dot{\varphi}^b - \dot{\overline{\psi}}^{\bar{a}}\varphi^b\right). \tag{70}$$

It fails to be positively definite, nevertheless, it seems to be physically interesting [1]. And finally, when $\alpha_5 \neq 0$, $\alpha_6 \neq 0$, Noether theorem tells us that the U(1)-constant of motion differs form the above ones by the term proportional to $\Gamma^{a\bar{b}}\dot{\Gamma}_{\bar{b}a}$. But now Γ is a dynamical variable and the last expression is not quadratic in $(\Gamma_{\bar{a}b}, \dot{\Gamma}_{\bar{a}b})$. Because of this there is no polarization procedure and no well-defined Noether-based scalar product at all.

4. Towards general systems

The main ideas of our nonlinearity model in quantum mechanics were formulated and presented in the simplest case of finite-level systems. Let us mention now about a more general situation. We concentrate mainly on the non-relativistic Schrödinger mechanics. It is true that on the very fundamental level of theoretical physics one is interested mainly in relativistic theory. Nevertheless, the non-relativistic counterparts are also interesting at least from the methodological point of view; they just enable us to understand deeper the peculiarity of relativistic theory. And of course they may be also directly useful physically in situations where in condensed matter theory, relativistic effects are not relevant, e.g., in superconductivity and superfluidity.

Galilei group and Galilei space-time are structurally much more complicated than Poincare (inhomogeneous Lorentz) group and Minkowski space-time. Because of this, the construction of Lagrangian for the Schrödinger theory need some comments and a few steps of reasoning.

One is rather used to start with wave-mechanical ideas of Schrödinger equation, i.e.,

$$i\hbar\frac{\partial\psi}{\partial t} = \mathbb{H}\psi \tag{71}$$

rather than with some yet unspecified precisely field theory on the Galilei space-time. The standard scalar product of Schrödinger amplitudes on the three-dimensional Euclidean space will be denoted as usually by

$$\langle\psi_1|\psi_2\rangle = \int \overline{\psi_1(x)}\psi_2(x)d_3x, \tag{72}$$

where, obviously, orthogonal Cartesian coordinates are meant. More generally, in curvilinear coordinates we would have

$$\langle\psi_1|\psi_2\rangle = \int \overline{\psi_1(q)}\psi_2(q)\sqrt{|g(3)|}d_3q, \tag{73}$$

where $|g(3)|$ is an abbreviation for the determinant of the matrix of Euclidean metric tensor $g(3)$, and q^i are generalized coordinates. The same formula is valid in the Riemann space, where all coordinates are "curvilinear". However, here we do not get into such details.

Hermitian conjugation of operators, $\mathbb{A} \mapsto \mathbb{A}^+$ is meant in the usual sense of $L^2(\mathbb{R}^3)$ with the above scalar product,

$$\langle\psi_1|\mathbb{A}\psi_2\rangle = \langle\mathbb{A}^+\psi_1|\psi_2\rangle, \tag{74}$$

obviously with some additional remarks concerning domains. The Hamilton operator is self-adjoint, $\mathbb{H}^{+} = \mathbb{H}$, also with some care concerning the domains.

The Schrödinger equation is again variational with the Lagrangian

$$\mathcal{L} = \mathcal{L}_T + \mathcal{L}_H, \tag{75}$$

where, $\mathcal{L}_T$, $\mathcal{L}_H$ denote respectively the terms depending linearly on first-order time derivatives and independent on them,

$$\mathcal{L}_T = \frac{i\hbar}{2}\left(\overline{\psi}\frac{\partial \psi}{\partial t} - \frac{\partial \overline{\psi}}{\partial t}\psi\right) = -\mathrm{Im}\left(\hbar\overline{\psi}\frac{\partial \psi}{\partial t}\right) = \mathrm{Re}\left(i\hbar\overline{\psi}\frac{\partial \psi}{\partial t}\right), \tag{76}$$

$$\mathcal{L}_H = -\overline{\psi}\left(\mathbb{H}\psi\right). \tag{77}$$

The total action is

$$I = I_T + I_H = \int \mathcal{L}_T dt d_3 x + \int \mathcal{L}_H dt d_3 x, \tag{78}$$

and the resulting variational derivative of I equals

$$\frac{\delta I}{\delta \overline{\psi}(t,x)} = -\mathbb{H}\psi + i\hbar\frac{\partial \psi}{\partial t}, \tag{79}$$

therefore, the stationary points are given indeed by the solutions of (71). In the very important special case of a material point moving in the potential field V,

$$\mathbb{H} = -\frac{\hbar^2}{2m}g^{ij}\frac{\partial^2}{\partial x^i \partial x^j} + V = -\frac{\hbar^2}{2m}\Delta + V, \tag{80}$$

it is convenient and instructive to rewrite the Lagrangian term $\mathcal{L}_H$ in a variationally equivalent form structurally similar to the Klein-Gordon Lagrangian,

$$\mathcal{L}'_H = -\frac{\hbar^2}{2m}g^{ij}\frac{\partial \overline{\psi}}{\partial x^i}\frac{\partial \psi}{\partial x^j} + V\overline{\psi}\psi, \tag{81}$$

which differs from the original one by a total divergence term. In particular, there are no artificial second derivatives in $\mathcal{L}'_H$. This resembles the situation one is faced with in General Relativity. For a free particle the total Lagrangian equals

$$\mathcal{L}' = \frac{i\hbar}{2}\left(\overline{\psi}\frac{\partial \psi}{\partial t} - \frac{\partial \overline{\psi}}{\partial t}\psi\right) - \frac{\hbar^2}{2m}g^{ij}\frac{\partial \overline{\psi}}{\partial x^i}\frac{\partial \psi}{\partial x^j}. \tag{82}$$

In this form, structurally as similar as possible to the Lagrangian for the Klein-Gordon field, the essential geometrical similarities and differences between Galilei and Poincare quantum symmetries are visible.

The well-known expressions for the probability density and probability current are obtained in non-relativistic quantum mechanics on the basis of some rather rough and intuitive statistical concepts. However, when the Schrödinger equation is interpreted within the framework of field theory on the Galilean space-time, those concepts appear as direct consequences of Noether theorem just as said above in the finite number of levels part. The Schrödinger Lagrangian for a free particle is invariant under the group $U(1)$ of global gauge transformations,

$$\psi \mapsto \exp\left(-\frac{i}{\hbar}e\chi\right)\psi, \tag{83}$$

where e denotes the coupling constant (elementary charge) and χ is the gauge parameter. Within the Kaluza-Klein formulation of electrodynamics of point charges, the variable χ and electric charge e are canonically conjugate quantities. The invariance under this gauge group implies the conservation law for the Galilean current $(j^t, \overline{j})$, where

$$j^t = \varrho = e\overline{\psi}\psi, \qquad j^a = \frac{e\hbar}{2im}\left(\overline{\psi}\frac{\partial\psi}{\partial x^a} - \frac{\partial\overline{\psi}}{\partial x^a}\psi\right). \tag{84}$$

The resulting continuity equation has the usual form:

$$\frac{\partial\varrho}{\partial t} + \frac{\partial j^a}{\partial x^a} = 0, \qquad \text{i.e.,} \qquad \frac{\partial\varrho}{\partial t} + \text{div}\overline{j} = 0. \tag{85}$$

Let us stress that all those formulas hold in rectilinear orthonormal coordinates. In general coordinates we would have to multiply the above expressions by $\sqrt{|g(3)|}$ so as to turn them respectively into scalar and contravariant vector densities. And so one does in a general Riemann space. Continuity equation is satisfied if ψ is a solution of the Schrödinger equation.

The functional

$$\psi \mapsto Q_t[\psi] = \int j^t d_3x = \int \varrho(t,x)d_3x \tag{86}$$

describes the total charge at the time instant t. On the basis of field equations it is independent on time,

$$\frac{d}{dt}Q_t[\psi] = 0. \tag{87}$$

This is the global law of the charge conservation. Let us observe that Q is a functional quadratic form of ψ. Its polarization reproduces the usual scalar product as a sesquilinear Hermitian form (up to a constant factor):

$$4e\langle\psi|\varphi\rangle = Q[\psi+\varphi] - Q[\psi-\varphi] - iQ[\psi+i\varphi] + iQ[\psi-i\varphi]. \tag{88}$$

In a consequence of charge conservation law $\langle\psi|\varphi\rangle$ is time-independent if ψ, φ are solutions of the same Schrödinger equation.

One can do the same for multiplets of scalar Schrödinger fields ψ^a, i.e., for the continuum of systems described in the previous section. Then we have

$$\mathcal{L}_T = \frac{i\hbar}{2}\Gamma_{\bar{a}b}\left(\overline{\psi}^{\bar{a}}\frac{\partial\psi^b}{\partial t} - \frac{\partial\overline{\psi}^{\bar{a}}}{\partial t}\psi^b\right), \tag{89}$$

$$\langle\psi|\varphi\rangle = \int \Gamma_{\bar{a}b}\overline{\psi}^{\bar{a}}\varphi^b\sqrt{|g(3)|}d_3q, \tag{90}$$

$$L_T = \mathrm{Re}\left\langle \psi|i\hbar\frac{\partial\psi}{\partial t}\right\rangle = -\mathrm{Im}\left\langle \psi|\hbar\frac{\partial\psi}{\partial t}\right\rangle, \tag{91}$$

$$\mathcal{L}_H = -\Gamma_{\bar{a}b}\overline{\psi}^{\bar{a}}\left(\mathbb{H}\psi\right)^b, \qquad L_H = -\langle\psi|\mathbb{H}\psi\rangle. \tag{92}$$

And again, for systems with scalar potentials, when

$$(\mathbb{H}\psi)^a = -\frac{\hbar^2}{2m}\Delta\psi^a + V^a{}_b\psi^b, \tag{93}$$

we can remove second derivatives as a divergence term and obtain the modified Lagrangian

$$\mathcal{L}'_H = -\frac{\hbar^2}{2m}\Gamma_{\bar{a}b}\frac{\partial\overline{\psi}^{\bar{a}}}{\partial x^i}\frac{\partial\psi^b}{\partial x^j}g(3)^{ij} - V_{\bar{a}b}\overline{\psi}^{\bar{a}}\psi^b, \tag{94}$$

where

$$V_{\bar{a}b} = \Gamma_{\bar{a}c}V^c{}_b. \tag{95}$$

We have used here Cartesian coordinates for simplicity. Using the operator of linear momentum,

$$\mathbb{P}_a = \frac{\hbar}{i}\frac{\partial}{\partial x^a}, \tag{96}$$

we can write

$$L_H = -\frac{1}{2m}\langle\mathbb{P}_a\psi|\mathbb{P}_b\psi\rangle\, g(3)^{ab} - \langle\psi|V\psi\rangle. \tag{97}$$

In analogy to (16), (17), (21) we can write the direct-nonlinearity Lagrangian as follows:

$$\begin{aligned}\mathcal{L} = {} & i\alpha_1\Gamma_{\bar{a}b}\left(\overline{\psi}^{\bar{a}}\dot{\psi}^b - \dot{\overline{\psi}}^{\bar{a}}\psi^b\right) + \alpha_2\Gamma_{\bar{a}b}\dot{\overline{\psi}}^{\bar{a}}\dot{\psi}^b \\ & + \alpha_3\Gamma_{\bar{a}b}\frac{\partial\overline{\psi}^{\bar{a}}}{\partial x^i}\frac{\partial\psi^b}{\partial x^j}g(3)^{ij} + \alpha_4 V_{\bar{a}b}\overline{\psi}^{\bar{a}}\psi^b - \mathcal{V}(\psi,\Gamma),\end{aligned} \tag{98}$$

where, e.g., the non-quadratic term $\mathcal{V}$ responsible for nonlinearity may be chosen in a quartic form (22). The combination of the α_2-, α_3-terms may be purely academic, as explained above, but it may also describe something like the Klein-Gordon phenomena, when α_2 and α_3 are appropriately suited. Indeed, at an appropriate ratio $\alpha_2 : \alpha_3$ the above expression for $\mathcal{L}$ becomes:

$$\mathcal{L} = i\alpha_1\Gamma_{\overline{a}b}\left(\overline{\psi}^{\overline{a}}\dot{\psi}^b - \dot{\overline{\psi}}^{\overline{a}}\psi^b\right) + \alpha_{231}\Gamma_{\overline{a}b}\frac{\partial\overline{\psi}^{\overline{a}}}{\partial x^\mu}\frac{\partial\psi^b}{\partial x^\nu}g^{\mu\nu}$$
$$- \alpha_{232}\Gamma_{\overline{a}b}\overline{\psi}^{\overline{a}}\psi^b + \alpha_4 V_{\overline{a}b}\overline{\psi}^{\overline{a}}\psi^b - \mathcal{V}(\psi,\Gamma), \qquad (99)$$

where g is the Minkowskian metric. For the notational simplicity we used Cartesian coordinates in this formula, to avoid the multiplying of $\mathcal{L}$ by $\sqrt{|g|}$. Of course, within relativistic framework it is rather artificial to superpose the g-Minkowskian terms with the non-relativistic α_1-term. The consequent relativistic theory should rather combine the Klein-Gordon and Dirac terms. Therefore, $\mathcal{L}$ should be then postulated as:

$$\mathcal{L} = i\alpha_{11}\Gamma^\mu{}_{\overline{a}b}\left(\overline{\psi}^{\overline{a}}\frac{\partial\psi^b}{\partial x^\mu} - \frac{\partial\overline{\psi}^{\overline{a}}}{\partial x^\mu}\psi^b\right) + \alpha_{231}\Gamma_{\overline{a}b}\frac{\partial\overline{\psi}^{\overline{a}}}{\partial x^\mu}\frac{\partial\psi^b}{\partial x^\nu}g^{\mu\nu}$$
$$- \alpha_{232}\Gamma_{\overline{a}b}\overline{\psi}^{\overline{a}}\psi^b + \alpha_4 V_{\overline{a}b}\overline{\psi}^{\overline{a}}\psi^b - \mathcal{V}(\psi,\Gamma). \qquad (100)$$

Here we do not fix entities and constants. As usual in the Dirac framework, Γ in the α_{231}-, α_{232}-terms is the sesquilinear Hermitian form of the neutral signature $(++--)$. And Γ^μ are sesquilinear Hermitian Dirac forms, i.e., raising their first index with the help of the reciprocal contravariant $\Gamma^{a\overline{b}}$, $\Gamma^{a\overline{c}}\Gamma_{\overline{c}b} = \delta^a{}_b$, one obtains the Dirac matrices $\gamma^{\mu a}{}_b = \Gamma^{a\overline{c}}\Gamma^\mu{}_{\overline{c}b}$ satisfying the anticommutation rules:

$$\gamma^\mu\gamma^\nu + \gamma^\nu\gamma^\mu = 2g^{\mu\nu}I_4. \qquad (101)$$

Strictly speaking, the quantities Γ_{ab} at α_{231}-, α_{232}-terms need not be the same. Nevertheless, in the fundamental quantum studies it is convenient to identify them. Let us mention, e.g., some ideas connected with the Dirac-Klein-Gordon equation appearing in certain problems of mathematical physics [2, 3]. In particular, it turns out that the SU$(2,2)$-ruled (conformally ruled) theory of spinorial geometrodynamics has a specially-relativistic limit based on Lagrangian similar to (100). In any case, the specially-relativistic theory based on the superposition of the Klein-Gordon and Dirac Lagrangians is interesting from the point of view of some peculiar kinship between pairs of fundamental particles. We mean pairs of fermions and pairs of quarks which occur in the standard model [9, 14, 15].

Let us go back to our general model (98), not necessarily relativistic one. Its nonlinearity is contained only in the non-quadratic potential term $\mathcal{V}(\psi,\Gamma)$. Lagrangian $\mathcal{L}$ and the action are local in the x-space when the Hamilton operator is a sum of the position- and momentum-type operators. However, in general the corresponding contribution to action is given by

$$I_\chi[\psi] = -\gamma\int\overline{\psi}^{\overline{a}}(t,x)\chi_{\overline{a}b}(x,y)\psi^b(t,y)dtdxdy, \qquad (102)$$

therefore,

$$\frac{\delta I_\chi}{\delta \overline{\psi}^{\overline{a}}(t,x)} = -\gamma \int \chi_{\overline{a}b}(x,y)\psi^b(t,y)dy. \quad (103)$$

This integral expression implies that our equations of motion will be integro-differential ones. If we seriously assume that also the scalar product is given by the double spatial integral, so that, e.g.,

$$I_1 = i\alpha \int \Gamma_{\overline{a}b}(x,y)\left(\overline{\psi}^{\overline{a}}(t,x)\dot{\psi}^b(t,y) - \dot{\overline{\psi}}^{\overline{a}}(t,x)\psi^b(t,y)\right) dtdxdy, \quad (104)$$

and in general

$$\Gamma_{\overline{a}b}(x,y) \neq \Gamma_{\overline{a}b}\delta(x-y), \quad (105)$$

then the variational derivative of I_1 also leads to the integro-differential equation, because

$$\frac{\delta I_1}{\delta \overline{\psi}^{\overline{a}}(t,x)} = 2i\alpha \int \Gamma_{\overline{a}b}(x,y)\dot{\psi}^b(t,y)dy. \quad (106)$$

The necessity of using integro-differential equations is embarrassing. However, the main difficulty appears when we wish to follow the finite-level systems dynamics in the general case. Namely, it was relatively easy to write formally something like the elements $\left[\Gamma^{a\overline{b}}\right]$ of the matrix reciprocal to $[\Gamma_{\overline{c}d}]$ for the finite-dimensional system. And one can try to follow this procedure for the dynamical scalar product in the general infinite-dimensional case. However, such a hybrid is not convincing. It would be then only "internal degrees of freedom" subject to the procedure of the dynamical scalar product. No doubt that this does not seem satisfactory. Let us write the dynamical scalar product of the Schrödinger-like quantum mechanics in the form:

$$\langle\psi|\varphi\rangle = \int \Gamma_{\overline{a}b}(x,y)\overline{\psi}^{\overline{a}}(x)\varphi^b(y)dxdy, \quad (107)$$

where $\overline{\Gamma_{\overline{a}b}(x,y)} = \Gamma_{\overline{b}a}(y,x)$. Assuming in addition the translational invariance, we have $\Gamma(x,y) = \Gamma(x-y)$, i.e.,

$$\langle\psi|\varphi\rangle = \int \Gamma_{\overline{a}b}(x-y)\overline{\psi}^{\overline{a}}(x)\varphi^b(y)dxdy. \quad (108)$$

A simplifying assumption would be factorization

$$\Gamma_{\overline{a}b}(x,y) = \Gamma_{\overline{a}b}\mathcal{K}(x,y) = \Gamma_{\overline{a}b}\mathcal{K}(x-y). \quad (109)$$

Let us remind that in the usual quantum mechanics without the dynamical scalar product, we have simply

$$\Gamma_{\overline{a}b}(x,y) = \Delta_{\overline{a}b}\delta(x-y), \quad (110)$$

where Δ is a positively definite algebraic scalar product for internal modes. In appropriately chosen basis we have simply $\Delta_{\overline{a}b} = \delta_{\overline{a}b}$. In the mentioned half-a-way approach, (110) is still valid but with the dynamical, time-dependent $\Delta_{\overline{a}b}$.

Much more reasonable, although incomparatively more difficult, would be a consequent approach based on the dynamical scalar product (107). Here we only mention some ideas. The main point is the construction of the full inverse of Γ in (107), with the time-dependent Γ, i.e., analytically $\Gamma^{a\overline{b}}(x,y;t)$ such that

$$\int \Gamma^{a\overline{c}}(x,u;t)\Gamma_{\overline{c}b}(u,y;t)du = \delta^{a}{}_{b}\delta(x-y). \tag{111}$$

Obviously, in general it is a rather very difficult problem to find explicitly the formula for that inverse. One can try to discretize it by choosing some appropriate finite (or perhaps countable) family Ω of vectors a_p such that

$$\Gamma^{a\overline{b}}(x,y;t) = \sum_{p\in\Omega} \Gamma^{a\overline{b}}(p;t)\delta(x-y+a_p). \tag{112}$$

This may be used as a basis for some discretization procedure like the finite-element method for finding $\Gamma^{a\overline{b}}(x,y;t)$.

In any case, everything said in the former section about the dynamical scalar product and about the failure of the absolute scalar product (72), (73) remains true in wave mechanics on the differential configuration manifold. Although it is true that in this case everything becomes much more complicated.

Acknowledgements

This paper partially contains results obtained within the framework of the research project N N501 049 540 financed from the Scientific Research Support Fund in 2011-2014 just as the previous contribution by Jan J. Sławianowski *Order of Time Derivatives in Quantum-Mechanical Equations* to the book *Measurements in Quantum Mechanics* [4]. The authors are greatly indebted to the Ministry of Science and Higher Education for this financial support.

Author details

Jan J. Sławianowski* and Vasyl Kovalchuk

* Address all correspondence to: jslawian@ippt.gov.pl

Institute of Fundamental Technological Research, Polish Academy of Sciences, Warsaw, Poland

References

[1] Sławianowski JJ, Kovalchuk V. Schrödinger and Related Equations as Hamiltonian Systems, Manifolds of Second-Order Tensors and New Ideas of Nonlinearity in Quantum Mechanics. Reports on Mathematical Physics 2010;65(1) 29–76.

[2] Sławianowski JJ, Kovalchuk V, Sławianowska A, Gołubowska B, Martens A, Rożko EE, Zawistowski ZJ. Affine Symmetry in Mechanics of Collective and Internal Modes. Part I. Classical Models. Reports on Mathematical Physics 2004;54(3) 373–427.

[3] Sławianowski JJ, Kovalchuk V, Sławianowska A, Gołubowska B, Martens A, Rożko EE, Zawistowski ZJ. Affine Symmetry in Mechanics of Collective and Internal Modes. Part II. Quantum Models. Reports on Mathematical Physics 2005;55(1) 1–45.

[4] Sławianowski JJ. Order of Time Derivatives in Quantum-Mechanical Equations. In: Pahlavani MR. (ed.) Measurements in Quantum Mechanics. Rijeka: INTECH; 2012. p.57–74.

[5] Doebner HD, Goldin GA. Introducing Nonlinear Gauge Transformations in a Family of Nonlinear Schrödinger Equations. Phys. Rev. A 1996;54 3764–3771.

[6] Doebner HD, Goldin GA, Nattermann P. Gauge Transformations in Quantum Mechanics and the Unification of Nonlinear Schrödinger Equations. J. Math. Phys. 1999;40 49–63; quant-ph/9709036.

[7] Kozlowski M, Marciak-Kozlowska J. From Quarks to Bulk Matter. USA: Hadronic Press; 2001.

[8] Marciak-Kozlowska J, Kozlowski M. *Schrödinger Equation for Nano-Science*, cond-mat/0306699.

[9] Sławianowski JJ. New Approach to the U(2,2)-Symmetry in Spinor and Gravitation Theory. Fortschr. Phys./Progress of Physics 1996;44(2) 105–141.

[10] Bogoliubov NN, Shirkov DV. Quantum Fields. Reading, Mass.: Benjamin/Cummings Pub. Co.; 1982.

[11] Krawietz A. Natürliche Geometrie der Deformationsprozesse. ZAMM 1979;59 T199–T200.

[12] Landau LD, Lifshitz EM. Quantum Mechanics. London: Pergamon Press; 1958.

[13] Messiah A. Quantum Mechanics. Amsterdam: North-Holland Publishing Company; 1965.

[14] Dvoeglazov VV. The Barut Second-Order Equation, Dynamical Invariants and Interactions. J. Phys. Conf. Ser. 2005;24 236–240; math-ph/0503008.

[15] Kruglov SI. On the Generalized Dirac Equation for Fermions with Two Mass-States. An. Fond. Louis de Broglie 2004;29 1005–1–16; quant-ph0408056.

Chapter 6

Schrödinger Equation and (Future) Quantum Physics

Miloš V. Lokajíček, Vojtěch Kundrát and
Jiří Procházka

Additional information is available at the end of the chapter

1. Introduction

The Schrödinger equation based on the Hamiltonian taken from the classical physics provides the solutions that may be correlated to the solutions of Hamilton equations or to their superpositions, if the additional assumptions introduced by Bohr have not been added; i.e., simple Hilbert space spanned on one set of Hamiltonian eigenfunctions and each vector representing pure state. It may describe, therefore, physical processes at microscopic as well as macroscopic levels; only the set of allowed states being partially limited. However, the given equation represents approximative phenomenological theory that is not able to explain the emergence of quantum states on the basis of Coulomb potential only; e.g., in the case of hydrogen atom such a state may arise evidently only if a repulsive short-ranged force (or some contact interaction) exists between electron and proton at low kinetic energy values. The dimensions and other properties of proton should play then important role in such a case. These characteristics may be derived from different experiments studying collisions between corresponding objects at different energies. They should be taken into account when the quantum physics is to pass from hitherto mere phenomenological description of physical evolution to causal realistic interpretation as it was common in classical physics. It concerns mainly the spontaneous transitions between different quantum states in atoms.

However, let us start with short summary of main points concerning the evolution of quantum mechanics theory during the 20th century. The Copenhagen quantum mechanics (based fundamentally on the Schrödinger equation [1] and proposed by Bohr [2] in 1927) has been taken as the only theory of microscopic physical processes till the end of the 20th century, even if it has involved a series of paradoxes corresponding in principle to internal contradictions contained in corresponding mathematical model. Already in 1933 Pauli [3] showed that the corresponding Hamiltonian was to exhibit continuous energy spectrum in the whole interval $E \in (-\infty, +\infty)$, which contradicted the existence of quantized energy states in closed systems. However, it has been possible to say that this fact did not raised any greater interest at that time.

More attention was devoted to the criticism of Einstein [4] in 1935, who showed with the help of a Gedankenexperiment that the given quantum theory required the existence of immediate interaction (or rather linkage) between two very distant matter objects, which was denoted by him as unacceptable on the basis of standard ontological experience. The given critique was refused by Bohr [5] having argued that this distant immediate interaction might exist between microscopic objects (at the difference to macroscopic phenomena). The world scientific community accepted fully Bohr's standpoint that was supported partially by the earlier argument of von Neumann [6] who refused the existence of any local (hidden) parameters in Copenhagen theory; the problem having been discussed several years ago. The argument of Grete Herrmann that the given conclusion of von Neumann was based on circular proof [7] was not taken into account, having remained practically unknown at that time.

A partial change occurred in 1952 when Bohm [8] showed that an additional (hidden) parameter existed already in the simplest solutions of Schrödinger equation. Two alternatives (Copenhagen mechanics and hidden-variable theory) were then considered in the microscopic region. The decision between them was expected to be done on the basis of experimental results when Bell [9] derived his inequalities in 1964 which were assumed to hold in the hidden-variable theory and not in the Copenhagen alternative. The corresponding experiments based in principle on the original proposal of Einstein were finished in 1982 and the inequalities of Bell were provably violated [10]. The Copenhagen quantum mechanics was then taken as the only physical theory valid for microscopic physical processes.

Only in the end of the last century it was shown for the first time that the given conclusion was based on the mistaking assumption and that the given inequalities did not hold in any quantum alternative based on Schrödinger equation [11]. In 2004 it was then shown by Rosinger [12] that Bell's inequalities contradicted also the inequalities derived by Boole (in 1854) for any probabilistic system [13, 14]:

$$max\,\{p_1, p_2, ..., p_n\} \;\leq\; P(A_1 \cup A_2 \cup .... \cup A_n) \;\leq\; min\,\{1, p_1 + p_2 + ... + p_n\}$$

$$max\,\{0, p_1 + p_2 + ... + p_n \;- n + 1\} \;\leq\; P(A_1 \cap A_2 \cap .... \cap A_n) \;\leq\; min\,\{1, p_1, p_2, ..., p_n\}$$

where the first relation concerns the probabilities p_j of possible alternative phenomena A_j and the other one concerns those of contemporary phenomena.

More detailed analysis of the Bell inequalities problem may be found in [15]. It has been possible to conclude that these inequalities have been valid only in the classical physics, not in any kind of quantum theory. They have been mistakenly applied to experiment where photon polarizations have been measured while this possibility has been excluded by the assumption on the basis of which they were derived.

The consequences following from this fact have been summarized in [16]; more complete summary of all problems of contemporary quantum mechanics has been then presented in

[17]. It has followed from these results that the Schrödinger equation itself may be denoted as common theory of all physical reality when the Hilbert space formed by its solutions has been adapted to given physical system (i.e., correspondingly extended) in contradiction to limiting and deforming conditions imposed by Bohr.

The given conclusion has followed practically from the fact that the Schrödinger equation may be derived for the set of statistical combinations of Hamilton equation solutions when the given set has been limited by a suitable condition (e.g., by Boltzmann statistics); see [18, 19]. It means that any Schrödinger function $\psi(x,t)$ may represent always a classical state or a statistical combination of such states; see also [20] (or already quoted [17]).

However, the Schrödinger equation (including Coulomb potential only) represents approximative phenomenological theory only. It is not able to explain any emergence of quantum energy states. In such a case some additional properties of individual objects forming a quantized system must be taken into account. E.g., in the case of hydrogen atom any quantum state could not come into being only on the basis of Coulomb force if some further repulsive or contact interaction did not exist between electron and proton that would depend necessarily on dimensions and structures of main constituents. The study of corresponding structures of individual objects should represent, therefore, inseparable part of contemporary quantum physics.

These structures may be derived mainly from the data gained in experiments concerning the collisions between corresponding particles. However, it is not sufficient to look for a phenomenological description of some measured values. More detailed collision models must be made use of in such a case. They must be able to study the dependence on impact parameter value; e.g., at least the so called eikonal model should be made use of. Some results obtained on the given realistic basis will be introduced in the following; it will be discussed how they may be made use of in solving the problems of quantum physics in the next future.

In Sec. 2 we shall start with discussing the significance of the Schrödinger equation, based on the fact that it may be derived in principle from Hamilton equations. The necessity of repulsive force in the emergence of quantum states in closed systems will be then discussed in Sec. 3. The eikonal model of elastic collisions between hadrons and some results obtained for elastic proton-proton collisions in the impact parameter space will be presented in Sec. 4. The problem of validity of optical theorem on which practically all contemporary elastic collision results have been based will be discussed in Sec. 5. In Sec. 6 new probabilistic ontological model enabling to study the existence of internal proton structures will be applied to experimental data and new results will be presented. In Sec. 7 some metaphysical consequences will be considered. Several open questions will be then mentioned in Sec. 8.

2. Schrödinger equation and Hilbert space structure

The evolution of a physical system consisting of different matter objects may be described with the help of Hamilton equations

$$\dot{q}_j = \frac{\partial H}{\partial p_j}, \quad \dot{p}_j = -\frac{\partial H}{\partial q_j} \tag{1}$$

where q_j and p_j are space coordinates and momenta of all individual objects; the corresponding Hamiltonian given by

$$H = \frac{p_j^2}{2m} + V(q_j) \tag{2}$$

represents the total energy of the given system.

Schrödinger equation may be then written as

$$i\hbar\frac{\partial\psi(x,t)}{\partial t} = H\psi(x,t) \tag{3}$$

where x represents the set of coordinates and q_j and p_j in the Hamiltonian are substituted by operators $q^j_{op} = x_j$ and $p^j_{op} = -i\hbar\frac{\partial}{\partial x_j}$. Time-dependent physical quantities are then defined as expectation values of corresponding operators

$$A(t) = \int \psi^*(x,t)\, A_{op}\, \psi(x,t)\, dx \tag{4}$$

where A_{op} is represented by corresponding combination of operators q^j_{op} and p^j_{op}.

It has been introduced in the preceding section that any solution $\psi(x,t)$ of Schrödinger equation may be identified with a solution of Hamilton equations or with a superposition of these classical solutions. Any function $\psi(x,t)$ at a given t may be then represented by a vector in the correspondingly constructed Hilbert space. This Hilbert space must consist of several mutually orthogonal Hilbert subspaces, each of them being spanned on the set of Hamiltonian eigenfunctions:

$$H\psi_E(x_j) = E\psi_E(x_j). \tag{5}$$

Two identical $\psi(x,t)$ functions exhibiting opposite time derivatives must belong always to different subspaces (i.e., incoming and outgoing states of evolving physical system must be taken as fundamentally different). The arrangements of total Hilbert spaces differ somewhat for continuous and discrete parts of Hamiltonian energy spectrum; corresponding details being found in [21, 22].

In the system of two free particles in their center-of-mass system, e.g., the Hilbert space must consist of two mutually orthogonal Hilbert subspaces (see also [23, 24]) being formed by incoming and outgoing states:

$$\mathcal{H} \equiv \{\Delta^- \oplus \Delta^+\}; \tag{6}$$

the given subspaces being mutually related with the help of evolution operator

$$U(t) = e^{-iHt}. \tag{7}$$

The individual vectors of evolution trajectory correspond then to the different expectation values of time operator T fulfilling the condition

$$i[H,T] = 1 \tag{8}$$

where the state corresponding to zero value of T may be arbitrarily chosen; being usually attributed to the state when mutual particle distance (impact parameter) is minimal. Eq. (8) defines the time unit for a corresponding physical system; or for any subsystem that evolves independently of the whole greater system.

Any time evolution trajectory may be characterized by the Hamiltonian expectation value E that is conserved during the whole evolution. It is, of course, also the expectation value of the angular momentum operator

$$M^2 = M_i M_i, \quad M_i = \varepsilon_{ijk} q^j_{op} p^k_{op} \tag{9}$$

commuting with the Hamiltonian

$$[M^2, H] = 0 \tag{10}$$

that is conserved. In principle it holds also $[M_i, H] = 0$ and $[M_i, M^2] = 0$; thus one of components M_i should be also conserved. However, its value depends on the orientation of the coordinate system. It is always possible to choose it so that the expectation value of given M_i is maximal. It means that any evolution trajectory of a given physical system is defined in addition to energy E also by the value of M^2; and in principle also by the sign of corresponding M_i.

As to the closed physical systems they are standardly taken as corresponding to the discrete part of Hamiltonian spectrum. In such a case the Hilbert space should be at least doubled in comparison to Eq. (6) as two different kinds of incoming and outgoing states exist; for more details see [21, 22]. Some other problems concerning closed physical systems (e.g., their emergence) will be discussed in the next section.

Let us return, however, to the problem of free two-particle system. It is necessary to mention one additional possibility when two incoming particles having corresponding kinetic energy may form an unstable object that decays after a short time. In such a case the Hilbert space may be further extended:

$$\mathcal{H} \equiv \{\Delta^- \oplus \Theta \oplus \Delta^+\} \tag{11}$$

where Θ may represent the object (or physical system) arisen by merging of two incoming particles (belonging to Δ^- subspace). This new object (or system) may be stable or (according to available free energy) unstable decaying into a state lying in Δ^+ or creating a further kind of physical systems or objects. The terms representing corresponding transition probabilities between different Hilbert subspaces must be added to earlier (classical) potential between two particles.

In this case it is, of course, necessary to respect always that the evolution goes in irreversible way. Evolution of the pair of colliding particles is described in the beginning in Δ^-, particles going nearer one to the other (incoming). In dependence on impact parameter value they are scattered by mutual potential and continue in Δ^+ as outgoing pair. However, in the case of small impact parameter values (at corresponding energy values) they may form also one common object being represented by a vector in Hilbert subspace Θ. The corresponding evolution may be described, of course, at the present on phenomenological level only as we do not know actual internal dynamics of arising object.

The object represented by the subspace Θ may be stable (e.g., creation of atom from nucleus and electron) or unstable (decaying in the same particle pair or in another one). At much higher energy values many other particles may be, of course, formed. In such a case the subspace Θ is to be substituted by a more complex system of products (or sums) of corresponding subspaces describing further evolution of separated physical systems.

While the simple subspaces (in which the interacting particles remain stable) may be in principle described with the help of corresponding Schrödinger equations the other processes require to be characterized by additional probabilities between concrete states in individual subspaces. There is not any interference between amplitudes from different orthogonal subspaces.

The individual (stable as well as unstable) objects represent closed physical systems that are characterized by some quantum physical values. Each object has some internal dynamics (eventually, exists in some different internal states - stable or unstable). And just these questions represent evidently one of the main problems of the future quantum physics.

3. Closed systems and quantum states

The idea of quantum states has been based on experimental data concerning the measured light spectra emitted by excited atoms, as it was formulated in two phenomenological postulates of N. Bohr [25]. These spectra have been correlated to transitions between different quantum energy levels. The existence of quantum states have been then derived with the help of Schrödinger equation containing Coulomb potential.

However, it is necessary to call attention to the fact that the Schrödinger equation provides an approximative phenomenological description of quantum phenomena only. It predicts and admits the existence of quantum states but it cannot explain at all how such a quantum state may arise when two corresponding objects (forming then the closed system) are mutually attracted and go always nearer one to the other. It is evident that the emergence of quantum state is not possible without the conjoint existence of short-ranged repulsive (or contact) force acting between these objects, too. One should expect that both the corresponding potentials (forces) are to be responsible for the final effect.

In the standard approach the corresponding quantum states are represented by special trajectories of electrons orbiting around a nucleus. It means that the quantum energy value should be correlated to a special value of angular momentum. It should be correlated also to dimensions of the physical system in the given state (and also to the dimensions of individual objects), which does not seem to have been analyzed sufficiently until now. It relates probably to the emergence of quantum states, which represents open question, too.

Let us start with this last problem in the case of hydrogen atom consisting of one electron and one proton. A stable hydrogen atom should arise always when a slowly moving electron appears in the neighborhood of a proton. The electron is attracted to a proton and it is evident that the given atom might be hardly formed if a repulsive (short-ranged) force between electron and proton did not exist. Some contact linkage of these two objects (e.g., some adhesive force between them) should be also taken into consideration.

At the present any force is being interpreted as the result of some potential. In such a case it might be expected that corresponding quantum states correspond to the distance when both these potentials compensate:

$$V(x) = V_{Coul}(x) + V_{rep}(x) \cong 0 .$$

The corresponding quantum states should then depend on the shapes of these two potentials. And we should ask how they might be influenced, e.g., by the proton or by its internal states that might be changeable as one must expect for the proton to exhibit some internal dynamics. The other question concerns then the problem how the energy of emitted photon is determined to correspond to the difference of corresponding quantum energy values and further which atom constituent emits the given photon.

It seems, therefore, that to expect for the quantum states to be mediated by some potential acting at the distance may be hardly convenient. The other possibility, i.e., the existence of weak adhesive force between electron and proton, might be more acceptable. In such a case the properties of hydrogen atom should follow from the properties of proton and its internal dynamics. The changes in the given adhesive linkage might be then responsible also for the energy of emitted photons.

In both the cases it is, of course, necessary to expect that the quantum states of hydrogen atom are to be fundamentally dependent on the dimensions and structure of proton. One must, therefore, assemble and analyze all corresponding characteristics that may be derived from available experimental data.

The characteristics of proton may be studied mainly with the help of experiments concerning the collisions between electrons and protons or between two protons. Unfortunately, in the corresponding analyzes only some mathematical models are usually used that describe some average phenomenological characteristics of given particles, which can hardly contribute to understanding the discussed problems of quantum physics. The models respecting the size and structures of individual particles must be made use of.

In the last time we have studied the elastic proton-proton collisions at higher values of collision energy using the eikonal model where the probability of different processes in the dependence on impact parameter values may be derived [26]; some conclusions differing from earlier ones having been obtained. First of all it has been demonstrated that there is not any reason for arguing that the elastic processes should be central, i.e., existing at very small (even zero) values of impact parameter. When any limiting condition (facilitating the calculations) has not been imposed the elastic collisions may be interpreted as peripheral in full agreement with ontological interpretation of microscopic objects; see more details in the next section.

4. Proton-proton collisions in impact parameter space

In this section we shall present some results of elastic proton-proton collisions in the impact parameter space obtained earlier. The corresponding experimental data are represented by elastic differential cross section that is in the given case given by two different mutual interactions: Coulomb and strong ones. If the influence of proton spins is neglected one measures and establishes the dependence $\frac{d\sigma^{C+N}(s,t)}{dt}$ where s is the square of center-of-mass energy and t is the square of center-of-mass four-momentum transfer (it is zero or negative and it is a function of scattering angle).

It is not possible to measure the influence of individual interactions separately. In the standard theoretical framework any collision process is regarded as fully described provided its complex scattering amplitude is given. The measured differential cross section is written as

$$\frac{d\sigma^{C+N}(s,t)}{dt} = \frac{\pi}{sp^2}\left|F^{C+N}(s,t)\right|^2 \tag{12}$$

where p is the three-momentum of incident nucleon in the center-of-mass system ($s = 4(p^2 + m^2)$; m being mass of a proton); natural units have been used: $\hbar = c = 1$. In the case of only hadronic (resp. Coulomb) elastic scattering it is, therefore, necessary to know amplitude $F^N(s,t)$ (resp. $F^C(s,t)$). And one must ask how to express complete amplitude $F^{C+N}(s,t)$ with the help of individual amplitudes $F^N(s,t)$ and $F^C(s,t)$.

Formulas still standardly used for complete elastic scattering amplitude $F^{C+N}(s,t)$ were derived under several very limiting assumptions. One of such commonly used formulas has been the simplified formula of West and Yennie (WY) published in 1968 [27] which was derived only for very low values of $|t|$ under assumption that the modulus of $F^N(s,t)$ was purely exponential in t and the phase of $F^N(s,t)$ was t-independent; both these assumptions were supposed to be valid in the whole region of kinematically allowed values of t at that time. Detailed discussion concerning both theoretical and experimental problems following from drastic limitation involved in this formula may be found, e.g., in [28].

The question has been raised whether it is possible to derive more general formula *without* any a priory limitation on $F^{C+N}(s,t)$, which could be used for more relevant analysis of experimental data. It has been possible to remove the most of earlier limitations if eikonal model has been applied to. According to [26] (in 1994) it has been possible to derive on the basis of the eikonal model more general formula for the complete elastic amplitude for *any* s and t with the accuracy up to terms linear in α and *without* any a priory restriction on hadronic amplitude $F^N(s,t)$.

If this hadronic amplitude $F^N(s,t)$ is given then some physically significant quantities may be calculated from it. For example, one may calculate distribution functions of total, elastic and inelastic hadronic collisions in the impact parameter space and if they are determined then one may evaluate mean-squares of impact parameter for corresponding processes. The mean-square of impact parameter in the case of elastic processes may be calculated from corresponding elastic distribution function $D_{el}(s,b)$ as

$$\langle b^2(s)\rangle_{el} = \frac{\int_0^\infty bdb\, b^2 D_{el}(s,b)}{\int_0^\infty bdb\, D_{el}(s,b)}. \tag{13}$$

and similarly we may define also $\langle b^2(s)\rangle_{tot}$ and $\langle b^2(s)\rangle_{inel}$ in the case of total and inelastic hadronic collisions. However, in eikonal model the mean-squares of all these processes may be derived directly from the t-dependent elastic hadronic amplitude $F^N(s,t)$ without trying to establish the whole distribution functions, which is mathematically much more complicated; detailed discussion how to actually determine all distribution functions on the basis of corresponding experimental data can be found in [29]. For a given hadronic amplitude $F^N(s,t)$ one may thus calculate relatively easily root-mean-squares $\sqrt{\langle b^2(s)\rangle_{el}}$ and $\sqrt{\langle b^2(s)\rangle_{inel}}$ and compare both the values. If the value $\sqrt{\langle b^2(s)\rangle_{el}}$ is lesser than $\sqrt{\langle b^2(s)\rangle_{inel}}$ then it would mean that elastic hadronic processes should be realized in average at lower impact parameter values than inelastic processes; the protons should be rather "transparent" which might be hardly acceptable from the ontological point of view. We may denote this situation as "central" behavior of elastic scattering. If the value $\sqrt{\langle b^2(s)\rangle_{el}}$ is greater than $\sqrt{\langle b^2(s)\rangle_{inel}}$ then we denote situation as "peripheral" behavior of elastic scattering; the situation being in agreement with usual interpretation of two matter object collisions.

As already mentioned the corresponding proton-proton collision data have been interpreted with the help of rather simplified phenomenological mathematical models in the past. Some conclusions have been rather surprising. Especially, when it has been concluded that a rather great ratio of elastic processes has corresponded to purely central collisions (i.e., protons could scatter elastically even at impact parameter $b = 0$). This kind of "transparency" of protons has been denoted already in 1979 in [30] as a "puzzle". In 1981 it has been then shown that the corresponding result has depended mainly on the t-dependence of the phase of elastic hadronic scattering amplitude, see [31]. The mentioned central behavior has been derived when the t-dependence of the phase has been strongly limited. And it has been shown that if the modulus is purely exponential and hadronic phase is practically t-independent in the whole region of kinematically allowed values of t (at arbitrary collision energy $\sqrt{s}$) one obtains necessarily the mentioned central behavior of elastic hadronic collisions in the impact parameter space. These two (over)simplified assumptions have been commonly used in many analysis of corresponding experimental data (they are included also in the simplified WY formula).

Consequently, it has been very interesting to put the opposite question: How to modify the given mathematical model to obtain collision processes corresponding to usual ideas. The experimentally established elastic proton-proton differential cross section at energy of $\sqrt{s} = 53$ GeV has been analyzed in the whole measured t-range with the help of more general eikonal formula used for complete scattering amplitude in [26]. It was possible to obtain acceptable fits for different t-dependencies of the phase (according to chosen parametrization). Two quite different dependencies of hadronic phase have been then shown in the quoted paper. The first phase was the so-called "standard" phase (used, e.g., in [32] for interpretation of experimental data) and the second phase corresponded to natural peripheral elastic collisions. Both the hadronic phases are plotted in Fig. 1. The root-mean-squares

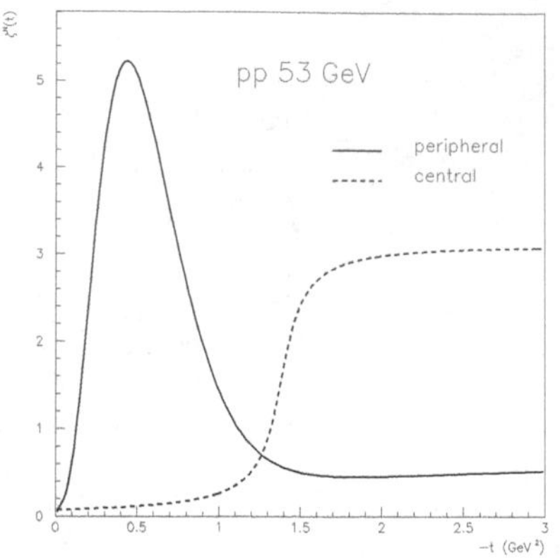

Figure 1. Two different hadronic phases $\zeta^N(s,t)$ defined by $F^N(s,t) = i\left|F^N(s,t)\right| e^{-i\zeta^N(s,t)}$ and fitted to corresponding experimental data at energy of $\sqrt{s} = 53$ GeV. Dashed line - standard hadronic phase admitting central elastic hadronic collisions; full line - peripheral phase.

corresponding to the two hadronic phases are then given in Table 1. The case with the standard hadronic phase leads to central elastic hadronic collisions. Similar result as in the case of proton-proton collisions was obtained also in the case of elastic antiproton-proton scattering at energy of 541 GeV (see [26, 29]). The phase which leads to peripheral behavior of hadronic proton-proton collisions has very similar t-dependence as that obtained earlier at lower collision energy of $\sqrt{s} = 23$ GeV in the already quoted older paper [31].

Data	Hadronic phase	Profile	$\sqrt{\langle b^2(s)\rangle_{tot}}$	$\sqrt{\langle b^2(s)\rangle_{el}}$	$\sqrt{\langle b^2(s)\rangle_{inel}}$
pp 53 GeV	peripheral	peripheral	1.028	1.803	0.772
pp 53 GeV	standard	central	1.028	0.679	1.087
$\bar{p}p$ 541 GeV	peripheral	peripheral	1.140	2.205	0.609
$\bar{p}p$ 541 GeV	standard	central	1.140	0.756	1.220

Table 1. Values of root-mean-squares (in femtometers) obtained from analysis of experimental data with standard and peripheral hadronic phase.

Model	$\sqrt{\langle b^2(s)\rangle_{tot}}$	$\sqrt{\langle b^2(s)\rangle_{el}}$	$\sqrt{\langle b^2(s)\rangle_{inel}}$
Bourelly et al.	1.249	0.876	1.399
Petrov et al. (2P)	1.227	0.875	1.324
Petrov et al. (3P)	1.263	0.901	1.375
Block et al.	1.223	0.883	1.336
Islam et al.	1.552	1.048	1.659

Table 2. Values of root-mean-squares (in femtometers) predicted by different models of proton-proton collisions at collision energy of 14 TeV. All the models predicts $\sqrt{\langle b^2(s)\rangle_{el}} < \sqrt{\langle b^2(s)\rangle_{inel}}$, i.e., central behavior of elastic scattering.

The elastic and inelastic root-mean-squares have been calculated in [28] also for several relatively new phenomenological models proposed for elastic hadronic proton-proton scattering at energy of 14 TeV (planned ultimate LHC collision energy). In all these cases it has been found that $\langle b^2(s)\rangle_{el} < \langle b^2(s)\rangle_{inel}$, i.e., all these models correspond to central behavior of proton collisions, as shown in Table 2.

One of the results obtained in [26] was that the modulus of hadronic amplitude $|F^N(s,t)|$ was practically uniquely determined from corresponding measured elastic differential cross section while the hadronic phase remained quite undetermined, being only slightly limited due to Coulomb interaction. It should be determined on the basis of other physical requirements, not on the basis of some arbitrary assumptions. The results gained in elastic proton-proton collisions might be very helpful for nuclear and particle physics in general. However, practically in all current descriptions of proton-proton elastic scattering the validity of optical theorem which relates the imaginary part of hadronic amplitude to total hadronic cross section, has been taken as a basic assumption. The given problem will be discussed in the next section, as the given validity does not seem to be sufficiently reasoned.

As to the other problem mentioned in the end of preceding section (the use of mathematical models representing only average structure of colliding objects) it has been at least partially removed when the more general eikonal model has been applied to and the results in dependence on impact parameter values have been established. However, the main character of some average phenomenological characteristics obtained from hitherto phenomenological models remains, which can hardly contribute to understanding the results of contemporary quantum physics. A new more detailed model of proton-proton elastic collisions with realistic behavior of the collisions in the impact parameter space and respecting the sizes and structures of individual colliding particles will be formulated and applied in preliminary form to experimental data in Sec. 6.

5. Optical theorem and its validity

Practically all hitherto mathematical models of proton collision processes (including the eikonal model described in Sec. 4) have been based on the assumption of optical theorem validity. This theorem has been taken from optical approaches where the total cross section was correlated to the imaginary part of complex index of refraction (see, e.g., the description of the given problem in [35]) and the behavior of light was studied on wave basis. The optical theorem has been applied then also to elastic particle collisions when the wave behavior has been attributed to all physical objects.

The approaches of deriving the validity of optical theorem in particle collisions have been summarized recently, e.g., in [36]. The goal of these approaches has consisted in deducing that it has held for the total cross section

$$\sigma^{tot}(s) = \sigma^{el}(s) + \sigma^{inel}(s) = \frac{4\pi}{p\sqrt{s}}\Im F(s,t=0) \tag{14}$$

where the function $F(t)$ defines the elastic differential cross section

$$\frac{d\sigma}{dt} = \frac{\pi}{sp^2}|F(s,t)|^2. \tag{15}$$

It holds then also

$$F(s,t) = 2\pi\sqrt{s}\int_0^\infty b\,db\,J_0(b\sqrt{-t})\Gamma(b). \tag{16}$$

where $\Gamma(b)$ is the profile function in impact parameter space characterizing the probability of elastic processes in dependence on impact parameter values and on their initial distribution.

Until now there is not any reliable theory of elastic scattering and only different phenomenological models are practically available. The complex function

$$F(s,t) = i|F(s,t)|\, e^{-i\zeta(s,t)} \tag{17}$$

is being, therefore, derived usually from corresponding experimental data. However, from them only the modulus may be established while the phase $\zeta(t)$ remains quite undetermined. It is to be predicted on the basis of additional assumptions that may influence strongly the conclusions concerning the value of total cross section. Also the optical theorem has been practically always involved even if it has been derived on the basis of other additional assumptions that have not corresponded to actual characteristics of strong interactions as it was mentioned already earlier in [37].

In all attempts to derive the validity of optical theorem in particle collisions the standard approaches based on the Copenhagen quantum mechanics have been made use of; especially, the summation of amplitudes used commonly in S-matrix theory has been applied to. The states belonging to two mutually orthogonal Hilbert subspaces Δ^- and Δ^+ (see Sec. (2)), i.e., the states at time $\tau = -\infty$ and at time $\tau = +\infty$, were taken as lying in one common Hilbert space and corresponding amplitudes were superposed, which cannot be done if the ontological approach is to be respected (i.e., time evolution described correctly with the help of time-dependent Schröedinger equation). It may be then denoted as surprising that the optical theorem has been applied to strong interaction only and never considered in the case of Coulomb interaction, even if the given derivation approach does not relate to a kind of interaction.

The final effect in collision processes of two protons is to depend on the conditions in individual events characterized mainly by corresponding impact parameter value. However, only the Coulomb elastic scattering comes into account for greater values of b where short-ranged (or contact) strong interaction cannot exist while for lower values of b elastic interaction (Coulomb or hadronic) may exist together with inelastic one. Consequently, as to the strong interaction the particles may continue in original motion practically without any disturbance; it means that such states represent a special set of states in the subspace Δ^+ which cannot be added if the total cross section corresponding to strong interaction is considered. Only the states scattered elastically by strong interaction may be involved.

It has been mentioned that for a given collision energy $\sqrt{s}$ only the modulus of hadronic amplitude $F(s,t)$ may be established from experimental data and the hadronic phase $\zeta(s,t)$ may be practically arbitrary. It means that very different characteristics may be attributed to elastic particle collisions; see the problem of centrality and peripherality of elastic hadronic processes discussed in preceding section. The choice of phase t-dependence has been then fundamentally influenced also by the application of the optical theorem, mainly in the region of very low values of $|t|$ where the measurement cannot be done. Even the separation of Coulomb and strong interactions might be fundamentally influenced when the parametrization of the strong part of scattering amplitude has allowed only the differential cross section decreasing monotony from the maximum value at $t = 0$. In the case of short-ranged (contact) strong interactions it is necessary to admit that the maximum value may lie at non-zero value of t as it will be demonstrated in the next section.

6. Probabilistic ontological model of elastic proton-proton collisions

It follows from the preceding that the Schrödinger equation leads to the same results as classical physics. However, in the classical physics based fully on ontological approach the individual matter objects represent always only the source of corresponding potentials mediating corresponding interactions. It may be applied fully to electromagnetic or gravitational forces. The strong (and also weak) interactions differ, however, rather strongly from these interactions. The short-ranged effect of the latter ones indicates that these forces should be denoted rather as contact ones; being zero at any greater distance. It means that in elastic proton-proton collisions (especially, at higher energy values) there is not practically any actual interference between Coulomb and strong interactions. In any event only one of these interactions is effective according to impact parameter value and the corresponding probabilities of individual contributions to elastic differential cross section may be added. However, in such a case it is necessary to take into account possible proton dimensions in collision instant. According to contemporary knowledge protons consist evidently of other objects that must be in mutual interaction. It is also necessary to expect that the particles having special value of spin must exhibit some internal dynamics. We are to accept that the proton may exist in different spontaneously changeable internal states that may have also different shapes and dimensions (as it has been proposed earlier in [38]). And we shall attempt to test whether these dimensional characteristics may become evident in elastic collision characteristics.

To make use of the given idea in the analysis of elastic proton-proton processes let us assume that a colliding proton may exist in n different states, each being characterized by maximal possible dimension d_k $(d_1 \geq ... \geq d_k \geq ... \geq d_n)$. The corresponding probabilities that proton is in such a state in the moment of interaction are then p_k $(\sum_k p_k = 1)$.

If two such protons or other similar objects collide the different pairs of states may interact; the individual probabilities of such collision channels (or collision states) being equal to

$$r_{k,l} = \begin{cases} p_k p_l & \text{if } k = l \text{ or the colliding object are not identical} \\ 2p_k p_l & \text{if } k \neq l \text{ and the colliding object are identical, } k < l \end{cases} \quad (18)$$

The factor 2 in the last relation follows from the fact that for $k \neq l$ and identical colliding objects the cases with interchanged collision types k, l and l, k are the same and the corresponding probabilities may be summed but we need to introduce convention $k < l$ to count each *distinct* collision type only once.

6.1. Nuclear and Coulomb scattering

As to the strong interactions (taken as contact ones) the maximum effective impact parameter for which two protons may still interact hadronically in corresponding collisions will be then equal to

$$B_{k,l} = (d_k + d_l)/2. \quad (19)$$

The two indexes k and l in $r_{k,l}$ and $B_{k,l}$ may be for convenience substituted by only one index j using a one-to-one correspondence $(k,l) \leftrightarrow j$. In other words, if one object in a state k collide with an another one in a state l then this always implies unique collision type j and vice versa. It is obvious that $\sum_j r_j = 1$.

It is necessary to expect that for each j-th collision channel at any given impact parameter $b < B_j$ the value of scattering angle (or equivalently $|t|$) of two elastically (and hadronically only) scattered protons will belong to a limited interval of values beginning always with zero value. The corresponding frequency of individual values of scattering angle will go to a maximum and will diminish again to zero; the interval being reduced to one point at $b = B_j$. However, in the following very preliminary (and simplified) analysis of experimental data we have substituted the given interval by the corresponding mean value. Let us denote the corresponding function as $\bar{t}_j(b)$ which smoothly fall from zero to some lower negative values when b will change from B_j to lower values. The inverse function will be denoted as $\bar{b}_j(t)$.

We may write for partial elastic hadronic cross section corresponding to a j-th collision state following expression

$$\frac{d\sigma_j^N(t)}{dt} = 2\pi P_j^{el}\left(\bar{b}_j(t)\right)\bar{b}_j(t)\frac{d\bar{b}_j(t)}{dt} \tag{20}$$

where the factor $P_j^{el}(b)$ represents individual probability of elastic processes at corresponding impact parameter values b if cylindrical symmetry in impact parameter dependence has been assumed. The elastic hadronic differential cross section is then given by the sum of individual contributions from all the collision states j:

$$\frac{d\sigma^N(t)}{dt} = \sum_j r_j \frac{d\sigma_j^N(t)}{dt}. \tag{21}$$

Parameters r_j in the last relation represent the weights (probabilities) of individual collision channels.

The behavior of the given pair of two colliding protons depends then on probabilities p_k of individual states, their dimension values d_k and on two series of functions $P_j^{el}(b)$ and $\bar{b}_j(t)$ (or $\bar{t}_j(b)$). Their values or shapes are to be derived from corresponding experimental data. However, the function $P_j^{el}(b)$ may be expressed as the product of two functions

$$P_j^{el}(b) = P_j^{tot}(b)\,P_j^{rat}(b) \tag{22}$$

where $P_j^{tot}(b)$ is the probability of any mutual hadronic particle interaction (elastic as well as inelastic) at impact parameter b corresponding to a j-th collision type while $P_j^{rat}(b)$ represents the corresponding ratio of elastic hadronic processes from all possible hadronic interactions. Both the functions $P_j^{tot}(b)$ and $P_j^{rat}(b)$ are evidently monotonous. The former one is to be non-increasing function of impact parameter b while the latter one is non-decreasing in dependence on $b \in \langle 0, B_j \rangle$. The monotony of the functions brings very important simplification in the choice of parametrization of both the new functions; their values moving in the whole interval $\langle 0,1\rangle$. The probability of any inelastic process may be then defined as

$$P_j^{inel}(b) = P_j^{tot}(b) - P_j^{el}(b). \tag{23}$$

It must hold always $P_j^{tot}(b) = P_j^{el}(b) = P_j^{inel}(b) = 0$ for any $b \geq B_j$. This will allow us to integrate in corresponding cases over finite interval of impact parameter $\langle 0, B_j\rangle$ instead of infinite interval $\langle 0,\infty\rangle$. We will use this fact in the following quite frequently; it also simplifies numerical calculations of corresponding expressions.

The (integrated) elastic hadronic cross section may be written with the help of Eq. (20) and Eq. (21) as

$$\sigma^{el,N} = \int_{t_{min}}^{0} dt \frac{d\sigma^N(t)}{dt} = \sum_j r_j \sigma_j^{el,N} \tag{24}$$

where we have introduced elastic hadronic cross section just for j-th collision channel

$$\sigma_j^{el,N} = 2\pi \int_0^{B_j} db P_j^{el}(b) b. \tag{25}$$

In the given model it is possible to derive also total hadronic cross sections for individual collision types $\sigma_j^{tot,N}$ from elastic experimental data without adding any further assumption; it holds

$$\sigma_j^{tot,N} = 2\pi \int_0^{B_j} db P_j^{tot}(b) b. \tag{26}$$

The total hadronic cross section may be then written again as a sum of contributions from all the collision channels

$$\sigma^{tot,N} = \sum_j r_j \sigma_j^{tot,N} \tag{27}$$

And using the relation (23) the corresponding values for inelastic cross section may be established, too.

The mutual elastic collisions between two protons at smaller scattering angles (smaller momentum transfers $|t|$) are caused, however, not only by strong interactions that may be interpreted practically as contact but also by mutual Coulomb forces acting at distance. One can express then the experimentally measured elastic differential cross section as the sum of the two given interactions

$$\frac{d\sigma^{C+N}(t)}{dt} = \sum_j r_j \frac{d\sigma_j^N(t)}{dt} + \frac{d\sigma^C(t)}{dt} \tag{28}$$

where the Coulomb differential cross section will be established directly by fitting experimental data. It will not be determined with the help of formfactors from the standard Coulomb potential valid for pointlike particles as such approach does not correspond to reality. The infinite Coulomb elastic differential cross section for $t = 0$ may be theoretically obtained at infinite distance while in real experiments the zero value of t may exist inside narrow particle beam as combined effect from different surrounding scattering centers. The similar difference may concern, of course, also the frequencies for higher values of $|t|$ due to asymmetric positions of charged partons in individual protons.

In our model we have introduced some free parameters and some unknown functions which are to be determined from corresponding measured elastic differential cross section using formula (28). It is necessary to fit maximal dimensions d_k of all the (considered) hadron states and their corresponding probabilities p_k at the moment of collision. We also need to parametrize and then to fit three monotonic functions $P_j^{tot}(b)$, $P_j^{rat}(b)$ and $\bar{b}_j(t)$ (or $\bar{t}_j(b)$) for each corresponding j-th hadronic collision channel. It is also the Coulomb interaction effect $\frac{d\sigma^C(t)}{dt}$ which needs to be determined from data. Several other very interesting and physically significant quantities may be then calculated from these parameters and functions as it has been shown in preceding.

6.2. Analysis of experimental data

We shall apply the given probabilistic model as we have already mentioned to the data obtained at ISR at CERN at the energy of 53 GeV [33, 34] (the same data as made use of in Sec. 4). We shall try to show that two proton states exhibiting the largest dimensions may be responsible for the part of differential elastic cross section data corresponding to $|t| \in (0., 1.25)$ GeV^2; see the corresponding part of experimental points shown in Fig. 2. As in the lower part of $|t|$ the density of measured points has been very great only one fifth of experimental points has been pictured in the interval $(0., 0.45)$ GeV^2.

If one proton in k-state (k=1,2) collides with another proton in l-state (l=1,2) we may define collision state j using a following one-to-one correspondence $(k,l) \leftrightarrow j$ ($k < l$): $(1,1) \leftrightarrow 1$, $(1,2) \leftrightarrow 2$ and $(2,2) \leftrightarrow 3$. We have thus three distinct collision types that will be responsible for the given part of elastic differential cross section.

It is, of course, necessary to parametrize suitably the corresponding functions used in the description of the given process. The following parameterizations of three monotonous functions $P_j^{tot}(b)$, $P_j^{rat}(b)$ and $\bar{b}_j(t)$ have been used for individual collision channels

$$P_j^{tot}(b) = \begin{cases} 1 & \text{if } 0 \le b \le \mu_{0,j} \\ \frac{e^{-\left(\mu_{1,j}(b-\mu_{0,j})\right)^{\mu_{2,j}}}(1+\mu_{3,j})}{1+\mu_{3,j}e^{-\left(\mu_{1,j}(b-\mu_{0,j})\right)^{\mu_{2,j}}}} & \text{if } \mu_{0,j} < b < B_j \\ 0 & \text{if } B_j \le b \end{cases} \quad (29)$$

$$P_j^{rat}(b) = \frac{e^{-\left(\nu_{0,j}(B_j-b)\right)^{\nu_{1,j}}}(1+\nu_{2,j})}{1+\nu_{2,j}e^{-\left(\nu_{0,j}(B_j-b)\right)^{\nu_{1,j}}}} \tag{30}$$

$$\bar{b}_j(t) = B_j\left(\frac{2}{\pi}\arccos\left[\left(\frac{|t|}{\eta_{0,j}}\right)^{1/\eta_{2,j}}\right]\right)^{1/\eta_{1,j}} \tag{31}$$

where $\mu_{0,j}, \mu_{1,j}, \mu_{2,j}, \mu_{3,j}; \ \nu_{0,j}, \nu_{1,j}, \nu_{2,j}; \ \eta_{0,j}, \eta_{1,j}, \eta_{2,j} \ \ (j = 1, 2, 3)$ are free parameters that are to be determined from experimental data together with parameters p_k and d_k $(k = 1, 2)$.

As it has been already mentioned also the Coulomb interaction effect is to be determined from corresponding experimental data. To enable precise fit the following parametrization has been chosen

$$\frac{d\sigma^C(t)}{dt} = \xi_0(1+(\xi_1|t|)^{\xi_2})\,e^{-\xi_3|t|} + \xi_4\,e^{-(\xi_5|t|)^{\xi_6}}\,\frac{1+\xi_7}{1+\xi_7\,e^{-(\xi_5|t|)^{\xi_6}}} \tag{32}$$

where ξ_i $(i = 0, .., 7)$ are additional free parameters that are to be derived from corresponding experimental data.

The number of free parameters may seem to be rather high. However, this is quite irrelevant at this stage of our research when quite new physical ideas are looked for and tested. The goal of our effort consists in principle in describing the characteristics of partial structures to a sufficient detail, to initiate further analysis with the help of other experiments.

The results of the corresponding fit are shown in Fig. 2. The probabilistic model can be fitted to experimental data practically in the whole considered interval of $|t| \in (0, 1.25)$ GeV^2. Hadronic differential cross sections $\frac{d\sigma_j^N(t)}{dt}$ given by Eq. (20) for all the three collision types are plotted in Fig. 2, too. Only their parts given by r_j parameters contribute to complete differential cross section $\frac{d\sigma^{C+N}(t)}{dt}$. The Coulomb differential cross section $\frac{d\sigma^C(t)}{dt}$ given by parametrization (32) is also shown in Fig. 2.

The following values of individual free parameters have been obtained on the basis of our analysis. The frequencies of two internal states considered in our fit to experimental data having the greatest dimensions are given by

$$p_1 = 0.48, \qquad p_2 = 0.39$$

and the corresponding dimensions are

$$d_1 = 2.50 \text{ fm}, \qquad d_2 = 2.29 \text{ fm}$$

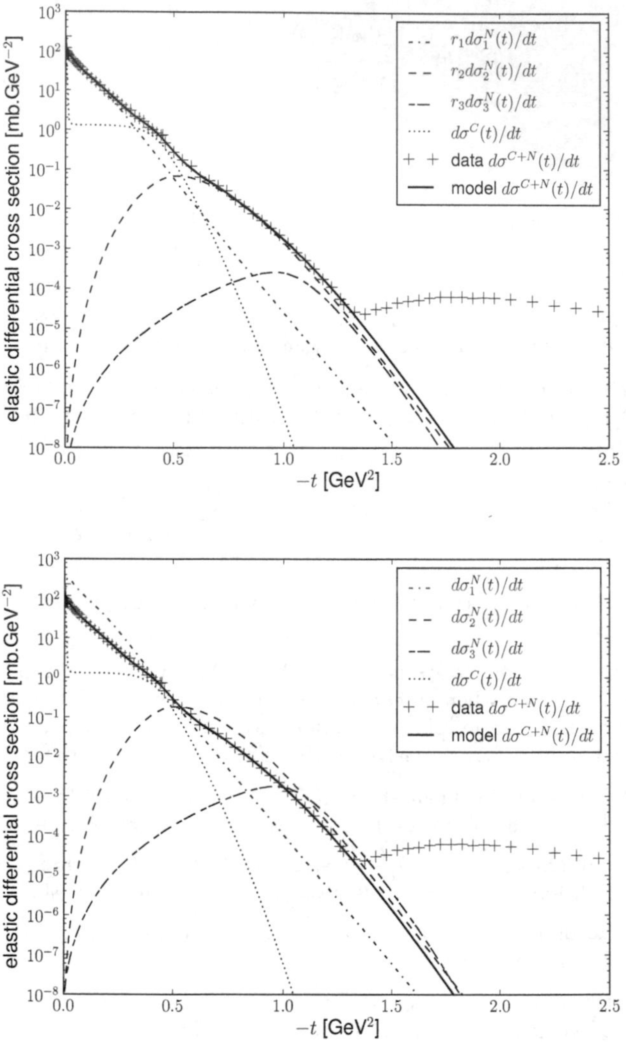

Figure 2. Differential elastic cross sections for proton-proton scattering at energy of 53 GeV; individual points - experimental data, full line - our probabilistic model fitted to experimental data, dashed line - Coulomb differential cross section $d\sigma^C(t)/dt$, three other lines - individual hadronic collision types (top - contributions $r_j d\sigma_j^N(t)/dt$ to complete differential cross $d\sigma^{C+N}(t)/dt$, bottom - individual differential cross sections $d\sigma_j^N(t)/dt$).

The functions $P_j^{tot}(b)$ and $P_j^{rat}(b)$ (see Eqs. (29) and (30)) representing the b-dependences of probabilities of total and elastic collisions together with function $\bar{t}_j(b)$ (inverse function of

$\bar{b}_j(t)$ given by Eq. (31)) are shown in Fig. 3. The values of free parameters in the functions describing the elastic scattering caused by strong interaction are shown in Table 3 for all the three considered collision states.

		j	1	2	3
$P_j^{tot}(b)$	$\mu_{0,j}$	[fm]	0.70	0.59	0.50
	$\mu_{1,j}$	[fm^{-1}]	0.95	2.35	4.34
	$\mu_{2,j}$	[1]	5.90	2.55	1.69
	$\mu_{3,j}$	[1]	193.	995.	15.
$P_j^{rat}(b)$	$\nu_{0,j}$	[fm^{-1}]	2.03	2.40	1.80
	$\nu_{1,j}$	[1]	4.34	3.38	5.65
	$\nu_{2,j}$	[1]	20.7	144.	3742.
$\bar{b}_j(t)$	$\eta_{0,j}$	[GeV^2]	24.8	26.8	35.3
	$\eta_{1,j}$	[1]	0.97	0.44	0.47
	$\eta_{2,j}$	[1]	5.15	2.63	2.85

Table 3. Values of free parameters of monotonous functions $P_j^{tot}(b)$, $P_j^{rat}(b)$ and $\bar{b}_j(t)$ for all the three collision types given by parametrizations (29), (30) and (31).

The additional effect caused by electric charges is then characterized by function (32) that contains some further free parameters. Their values obtained by fitting the given experimental data are

$$\xi_0 = 491 \text{ mb.GeV}^{-2}, \;\; \xi_1 = 265 \text{ GeV}^{-2}, \;\; \xi_2 = 3.70, \;\; \xi_3 = 742\text{GeV}^{-2},$$

$$\xi_4 = 1.32 \text{ mb.GeV}^{-2}, \;\; \xi_5 = 6.14 \text{ GeV}^{-2}, \;\; \xi_6 = 1.70, \;\; \xi_7 = 106\,.$$

Once we have determined all the free parameters and unknown functions from the fit we may calculate several other physically significant quantities. Let us start with quantities which we can calculate for all the three individual collision states considered in our fit.

The parameters r_j (see Eq. (18)) determining the contribution of individual channels (their probabilities) are in Table 4. Corresponding maximal effective impact parameters B_j for which protons might still interact via hadronic interaction calculated from Eq. (19) are shown, too; they are around 2.4 fm slightly different for each collision type j. Further total, elastic and inelastic hadronic cross sections having been calculated for all three collision states j are introduced in Table 4.

According to our very very rough model a colliding proton may be in one of two considered internal states with probability $p_1 + p_2 = 0.86$. It means that in 14% of cases the proton is to be in different internal states with different maximal dimensions. The proton collisions go then in the given three channels with the probability $\sum_{j=1}^{3} r_j = p_1^2 + 2p_1p_2 + p_2^2 = 0.75$. The additional internal states may be responsible for measured differential cross section outside our considered t-range (for $|t| > 1.25$ GeV^2); partially also in combination with two already considered states, which is in agreement with preliminary tests done already by us, too. It means that the actual total and inelastic cross sections will be higher than the values

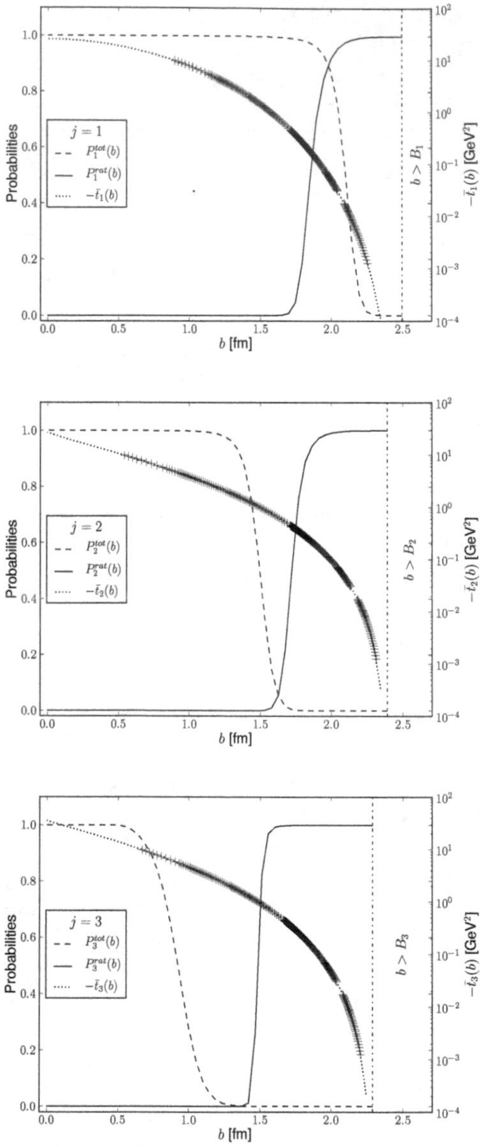

Figure 3. Functions $P_j^{tot}(b)$ and $P_j^{rat}(b)$ with opposite monotony (dashed and full lines) and functions $|\bar{t}_j(b)|$ (dotted line) for individual collision types (top $j = 1$, middle $j = 2$, bottom $j = 3$). Individual points lying on $|\bar{t}_j(b)|$ curves correspond to values of impact parameters b at experimentally established values of t calculated using functions $\bar{b}_j(t)$.

j		1	2	3	$\sum_{j=1}^{3}$
k,l		1,1	1,2	2,2	
r_j	[1]	0.23	0.37	0.15	0.75
B_j	[fm]	2.50	2.39	2.29	-
$\sigma_j^{tot,N}$	[mb]	137.	69.7	27.1	-
$\sigma_j^{el,N}$	[mb]	28.1	6.2×10^{-2}	7.2×10^{-4}	-
σ_j^{inel}	[mb]	109.	69.7	27.1	-
$r_j\sigma_j^{tot,N}$	[mb]	31.5	25.7	4.0	61.2
$r_j\sigma_j^{el,N}$	[mb]	6.5	2.3×10^{-2}	1.1×10^{-4}	6.5
$r_j\sigma_j^{inel}$	[mb]	25.0	25.7	4.0	54.7

Table 4. Values of some physically significant quantities obtained as a result of the probabilistic ontological model fitted to corresponding experimental data at energy of 53 GeV.

introduced in Table 4: $\sigma^{tot,N} > \sum_{j=1}^{3} r_j\sigma_j^{tot,N} = 61.2$ mb (resp. $\sigma^{inel,N} > \sum_{j=1}^{3} r_j\sigma_j^{inel,N} = 54.7$ mb), while the increase of elastic cross section may be neglected. The given values may be, of course, influenced by the very rough approximation neglecting the distribution of momentum transverses at any impact parameter value.

It is evident that according to our probabilistic model the Coulomb effect appears as significant till $|t| \cong 0.5\ \text{GeV}^2$, which might throw quite new light to the structure of charge distribution in individual protons and contribute fundamentally to our understanding of the internal structure of hadron objects.

The presented model has been based, of course, on one very simplifying and rough assumption ($\bar{t}_j(b)$ - relating always only one momentum value to each impact parameter), which might significantly influence the corresponding numerical results. We are working already on the model that will respect the existence of momentum transfer spectrum for any impact parameter value, which should allow much more realistic fit of experimental data on the given basis. Also the fitting of experimental data at other energy values will be performed; the fits for higher values of $|t|$ will be done, too.

7. Contemporary quantum physics and metaphysics

We have mentioned already in the preceding that the new results require for the physicists to return to ontological basis the classical physics was based on. This basic approach changed fundamentally in the beginning of modern period when Descartes formulated his mathematically-philosophical concept in which any linkage between human knowledge and ontological approach to matter world was practically excluded. Any knowledge of matter existence started to be based fully on human thinking. Also the participation of human senses was excluded at that time, which was criticized by some philosophers. It led then to the positivism that influenced fully the thinking of European society in the course of the 19th century.

In the middle of the 19th century the given thinking dominated also scientific approaches and scientific knowledge. Boltzmann started it when he denoted one phenomenological characteristics of the system consisting of a great number of particles as natural law. It is possible to say that the given way was accomplished when Bohr formulated his Copenhagen quantum mechanics in 1927. He started from Schrödinger equation proposed in 1925 which itself was fully acceptable as the same results followed from it as from the classical concept of Galileo and Newton (only the set of corresponding states was smaller); see short paragraph (with corresponding quotations) in Sec. 1. However, Bohr deformed significantly physical conclusions (following from proper Schröedinger equation) by adding further very strong assumptions without any actual reason, which influenced fundamentally the evolution of quantum physics in the 20th century.

The corresponding (quantum) era of physical research started in principle at the break of the 19th and 20th centuries when black-body radiation was being intensively studied and at the same time new phenomena at the level of microscopic world were discovered (X and gamma radiations and electron (1895-1897)); and further, quantum energy transfer (1900) and photon existence (1904) were predicted. A broad space was opened for formulation of new hypotheses that could not be tested directly but only on the basis of indirect effects established with the help of macroscopic measuring devices. In such a situation the phenomenological models seemed to be very advantageous.

However, to understand the possibility of scientific knowledge it is necessary to realize what are the approaches of scientific research. As to the region of physical research it is possible to say that it is based on falsification approach. The basic step consists then in formulating some more general statements or unifying hypotheses with the help of our reason on the basis of observation and measurement; with the help of the approach making use of logical induction, or intuition. The goal of further approaches and analyses consists then in deriving all possible consequences that follow from a given hypothesis (or a set of hypotheses). As such hypotheses start always from a limited set of our pieces of knowledge it is clear in some cases immediately that they cannot sustain in further considerations. Generally, it is necessary to derive for any hypothesis all possible consequences with the help of logical deduction and to analyze, whether some logical contradictions between individual consequences do not exist, and further to compare these consequences to all possible observations of natural phenomena. If a contradiction is found the given hypothesis (or a set of hypotheses) must be refused, resp. modified, so as the given contradiction might be removed. If one does not come after sufficient falsifying effort to any contradiction the given set of hypotheses may be denoted as plausible; and it is possible to start from it in further considerations about the natural laws concerning the world and human being.

It follows from the preceding that the falsification approach represents important and practically basic knowledge method based on human reason, as K. Popper (1902-94) presented in the first half of the 20th century. One is never justified to denote our reason knowledge for a verified truth, as one can never grant that in following steps a logical contradiction or a contradiction to nature observation will not appear. On the other side one must accept any non-falsified hypothesis as plausible, even if it is in contradiction to another plausible hypothesis. All these statements or hypotheses must be fully tolerated. The decision concerning the preference of one of them must be left to other logical and experimental analyses. On the other side in contradiction to contemporary requirements of some human groups any falsified statement cannot be tolerated and must be decisively refused.

The preceding arguments have thrown new light also to the principle of falsifiability propagated in the region of physical research in the last century. The requirement of falsification tests to any statement has been interpreted as the possibility to prove the validity of the given statement, which contradicts the possibilities of our reason knowledge. At least some positive test should be necessary to admit the given statement as plausible; however, it can never represent any proof of its validity.

In the 20th century purely phenomenological view was taken as the knowledge basis and any ontological aspects have not been respected. Evidently mutually contradicting assumptions were being applied especially to elastic collision processes. It follows from these facts that the corresponding metaphysical view must be based on ontological view respecting fully our experience with matter nature.

8. Some open questions

One of the most important pieces of knowledge introduced in preceding sections is the fact that there is fundamental difference between Coulomb forces and strong (nuclear) interaction; the former acting at a distance and the latter representing a contact force. To interpret this force as the effect of some potential may be misleading. And it is the task of contemporary research to find some new approaches how to describe successfully the given phenomenon. The analyses of elastic collisions between the particles exhibiting corresponding interaction may be very helpful in this direction. It is even possible to say that the further progress of future quantum physics is strongly bound to the problem of particle structure. The existence and emergence of quantum states where not only the values of energy but also the values of angular momentum are quantized may be hardly understood without the progress in this region.

It follows from the preceding that it will be probably necessary to distinguish between two kinds of quantum states: one relating to the quantum states of atoms and the other relating to existence of individual matter objects characterized by special values of rest energy and angular momentum. The former relates to the mutual properties (interaction) of electrons and protons and the latter to mutual interactions of strongly interacting objects. It has been shown that the Coulomb interaction itself cannot explain the emergence of quantum states and that some additional interaction must exist between protons (nucleons) and electrons. And one should ask whether a kind of weak *contact* forces does not exist in addition to strong (nuclear) contact ones.

The existence of contact forces has opened a new question: How to describe the given situation with the help of corresponding mathematical models. It is evident that it will be necessary to limit the solutions of Schrödinger equation to subsystems exhibiting continuous evolution. This approach should be used in individual subsystems where no sudden changes (caused by contact forces) are occurring; and to describe the effect of contact forces probably as the passage to another subsystem being described again as developing continuously. The given situation may be represented in the Hilbert space consisting of individual mutually orthogonal subspaces defined with the help of corresponding Schrödinger equation describing respective continuous evolution.

The representation of physical processes in a suitable Hilbert space is surely very helpful. It means, of course, that we must represent the states corresponding to quantities conserving during the whole time evolution in mutually orthogonal subspaces. It concerns mainly two quantities: energy and angular momentum. The corresponding trajectories should always belong to mutually orthogonal subspaces. It may concern also the representation of states characterized by impact parameter values in the case of collision processes.

In any case it is possible to say that we have pass to the other ontological level that differs from that the classical physics has been based on and opens quite new questions. It is also the fact that these basic particles are to be characterized by special values of spins, testifying that these particles exhibit special internal dynamics which may lead to the existence of their different internal states. It has been demonstrated that these characteristics may be studied with the help of corresponding elastic collisions processes.

The given results might then help in revealing the internal dynamics running in corresponding objects. It is evident that practically all particles existing in the nature must be taken as complex quantum objects. The new results may contribute in looking for the characteristics of constituents of which the given particles consist and which are responsible for their structures. At the present the main attention is to be devoted to the protons and other hadrons; especially the reasons for their stability or instability should be analyzed.

From the presented results it is also possible to conclude that the contemporary quark theory having been proposed on the basis of phenomenological description of corresponding physical situation may hardly represent suitable basis to the given goal. It is, of course, necessary to expect that some basic objects (some "quasi-quarks") should exist that should exhibit some very strong (superstrong) contact interaction. This interaction kind might be responsible for forming some conglomerates equivalent to the so called partons corresponding in principle to experimentally observable hadrons. These partons might form changeable structures (being held mutually together by standard strong forces) of individual hadrons while they might be kicked out from the given object when sufficient amount of kinetic energy has been furnished to them in the corresponding collision event.

The preceding considerations have been based on the ontological approach respecting the basic matter properties when it has been shown that the past refusal of ontology led to key mistakes in the physics of the 20th century. One must be very careful in formulating and testing different hypotheses when some quite new properties of nanoscopic objects should be taken into account (e.g., the existence of contact forces). However, the ontological basis (including causal sequentiality) in the description of matter world should not be abandoned.

9. Conclusion

Even if the Schrödinger equation might represent in principle basic theory of the whole matter reality it corresponds to approximate phenomenological description only. Especially, it is not possible to explain at all how a quantum state may emerge. E.g., it is evident that the hydrogen atom arises always when slowly moving electron and proton meet and are attracted mutually by Coulomb force, which cannot be theoretically reasoned. It may occurred only if a kind of repulsive force between two given objects or a kind of impenetrability of proton having certain dimensions is to be involved. It means that the structure of protons (and other hadrons) represents indivisible part of the contemporary quantum physics.

Consequently, the concept of the hydrogen atom formed by proton and electron orbiting around should be taken hardly as acceptable. The adhesive merging of both the constituents must be regarded as much more probable concept. Here, of course, the existence of different internal proton structures should be responsible for divers quantum states of hydrogen atoms. It is necessary to look for the experiments how to get corresponding pieces of knowledge.

It has been demonstrated that some characteristics of these different proton structures might be derived from the data of elastic two-proton collisions. However, the earlier antiontological proton properties have been obtained on the basis of phenomenological models where some arbitrary additional assumptions have been added. The standard ontological characteristics have been obtained when the eikonal model enabling to test the dependence on impact parameter value has been made use of. When this ontological model has been further generalized (and corresponding probabilities have been derived directly from measured values) it has been shown that the momentum transfer dependence of differential cross section may be reconstructed as the consequence of colliding protons exhibiting a series of structures of different external dimensions. It opens new way how the existence of quantum atom states might be interpreted on much more realistic basis.

It represents also important argument that the purely phenomenological approach to physical reality should be abandoned as quite insufficient. It is necessary to return to ontological approach on which all successes of the classical physics were based and from which practically all world civilization emerged. Some interesting orientation results have been already obtained with the help of the mentioned very rough model. Now the attention is to be devoted to its generalization to correspond fully to all ontological requirements and not to involve any unphysical limitation.

The given results should contribute mainly to understand better the existence of atom quantum states; and how they may be influenced by proton structure. It is necessary to analyze corresponding data from experiments that might help in this direction; mainly elastic collision experiments may be very helpful. However, it is not more possible to look for a phenomenological description of measured values only, but for the interpretation of corresponding processes on ontological basis.

However, it may be also helpful to answer the question how it was possible that the Copenhagen alternative was influencing scientific thinking in greater part of the past century. It followed from the fact that two different kinds of quantum physics have existed; one based on the Copenhagen quantum mechanics and looking for the support of quantum paradoxes and the other one solving in principle successfully different physical and technological problems on the basis of standard Schrödinger approach (no additional assumption having been added - without mentioning it explicitly). It is possible to say that it followed from the fact that the scientific thinking in the modern period was fundamentally influenced by mathematical philosophy of Descartes refusing ontological approach. It was also the reason why also Einstein's criticism based on ontological argument has been refused by scientific community. Our main contemporary task consists in devoting more attention to the ontological properties of physical objects.

Author details

Miloš V. Lokajíček*,
Vojtěch Kundrát and Jiří Procházka

* Address all correspondence to: lokaj@fzu.cz; kundrat@fzu.cz; prochazka@fzu.cz

Institute of Physics of the AS CR, Prague, Czech Republic

References

[1] E. Schrödinger: Quantisierung als Eigenwertproblem; Ann. Phys. 79, 361-76; 489-527; 80, 437-90; 81, 109-39 (1926).

[2] N. Bohr: The quantum postulate and the development of atomic theory; Nature 121, 580-90 (1928).

[3] W. Pauli: Die allgemeinen Prinzipien der Wellenmechanik; Handbuch der Physik XXIV, Springer, Berlin, p. 140 (1933).

[4] A. Einstein, B. Podolsky, N. Rosen: Can quantum-mechanical description of physical reality be considered complete?; Phys. Rev. 47, 777-80 (1935).

[5] N. Bohr: Can quantum-mechanical description of physical reality be considered complete?; Phys. Rev. 48, 696-702 (1935).

[6] J. von Neumann: Mathematische Grundlagen der Quantenmechanik; Springer (1932).

[7] Grete Herrmann: Die Naturphilosophischen Grundlagen der Quantenmechanik; Abhandlungen der Fries'schen Schule 6, 75-152 (1935).

[8] D. Bohm: A suggested interpretation of the quantum theory in terms of "hidden variables"; Phys. Rev. 85, 180-93 (1952).

[9] J. S. Bell: On the Einstein Podolsky Rosen paradox; Physics 1, 195-200 (1964).

[10] A. Aspect, P. Grangier, G. Roger: Experimental realization of Einstein-Podolsky-Rosen-Bohm Gedankenexperiment: A new violation of Bell's inequalities; Phys. Rev. Lett. 49, 91-4 (1982).

[11] M. V. Lokajíček: Locality problem, Bell's inequalities and EPR experiments; /arXiv:quant-ph/9808005 (1998).

[12] E. E. Rosinger: George Boole and the Bell inequalities; /arXiv:quant-ph/0406004.

[13] G. Boole: On the Theory of Probabilities; Philos. Trans.; R. Soc., London 152, 225-52 (1862).

[14] N. Vorob'ev: Theor. Probab. Appl. 7, 147 (1952).

[15] M. V. Lokajíček: The assumption in Bell's inequalities and entanglement problem; J. Comp. Theor. Nanosci. (accepted for publication), /arXiv:1108.0922.

[16] M. V. Lokajíček: Einstein-Bohr controversy and theory of hidden variables; NeuroQuantology (section: Basics of Quantum Physics) 8 (2010), issue 4, 638-45; see also /arXiv:1004.3005[quant-ph].

[17] M. V. Lokajíček: Einstein-Bohr controversy after 75 years, its actual solution and consequences; *Some Applications of Quantum Mechanics* (ed. M. R. Pahlavani), InTech Publisher (February 2012), 409-24.

[18] U. Hoyer: Synthetische Quantentheorie; Georg Olms Verlag, Hildesheim (2002).

[19] H. Ioannidou: A new derivation of Schrödinger equation; Lett. al Nuovo Cim. 34, 453-8 (1982).

[20] M. V. Lokajíček: Schrödinger equation, classical physics and Copenhagen quantum mechanics; New Advances in Physics 1, No. 1, 69-77 (2007); see also /arxiv/quant-ph/0611176.

[21] P. Kundrát, M. Lokajíček: Three-dimensional harmonic oscillator and time evolution in quantum mechanics; Phys. Rev. A 67, art. 012104 (2003).

[22] P. Kundrát, M. Lokajíček: Irreversible time flow and Hilbert space structure; New Research in Quantum Physics (eds. Vl. Krasnoholovets, F. Columbus), Nova Science Publishers, Inc., pp. 17-41 (2004).

[23] P.D.Lax, R.S.Phillips: Scattering theory; Academic Press (1967).

[24] P.D.Lax, R.S.Phillips: Scattering theory for automorphic functions; Princeton (1976).

[25] N. Bohr: On the constitution of atoms and molecules; Philosophical Magazine 26, 1-24 (1913).

[26] V. Kundrát, M. V. Lokajíček: High-energy elastic scattering amplitude of unpolarized and charged hadrons; Z. Phys. C 63, 619-29 (1994).

[27] G. B. West and D. R. Yennie: Coulomb Interference in High-Energy Scattering; Phys. Rev. 172, 1413 (1968).

[28] J. Kašpar et al.: Phenomenological models of elastic nucleon scattering and predictions for LHC; Nucl. Phys. B 843, 84 (2011).

[29] V. Kundrát, M. Lokajíček and D. Krupa: Impact parameter structure derived from elastic collisions; Physics Letters B 544, 132 (2002).

[30] G. Giacomelli, M. Jacob: Physics at the CERN-ISR; Phys. Rep. 55, 1 (1979).

[31] V. Kundrát, M. Lokajíček Jr., M. V. Lokajíček: Are elastic collisions central or peripheral?; Czech. J. Phys. B 31, 1334 (1981).

[32] J. L. Bailly et al. (EHS-RCBC Collaboration): An impact parameter analysis of proton-proton elastic and inelastic interactions at 360 GeV/c; Z. Phys. C 37, 7 (1987).

[33] J. Bystricky et al., in Nucleon-nucleon and kaon-nucleon scattering edited by H. Schopper(Landolt-Börnstein Series, Vol. 1), (Springer, Berlin, 1980).

[34] M. K. Carter, P. D. B. Collins and M. R. Whalley, Compilation of Nucleon-Nucleon and Nucleon-Antinucleon Elastic Scattering Data, RAL-86-002, preprint.

[35] R. G. Newton: Optical theorem and beyond; Am. J. Phys. 44, 639-42 (1976).

[36] V. Barone, E. Predazzi: High-energy particle diffraction; Springer-Verlag (2002).

[37] M. V. Lokajíček, V. Kundrát: Optical theorem and elastic nucleon scattering; /arXiv:0906.3961 (see also Proc. of 13th Int. Conf., Blois Workshop; /arXiv:1002.3527 [hep-ph]).

[38] M. V. Lokajíček, V. Kundrát: Elastic pp scattering and the internal structure of colliding protons; (2009) /arXiv:0909.3199[hep-ph].

Chapter 7

Quantum Damped Harmonic Oscillator

Kazuyuki Fujii

Additional information is available at the end of the chapter

1. Introduction

In this chapter we introduce a toy model of Quantum Mechanics with Dissipation. Quantum Mechanics with Dissipation plays a crucial role to understand real world. However, it is not easy to master the theory for undergraduates. The target of this chapter is eager undergraduates in the world. Therefore, a good toy model to understand it deeply is required.

The quantum damped harmonic oscillator is just such a one because undergraduates must use (master) many fundamental techniques in Quantum Mechanics and Mathematics. That is, harmonic oscillator, density operator, Lindblad form, coherent state, squeezed state, tensor product, Lie algebra, representation theory, Baker–Campbell–Hausdorff formula, etc.

They are "jewels" in Quantum Mechanics and Mathematics. If undergraduates master this model, they will get a powerful weapon for Quantum Physics. I expect some of them will attack many hard problems of Quantum Mechanics with Dissipation.

The contents of this chapter are based on our two papers [3] and [6]. I will give a clear and fruitful explanation to them as much as I can.

2. Some preliminaries

In this section let us make some reviews from Physics and Mathematics within our necessity.

2.1. From physics

First we review the solution of classical damped harmonic oscillator, which is important to understand the text. For this topic see any textbook of Mathematical Physics.

The differential equation is given by

$$\ddot{x} + \omega^2 x = -\gamma \dot{x} \iff \ddot{x} + \gamma \dot{x} + \omega^2 x = 0 \quad (\gamma > 0) \tag{2.1}$$

where $x = x(t)$, $\dot{x} = dx/dt$ and the mass m is set to 1 for simplicity. In the following we treat only the case $\omega > \gamma/2$ (the case $\omega = \gamma/2$ may be interesting).

Solutions (with complex form) are well–known to be

$$x_{\pm}(t) = e^{-\left(\frac{\gamma}{2} \pm i\sqrt{\omega^2 - (\frac{\gamma}{2})^2}\right)t},$$

so the general solution is given by

$$\begin{aligned} x(t) = \alpha x_{+}(t) + \bar{\alpha} x_{-}(t) &= \alpha e^{-\left(\frac{\gamma}{2} + i\sqrt{\omega^2 - (\frac{\gamma}{2})^2}\right)t} + \bar{\alpha} e^{-\left(\frac{\gamma}{2} - i\sqrt{\omega^2 - (\frac{\gamma}{2})^2}\right)t} \\ &= \alpha e^{-\left(\frac{\gamma}{2} + i\omega\sqrt{1 - (\frac{\gamma}{2\omega})^2}\right)t} + \bar{\alpha} e^{-\left(\frac{\gamma}{2} - i\omega\sqrt{1 - (\frac{\gamma}{2\omega})^2}\right)t} \end{aligned} \tag{2.2}$$

where α is a complex number. If $\gamma/2\omega$ is small enough we have an approximate solution

$$x(t) \approx \alpha e^{-(\frac{\gamma}{2} + i\omega)t} + \bar{\alpha} e^{-(\frac{\gamma}{2} - i\omega)t}. \tag{2.3}$$

Next, we consider the quantum harmonic oscillator. This is well–known in textbooks of Quantum Mechanics. As standard textbooks of Quantum Mechanics see [2] and [11] ([2] is particularly interesting).

For the Hamiltonian

$$H = H(q, p) = \frac{1}{2}(p^2 + \omega^2 q^2) \tag{2.4}$$

where $q = q(t),\ p = p(t)$, the canonical equation of motion reads

$$\dot{q} \equiv \frac{\partial H}{\partial p} = p, \quad \dot{p} \equiv -\frac{\partial H}{\partial q} = -\omega^2 q.$$

From these we recover the equation

$$\ddot{q} = -\omega^2 q \iff \ddot{q} + \omega^2 q = 0.$$

See (2.1) with $q = x$ and $\lambda = 0$.

Next, we introduce the Poisson bracket. For $A = A(q, p),\ B = B(q, p)$ it is defined as

$$\{A, B\}_c \equiv \frac{\partial A}{\partial q}\frac{\partial B}{\partial p} - \frac{\partial A}{\partial p}\frac{\partial B}{\partial q} \tag{2.5}$$

where $\{,\}_c$ means classical. Then it is easy to see

$$\{q, q\}_c = 0, \quad \{p, p\}_c = 0, \quad \{q, p\}_c = 1. \tag{2.6}$$

Now, we are in a position to give a quantization condition due to Dirac. Before that we prepare some notation from algebra.

Square matrices A and B don't commute in general, so we need the commutator

$$[A, B] = AB - BA.$$

Then Dirac gives an abstract correspondence $q \longrightarrow \hat{q}, \quad p \longrightarrow \hat{p}$ which satisfies the condition

$$[\hat{q}, \hat{q}] = 0, \quad [\hat{p}, \hat{p}] = 0, \quad [\hat{q}, \hat{p}] = i\hbar \mathbf{1} \tag{2.7}$$

corresponding to (2.6). Here $\hbar$ is the Plank constant, and $\hat{q}$ and $\hat{p}$ are both Hermite operators on some Fock space (a kind of Hilbert space) given in the latter and $\mathbf{1}$ is the identity on it. Therefore, our quantum Hamiltonian should be

$$H = H(\hat{q}, \hat{p}) = \frac{1}{2}(\hat{p}^2 + \omega^2 \hat{q}^2) \tag{2.8}$$

from (2.4). Note that a notation H instead of $\hat{H}$ is used for simplicity. From now on we consider a complex version. From (2.4) and (2.8) we rewrite like

$$H(q, p) = \frac{1}{2}(p^2 + \omega^2 q^2) = \frac{\omega^2}{2}(q^2 + \frac{1}{\omega^2}p^2) = \frac{\omega^2}{2}(q - \frac{i}{\omega}p)(q + \frac{i}{\omega}p)$$

and

$$\begin{aligned} H(\hat{q}, \hat{p}) &= \frac{\omega^2}{2}(\hat{q}^2 + \frac{1}{\omega^2}\hat{p}^2) = \frac{\omega^2}{2}\left\{(\hat{q} - \frac{i}{\omega}\hat{p})(\hat{q} + \frac{i}{\omega}\hat{p}) - \frac{i}{\omega}[\hat{q}, \hat{p}]\right\} \\ &= \frac{\omega^2}{2}\left\{(\hat{q} - \frac{i}{\omega}\hat{p})(\hat{q} + \frac{i}{\omega}\hat{p}) + \frac{\hbar}{\omega}\right\} = \omega\hbar\left\{\frac{\omega}{2\hbar}(\hat{q} - \frac{i}{\omega}\hat{p})(\hat{q} + \frac{i}{\omega}\hat{p}) + \frac{1}{2}\right\} \end{aligned}$$

by use of (2.7), and if we set

$$a^\dagger = \sqrt{\frac{\omega}{2\hbar}}(\hat{q} - \frac{i}{\omega}\hat{p}), \quad a = \sqrt{\frac{\omega}{2\hbar}}(\hat{q} + \frac{i}{\omega}\hat{p}) \tag{2.9}$$

we have easily

$$[a, a^\dagger] = \frac{\omega}{2\hbar}[\hat{q} + \frac{i}{\omega}\hat{p}, \hat{q} - \frac{i}{\omega}\hat{p}] = \frac{\omega}{2\hbar}\left\{-\frac{2i}{\omega}[\hat{q}, \hat{p}]\right\} = \frac{\omega}{2\hbar}\left\{-\frac{2i}{\omega} \times i\hbar\right\} = \mathbf{1}$$

by use of (2.7). As a result we obtain a well–known form

$$H = \omega\hbar(a^\dagger a + \frac{1}{2}), \quad [a, a^\dagger] = \mathbf{1}. \tag{2.10}$$

Here we used an abbreviation $1/2$ in place of $(1/2)\mathbf{1}$ for simplicity.

If we define an operator $N = a^{\dagger}a$ (which is called the number operator) then it is easy to see the relations

$$[N, a^{\dagger}] = a^{\dagger}, \quad [N, a] = -a, \quad [a, a^{\dagger}] = \mathbf{1}. \tag{2.11}$$

For the proof a well–known formula $[AB, C] = [A, C]B + A[B, C]$ is used. Note that $aa^{\dagger} = a^{\dagger}a + [a, a^{\dagger}] = N + 1$. The set $\{a^{\dagger}, a, N\}$ is just a generator of Heisenberg algebra and we can construct a Fock space based on this. Let us note that a, $a^{\dagger}$ and N are called the annihilation operator, creation one and number one respectively.

First of all let us define a vacuum $|0\rangle$. This is defined by the equation $a|0\rangle = 0$. Based on this vacuum we construct the n state $|n\rangle$ like

$$|n\rangle = \frac{(a^{\dagger})^n}{\sqrt{n!}}|0\rangle \quad (0 \le n).$$

Then we can easily prove

$$a^{\dagger}|n\rangle = \sqrt{n+1}|n+1\rangle, \quad a|n\rangle = \sqrt{n}|n-1\rangle, \quad N|n\rangle = n|n\rangle \tag{2.12}$$

and moreover can prove both the orthogonality condition and the resolution of unity

$$\langle m|n\rangle = \delta_{mn}, \quad \sum_{n=0}^{\infty} |n\rangle\langle n| = \mathbf{1}. \tag{2.13}$$

For the proof one can use for example

$$a^2(a^{\dagger})^2 = a(aa^{\dagger})a^{\dagger} = a(N+1)a^{\dagger} = (N+2)aa^{\dagger} = (N+2)(N+1)$$

by (2.11), therefore we have

$$\langle 0|a^2(a^{\dagger})^2|0\rangle = \langle 0|(N+2)(N+1)|0\rangle = 2! \Longrightarrow \langle 2|2\rangle = 1.$$

The proof of the resolution of unity may be not easy for readers (we omit it here).

As a result we can define a Fock space generated by the generator $\{a^{\dagger}, a, N\}$

$$\mathcal{F} = \left\{ \sum_{n=0}^{\infty} c_n|n\rangle \in \mathbf{C}^{\infty} \mid \sum_{n=0}^{\infty} |c_n|^2 < \infty \right\}. \tag{2.14}$$

This is just a kind of Hilbert space. On this space the operators (= infinite dimensional matrices) $a^{\dagger}$, a and N are represented as

$$a = \begin{pmatrix} 0 & 1 & & & \\ & 0 & \sqrt{2} & & \\ & & 0 & \sqrt{3} & \\ & & & 0 & \ddots \\ & & & & \ddots \end{pmatrix}, \quad a^\dagger = \begin{pmatrix} 0 & & & & \\ 1 & 0 & & & \\ & \sqrt{2} & 0 & & \\ & & \sqrt{3} & 0 & \\ & & & \ddots & \ddots \end{pmatrix},$$

$$N = a^\dagger a = \begin{pmatrix} 0 & & & & \\ & 1 & & & \\ & & 2 & & \\ & & & 3 & \\ & & & & \ddots \end{pmatrix} \tag{2.15}$$

by use of (2.12).

Note We can add a phase to $\{a, a^\dagger\}$ like

$$b = e^{i\theta}a, \quad b^\dagger = e^{-i\theta}a^\dagger, \quad N = b^\dagger b = a^\dagger a$$

where θ is constant. Then we have another Heisenberg algebra

$$[N, b^\dagger] = b^\dagger, \quad [N, b] = -b, \quad [b, b^\dagger] = \mathbf{1}.$$

Next, we introduce a coherent state which plays a central role in Quantum Optics or Quantum Computation. For $z \in \mathbf{C}$ the coherent state $|z\rangle \in \mathcal{F}$ is defined by the equation

$$a|z\rangle = z|z\rangle \quad \text{and} \quad \langle z|z\rangle = 1.$$

The annihilation operator a is not hermitian, so this equation is never trivial. For this state the following three equations are equivalent :

$$\begin{cases} (1) \quad a|z\rangle = z|z\rangle \text{ and } \langle z|z\rangle = 1, \\ (2) \quad |z\rangle = e^{za^\dagger - \bar{z}a}|0\rangle, \\ (3) \quad |z\rangle = e^{-\frac{|z|^2}{2}} \sum_{n=0}^{\infty} \frac{z^n}{\sqrt{n!}}|n\rangle. \end{cases} \tag{2.16}$$

The proof is as follows. From (1) to (2) we use a popular formula

$$e^A B e^{-A} = B + [A, B] + \frac{1}{2!}[A, [A, B]] + \cdots$$

$(A, B$: operators) to prove

$$e^{-(za^\dagger - \bar{z}a)} a e^{za^\dagger - \bar{z}a} = a + z.$$

From (2) to (3) we use the Baker-Campbell-Hausdorff formula (see for example [17])

$$e^A e^B = e^{A+B+\frac{1}{2}[A,B]+\frac{1}{6}[A,[A,B]]+\frac{1}{6}[B,[A,B]]+\cdots}.$$

If $[A,[A,B]] = 0 = [B,[A,B]]$ (namely, $[A,B]$ commutes with both A and B) then we have

$$e^A e^B = e^{A+B+\frac{1}{2}[A,B]} = e^{\frac{1}{2}[A,B]} e^{A+B} \Longrightarrow e^{A+B} = e^{-\frac{1}{2}[A,B]} e^A e^B. \tag{2.17}$$

In our case the condition is satisfied because of $[a, a^\dagger] = 1$. Therefore we obtain a (famous) decomposition

$$e^{za^\dagger - \bar{z}a} = e^{-\frac{|z|^2}{2}} e^{za^\dagger} e^{-\bar{z}a}. \tag{2.18}$$

The remaining part of the proof is left to readers.

From the equation (3) in (2.16) we obtain the resolution of unity for coherent states

$$\int\int \frac{dxdy}{\pi} |z\rangle\langle z| = \sum_{n=0}^{\infty} |n\rangle\langle n| = \mathbf{1} \quad (z = x + iy). \tag{2.19}$$

The proof is reduced to the following formula

$$\int\int \frac{dxdy}{\pi} e^{-|z|^2} \bar{z}^m z^n = n!\, \delta_{mn} \quad (z = x + iy).$$

The proof is left to readers. See [14] for more general knowledge of coherent states.

2.2. From mathematics

We consider a simple matrix equation

$$\frac{d}{dt} X = AXB \tag{2.20}$$

where

$$X = X(t) = \begin{pmatrix} x_{11}(t) & x_{12}(t) \\ x_{21}(t) & x_{22}(t) \end{pmatrix}, \quad A = \begin{pmatrix} a_{11} & a_{12} \\ a_{21} & a_{22} \end{pmatrix}, \quad B = \begin{pmatrix} b_{11} & b_{12} \\ b_{21} & b_{22} \end{pmatrix}.$$

A standard form of linear differential equation which we usually treat is

$$\frac{d}{dt}\mathbf{x} = C\mathbf{x}$$

where $\mathbf{x} = \mathbf{x}(t)$ is a vector and C is a matrix associated to the vector. Therefore, we want to rewrite (2.20) into a standard form.

For the purpose we introduce the Kronecker product of matrices. For example, it is defined as

$$\begin{aligned} A \otimes B &= \begin{pmatrix} a_{11} & a_{12} \\ a_{21} & a_{22} \end{pmatrix} \otimes B \equiv \begin{pmatrix} a_{11}B & a_{12}B \\ a_{21}B & a_{22}B \end{pmatrix} \\ &= \begin{pmatrix} a_{11}b_{11} & a_{11}b_{12} & a_{12}b_{11} & a_{12}b_{12} \\ a_{11}b_{21} & a_{11}b_{22} & a_{12}b_{21} & a_{12}b_{22} \\ a_{21}b_{11} & a_{21}b_{12} & a_{22}b_{11} & a_{22}b_{12} \\ a_{21}b_{21} & a_{21}b_{22} & a_{22}b_{21} & a_{22}b_{22} \end{pmatrix} \end{aligned} \tag{2.21}$$

for A and B above. Note that recently we use the tensor product instead of the Kronecker product, so we use it in the following. Here, let us list some useful properties of the tensor product

$$\begin{aligned} &(1) \quad (A_1 \otimes B_1)(A_2 \otimes B_2) = A_1A_2 \otimes B_1B_2, \\ &(2) \quad (A \otimes E)(E \otimes B) = A \otimes B = (E \otimes B)(A \otimes E), \\ &(3) \quad e^{A\otimes E + E\otimes B} = e^{A\otimes E}e^{E\otimes B} = (e^A \otimes E)(E \otimes e^B) = e^A \otimes e^B, \\ &(4) \quad (A \otimes B)^\dagger = A^\dagger \otimes B^\dagger \end{aligned} \tag{2.22}$$

where E is the unit matrix. The proof is left to readers. [9] is recommended as a general introduction.

Then the equation (2.20) can be written in terms of components as

$$\begin{cases} \frac{dx_{11}}{dt} = a_{11}b_{11}x_{11} + a_{11}b_{21}x_{12} + a_{12}b_{11}x_{21} + a_{12}b_{21}x_{22}, \\ \frac{dx_{12}}{dt} = a_{11}b_{12}x_{11} + a_{11}b_{22}x_{12} + a_{12}b_{12}x_{21} + a_{12}b_{22}x_{22}, \\ \frac{dx_{21}}{dt} = a_{21}b_{11}x_{11} + a_{21}b_{21}x_{12} + a_{22}b_{11}x_{21} + a_{22}b_{21}x_{22}, \\ \frac{dx_{22}}{dt} = a_{21}b_{12}x_{11} + a_{21}b_{22}x_{12} + a_{22}b_{12}x_{21} + a_{22}b_{22}x_{22} \end{cases}$$

or in a matrix form

$$\frac{d}{dt}\begin{pmatrix} x_{11} \\ x_{12} \\ x_{21} \\ x_{22} \end{pmatrix} = \begin{pmatrix} a_{11}b_{11} & a_{11}b_{21} & a_{12}b_{11} & a_{12}b_{21} \\ a_{11}b_{12} & a_{11}b_{22} & a_{12}b_{12} & a_{12}b_{22} \\ a_{21}b_{11} & a_{21}b_{21} & a_{22}b_{11} & a_{22}b_{21} \\ a_{21}b_{12} & a_{21}b_{22} & a_{22}b_{12} & a_{22}b_{22} \end{pmatrix} \begin{pmatrix} x_{11} \\ x_{12} \\ x_{21} \\ x_{22} \end{pmatrix}.$$

If we set

$$X = \begin{pmatrix} x_{11} & x_{12} \\ x_{21} & x_{22} \end{pmatrix} \Longrightarrow \widehat{X} = (x_{11}, x_{12}, x_{21}, x_{22})^T$$

where T is the transpose, then we obtain a standard form

$$\frac{d}{dt}X = AXB \Longrightarrow \frac{d}{dt}\widehat{X} = (A \otimes B^T)\widehat{X} \tag{2.23}$$

from (2.21).

Similarly we have a standard form

$$\frac{d}{dt}X = AX + XB \Longrightarrow \frac{d}{dt}\widehat{X} = (A \otimes E + E \otimes B^T)\widehat{X} \tag{2.24}$$

where $E^T = E$ for the unit matrix E.

From these lessons there is no problem to generalize (2.23) and (2.24) based on 2×2 matrices to ones based on any (square) matrices or operators on $\mathcal{F}$. Namely, we have

$$\begin{cases} \frac{d}{dt}X = AXB \Longrightarrow \frac{d}{dt}\widehat{X} = (A \otimes B^T)\widehat{X}, \\ \frac{d}{dt}X = AX + XB \Longrightarrow \frac{d}{dt}\widehat{X} = (A \otimes I + I \otimes B^T)\widehat{X}. \end{cases} \tag{2.25}$$

where I is the identity E (matrices) or $\mathbf{1}$ (operators).

3. Quantum damped harmonic oscillator

In this section we treat the quantum damped harmonic oscillator. As a general introduction to this topic see [1] or [16].

3.1. Model

Before that we introduce the quantum harmonic oscillator. The Schrödinger equation is given by

$$i\hbar\frac{\partial}{\partial t}|\Psi(t)\rangle = H|\Psi(t)\rangle = \left(\omega\hbar(N + \frac{1}{2})\right)|\Psi(t)\rangle$$

by (2.10) (note $N = a^\dagger a$). In the following we use $\frac{\partial}{\partial t}$ instead of $\frac{d}{dt}$.

Now we change from a wave–function to a density operator because we want to treat a mixed state, which is a well–known technique in Quantum Mechanics or Quantum Optics.

If we set $\rho(t) = |\Psi(t)\rangle\langle\Psi(t)|$, then a little algebra gives

$$i\hbar\frac{\partial}{\partial t}\rho = [H,\rho] = [\omega\hbar N,\rho] \implies \frac{\partial}{\partial t}\rho = -i[\omega N,\rho]. \tag{3.1}$$

This is called the quantum Liouville equation. With this form we can treat a mixed state like

$$\rho = \rho(t) = \sum_{j=1}^{N} u_j|\Psi_j(t)\rangle\langle\Psi_j(t)|$$

where $u_j \geq 0$ and $\sum_{j=1}^{N} u_j = 1$. Note that the general solution of (3.1) is given by

$$\rho(t) = e^{-i\omega Nt}\rho(0)e^{i\omega Nt}.$$

We are in a position to state the equation of quantum damped harmonic oscillator by use of (3.1).

Definition The equation is given by

$$\frac{\partial}{\partial t}\rho = -i[\omega a^{\dagger}a,\rho] - \frac{\mu}{2}\left(a^{\dagger}a\rho + \rho a^{\dagger}a - 2a\rho a^{\dagger}\right) - \frac{\nu}{2}\left(aa^{\dagger}\rho + \rho aa^{\dagger} - 2a^{\dagger}\rho a\right) \tag{3.2}$$

where μ, ν ($\mu > \nu \geq 0$) are some real constants depending on the system (for example, a damping rate of the cavity mode)[1].

Note that the extra term

$$-\frac{\mu}{2}\left(a^{\dagger}a\rho + \rho a^{\dagger}a - 2a\rho a^{\dagger}\right) - \frac{\nu}{2}\left(aa^{\dagger}\rho + \rho aa^{\dagger} - 2a^{\dagger}\rho a\right)$$

is called the Lindblad form (term). Such a term plays an essential role in **Decoherence**.

3.2. Method of solution

First we solve the Lindblad equation :

$$\frac{\partial}{\partial t}\rho = -\frac{\mu}{2}\left(a^{\dagger}a\rho + \rho a^{\dagger}a - 2a\rho a^{\dagger}\right) - \frac{\nu}{2}\left(aa^{\dagger}\rho + \rho aa^{\dagger} - 2a^{\dagger}\rho a\right). \tag{3.3}$$

Interesting enough, we can solve this equation completely.

[1] The aim of this chapter is not to drive this equation. In fact, its derivation is not easy for non–experts, so see for example the original papers [15] and [10], or [12] as a short review paper

Let us rewrite (3.3) more conveniently using the number operator $N \equiv a^\dagger a$

$$\frac{\partial}{\partial t}\rho = \mu a \rho a^\dagger + \nu a^\dagger \rho a - \frac{\mu+\nu}{2}(N\rho + \rho N + \rho) + \frac{\mu-\nu}{2}\rho \tag{3.4}$$

where we have used $aa^\dagger = N + \mathbf{1}$.

From here we use the method developed in Section 2.2. For a matrix $X = (x_{ij}) \in M(\mathcal{F})$ over $\mathcal{F}$

$$X = \begin{pmatrix} x_{00} & x_{01} & x_{02} & \cdots \\ x_{10} & x_{11} & x_{12} & \cdots \\ x_{20} & x_{21} & x_{22} & \cdots \\ \vdots & \vdots & \vdots & \ddots \end{pmatrix}$$

we correspond to the vector $\widehat{X} \in \mathcal{F}^{\dim_{\mathbb{C}}\mathcal{F}}$ as

$$X = (x_{ij}) \longrightarrow \widehat{X} = (x_{00}, x_{01}, x_{02}, \cdots; x_{10}, x_{11}, x_{12}, \cdots; x_{20}, x_{21}, x_{22}, \cdots; \cdots\cdots)^T \tag{3.5}$$

where T means the transpose. The following formulas

$$\widehat{AXB} = (A \otimes B^T)\widehat{X}, \quad (\widehat{AX + XB}) = (A \otimes \mathbf{1} + \mathbf{1} \otimes B^T)\widehat{X} \tag{3.6}$$

hold for $A, B, X \in M(\mathcal{F})$, see (2.25).

Then (3.4) becomes

$$\begin{aligned} \frac{\partial}{\partial t}\widehat{\rho} &= \left\{\mu a \otimes (a^\dagger)^T + \nu a^\dagger \otimes a^T - \frac{\mu+\nu}{2}(N \otimes \mathbf{1} + \mathbf{1} \otimes N + \mathbf{1} \otimes \mathbf{1}) + \frac{\mu-\nu}{2}\mathbf{1} \otimes \mathbf{1}\right\}\widehat{\rho} \\ &= \left\{\frac{\mu-\nu}{2}\mathbf{1} \otimes \mathbf{1} + \nu a^\dagger \otimes a^\dagger + \mu a \otimes a - \frac{\mu+\nu}{2}(N \otimes \mathbf{1} + \mathbf{1} \otimes N + \mathbf{1} \otimes \mathbf{1})\right\}\widehat{\rho} \end{aligned} \tag{3.7}$$

where we have used $a^T = a^\dagger$ from the form (2.15), so that the solution is formally given by

$$\widehat{\rho}(t) = e^{\frac{\mu-\nu}{2}t} e^{t\left\{\nu a^\dagger \otimes a^\dagger + \mu a \otimes a - \frac{\mu+\nu}{2}(N \otimes \mathbf{1} + \mathbf{1} \otimes N + \mathbf{1} \otimes \mathbf{1})\right\}} \widehat{\rho}(0). \tag{3.8}$$

In order to use some techniques from Lie algebra we set

$$K_3 = \frac{1}{2}(N \otimes \mathbf{1} + \mathbf{1} \otimes N + \mathbf{1} \otimes \mathbf{1}), \quad K_+ = a^\dagger \otimes a^\dagger, \quad K_- = a \otimes a \quad \left(K_- = K_+^\dagger\right) \tag{3.9}$$

then we can show the relations

$$[K_3, K_+] = K_+, \quad [K_3, K_-] = -K_-, \quad [K_+, K_-] = -2K_3.$$

This is just the $su(1,1)$ algebra. The proof is very easy and is left to readers.

The equation (3.8) can be written simply as

$$\widehat{\rho}(t) = e^{\frac{\mu-\nu}{2}t} e^{t\{\nu K_+ + \mu K_- - (\mu+\nu)K_3\}} \widehat{\rho}(0), \tag{3.10}$$

so we have only to calculate the term

$$e^{t\{\nu K_+ + \mu K_- - (\mu+\nu)K_3\}}, \tag{3.11}$$

which is of course not simple. Now the disentangling formula in [4] is helpful in calculating (3.11).

If we set $\{k_+, k_-, k_3\}$ as

$$k_+ = \begin{pmatrix} 0 & 1 \\ 0 & 0 \end{pmatrix}, \quad k_- = \begin{pmatrix} 0 & 0 \\ -1 & 0 \end{pmatrix}, \quad k_3 = \frac{1}{2}\begin{pmatrix} 1 & 0 \\ 0 & -1 \end{pmatrix} \quad \left(k_- \neq k_+^{\dagger}\right) \tag{3.12}$$

then it is very easy to check the relations

$$[k_3, k_+] = k_+, \quad [k_3, k_-] = -k_-, \quad [k_+, k_-] = -2k_3.$$

That is, $\{k_+, k_-, k_3\}$ are generators of the Lie algebra $su(1,1)$. Let us show by $SU(1,1)$ the corresponding Lie group, which is a typical noncompact group.

Since $SU(1,1)$ is contained in the special linear group $SL(2;\mathbf{C})$, we **assume** that there exists an infinite dimensional unitary representation $\rho : SL(2;\mathbf{C}) \longrightarrow U(\mathcal{F} \otimes \mathcal{F})$ (group homomorphism) satisfying

$$d\rho(k_+) = K_+, \quad d\rho(k_-) = K_-, \quad d\rho(k_3) = K_3.$$

From (3.11) some algebra gives

$$\begin{aligned} e^{t\{\nu K_+ + \mu K_- - (\mu+\nu)K_3\}} &= e^{t\{\nu d\rho(k_+) + \mu d\rho(k_-) - (\mu+\nu) d\rho(k_3)\}} \\ &= e^{d\rho(t(\nu k_+ + \mu k_- - (\mu+\nu)k_3))} \\ &= \rho\left(e^{t(\nu k_+ + \mu k_- - (\mu+\nu)k_3)}\right) \quad (\Downarrow \text{ by definition}) \\ &\equiv \rho\left(e^{tA}\right) \end{aligned} \tag{3.13}$$

and we have

$$\begin{aligned} e^{tA} &= e^{t\{\nu k_+ + \mu k_- - (\mu+\nu)k_3\}} \\ &= \exp\left\{t\begin{pmatrix} -\frac{\mu+\nu}{2} & \nu \\ -\mu & \frac{\mu+\nu}{2}\end{pmatrix}\right\} \\ &= \begin{pmatrix} \cosh\left(\frac{\mu-\nu}{2}t\right) - \frac{\mu+\nu}{\mu-\nu}\sinh\left(\frac{\mu-\nu}{2}t\right) & \frac{2\nu}{\mu-\nu}\sinh\left(\frac{\mu-\nu}{2}t\right) \\ -\frac{2\mu}{\mu-\nu}\sinh\left(\frac{\mu-\nu}{2}t\right) & \cosh\left(\frac{\mu-\nu}{2}t\right) + \frac{\mu+\nu}{\mu-\nu}\sinh\left(\frac{\mu-\nu}{2}t\right)\end{pmatrix}. \end{aligned}$$

The proof is based on the following two facts.

$$(tA)^2 = t^2\begin{pmatrix} -\frac{\mu+\nu}{2} & \nu \\ -\mu & \frac{\mu+\nu}{2}\end{pmatrix}^2 = t^2\begin{pmatrix} \left(\frac{\mu+\nu}{2}\right)^2 - \mu\nu & 0 \\ 0 & \left(\frac{\mu+\nu}{2}\right)^2 - \mu\nu\end{pmatrix} = \left(\frac{\mu-\nu}{2}t\right)^2\begin{pmatrix} 1 & 0 \\ 0 & 1\end{pmatrix}$$

and

$$e^X = \sum_{n=0}^{\infty}\frac{1}{n!}X^n = \sum_{n=0}^{\infty}\frac{1}{(2n)!}X^{2n} + \sum_{n=0}^{\infty}\frac{1}{(2n+1)!}X^{2n+1} \quad (X = tA).$$

Note that

$$\cosh(x) = \frac{e^x + e^{-x}}{2} = \sum_{n=0}^{\infty}\frac{x^{2n}}{(2n)!} \quad \text{and} \quad \sinh(x) = \frac{e^x - e^{-x}}{2} = \sum_{n=0}^{\infty}\frac{x^{2n+1}}{(2n+1)!}.$$

The remainder is left to readers.

The Gauss decomposition formula (in $SL(2;\mathbf{C})$)

$$\begin{pmatrix} a & b \\ c & d\end{pmatrix} = \begin{pmatrix} 1 & \frac{b}{d} \\ 0 & 1\end{pmatrix}\begin{pmatrix} \frac{1}{d} & 0 \\ 0 & d\end{pmatrix}\begin{pmatrix} 1 & 0 \\ \frac{c}{d} & 1\end{pmatrix} \quad (ad - bc = 1)$$

gives the decomposition

$$
e^{tA} = \begin{pmatrix} 1 & \frac{\frac{2\nu}{\mu-\nu}\sinh(\frac{\mu-\nu}{2}t)}{\cosh(\frac{\mu-\nu}{2}t)+\frac{\mu+\nu}{\mu-\nu}\sinh(\frac{\mu-\nu}{2}t)} \\ 0 & 1 \end{pmatrix} \times
$$

$$
\begin{pmatrix} \frac{1}{\cosh(\frac{\mu-\nu}{2}t)+\frac{\mu+\nu}{\mu-\nu}\sinh(\frac{\mu-\nu}{2}t)} & 0 \\ 0 & \cosh\left(\frac{\mu-\nu}{2}t\right)+\frac{\mu+\nu}{\mu-\nu}\sinh\left(\frac{\mu-\nu}{2}t\right) \end{pmatrix} \times
$$

$$
\begin{pmatrix} 1 & 0 \\ -\frac{\frac{2\mu}{\mu-\nu}\sinh(\frac{\mu-\nu}{2}t)}{\cosh(\frac{\mu-\nu}{2}t)+\frac{\mu+\nu}{\mu-\nu}\sinh(\frac{\mu-\nu}{2}t)} & 1 \end{pmatrix}
$$

and moreover we have

$$
\begin{aligned}
e^{tA} &= \exp\begin{pmatrix} 0 & \frac{\frac{2\nu}{\mu-\nu}\sinh(\frac{\mu-\nu}{2}t)}{\cosh(\frac{\mu-\nu}{2}t)+\frac{\mu+\nu}{\mu-\nu}\sinh(\frac{\mu-\nu}{2}t)} \\ 0 & 0 \end{pmatrix} \times \\
&\exp\begin{pmatrix} -\log\left(\cosh(\frac{\mu-\nu}{2}t)+\frac{\mu+\nu}{\mu-\nu}\sinh(\frac{\mu-\nu}{2}t)\right) & 0 \\ 0 & \log\left(\cosh(\frac{\mu-\nu}{2}t)+\frac{\mu+\nu}{\mu-\nu}\sinh(\frac{\mu-\nu}{2}t)\right) \end{pmatrix} \times \\
&\exp\begin{pmatrix} 0 & 0 \\ -\frac{\frac{2\mu}{\mu-\nu}\sinh(\frac{\mu-\nu}{2}t)}{\cosh(\frac{\mu-\nu}{2}t)+\frac{\mu+\nu}{\mu-\nu}\sinh(\frac{\mu-\nu}{2}t)} & 0 \end{pmatrix} \\
&= \exp\left(\frac{\frac{2\nu}{\mu-\nu}\sinh\left(\frac{\mu-\nu}{2}t\right)}{\cosh\left(\frac{\mu-\nu}{2}t\right)+\frac{\mu+\nu}{\mu-\nu}\sinh\left(\frac{\mu-\nu}{2}t\right)}k_+\right) \times \\
&\exp\left(-2\log\left(\cosh\left(\frac{\mu-\nu}{2}t\right)+\frac{\mu+\nu}{\mu-\nu}\sinh\left(\frac{\mu-\nu}{2}t\right)\right)k_3\right) \times \\
&\exp\left(\frac{\frac{2\mu}{\mu-\nu}\sinh\left(\frac{\mu-\nu}{2}t\right)}{\cosh\left(\frac{\mu-\nu}{2}t\right)+\frac{\mu+\nu}{\mu-\nu}\sinh\left(\frac{\mu-\nu}{2}t\right)}k_-\right).
\end{aligned}
$$

Since ρ is a group homomorphism ($\rho(XYZ) = \rho(X)\rho(Y)\rho(Z)$) and the formula $\rho\left(e^{Lk}\right) = e^{Ld\rho(k)}$ ($k = k_+, k_3, k_-$) holds we obtain

$$\rho\left(e^{tA}\right) \exp\left(\frac{\frac{2\nu}{\mu-\nu}\sinh\left(\frac{\mu-\nu}{2}t\right)}{\cosh\left(\frac{\mu-\nu}{2}t\right)+\frac{\mu+\nu}{\mu-\nu}\sinh\left(\frac{\mu-\nu}{2}t\right)} d\rho(k_+)\right) \times$$
$$\exp\left(-2\log\left(\cosh\left(\frac{\mu-\nu}{2}t\right)+\frac{\mu+\nu}{\mu-\nu}\sinh\left(\frac{\mu-\nu}{2}t\right)\right) d\rho(k_3)\right) \times$$
$$\exp\left(\frac{\frac{2\mu}{\mu-\nu}\sinh\left(\frac{\mu-\nu}{2}t\right)}{\cosh\left(\frac{\mu-\nu}{2}t\right)+\frac{\mu+\nu}{\mu-\nu}\sinh\left(\frac{\mu-\nu}{2}t\right)} d\rho(k_-)\right).$$

As a result we have the disentangling formula

$$e^{t\{\nu K_+ + \mu K_- - (\mu+\nu)K_3\}} = \exp\left(\frac{\frac{2\nu}{\mu-\nu}\sinh\left(\frac{\mu-\nu}{2}t\right)}{\cosh\left(\frac{\mu-\nu}{2}t\right)+\frac{\mu+\nu}{\mu-\nu}\sinh\left(\frac{\mu-\nu}{2}t\right)} K_+\right) \times$$
$$\exp\left(-2\log\left(\cosh\left(\frac{\mu-\nu}{2}t\right)+\frac{\mu+\nu}{\mu-\nu}\sinh\left(\frac{\mu-\nu}{2}t\right)\right) K_3\right) \times$$
$$\exp\left(\frac{\frac{2\mu}{\mu-\nu}\sinh\left(\frac{\mu-\nu}{2}t\right)}{\cosh\left(\frac{\mu-\nu}{2}t\right)+\frac{\mu+\nu}{\mu-\nu}\sinh\left(\frac{\mu-\nu}{2}t\right)} K_-\right) \tag{3.14}$$

by (3.13).

In the following we set for simplicity

$$E(t) = \frac{\frac{2\mu}{\mu-\nu}\sinh\left(\frac{\mu-\nu}{2}t\right)}{\cosh\left(\frac{\mu-\nu}{2}t\right)+\frac{\mu+\nu}{\mu-\nu}\sinh\left(\frac{\mu-\nu}{2}t\right)},$$
$$F(t) = \cosh\left(\frac{\mu-\nu}{2}t\right)+\frac{\mu+\nu}{\mu-\nu}\sinh\left(\frac{\mu-\nu}{2}t\right), \tag{3.15}$$
$$G(t) = \frac{\frac{2\nu}{\mu-\nu}\sinh\left(\frac{\mu-\nu}{2}t\right)}{\cosh\left(\frac{\mu-\nu}{2}t\right)+\frac{\mu+\nu}{\mu-\nu}\sinh\left(\frac{\mu-\nu}{2}t\right)}.$$

Readers should be careful of this "proof", which is a heuristic method. In fact, it is incomplete because we have assumed a group homomorphism. In order to complete it we want to show a disentangling formula like

$$e^{t\{\nu K_+ + \mu K_- - (\mu+\nu)K_3\}} = e^{f(t)K_+} e^{g(t)K_3} e^{h(t)K_-}$$

with unknowns $f(t)$, $g(t)$, $h(t)$ satisfying $f(0) = g(0) = h(0) = 0$. For the purpose we set

$$A(t) = e^{t\{\nu K_+ + \mu K_- - (\mu+\nu)K_3\}}, \quad B(t) = e^{f(t)K_+}e^{g(t)K_3}e^{h(t)K_-}.$$

For $t = 0$ we have $A(0) = B(0) =$ identity and

$$\dot{A}(t) = \{\nu K_+ + \mu K_- - (\mu + \nu)K_3\}A(t).$$

Next, let us calculate $\dot{B}(t)$. By use of the Leibniz rule

$$\begin{aligned}
\dot{B}(t) &= (\dot{f}K_+)e^{f(t)K_+}e^{g(t)K_3}e^{h(t)K_-} + e^{f(t)K_+}(\dot{g}K_3)e^{g(t)K_3}e^{h(t)K_-} + e^{f(t)K_+}e^{g(t)K_3}(\dot{h}K_-)e^{h(t)K_-} \\
&= \left\{\dot{f}K_+ + \dot{g}e^{fK_+}K_3e^{-fK_+} + \dot{h}e^{fK_+}e^{gK_3}K_-e^{-gK_3}e^{-fK_+}\right\}e^{f(t)K_+}e^{g(t)K_3}e^{h(t)K_-} \\
&= \left\{\dot{f}K_+ + \dot{g}(K_3 - fK_+) + \dot{h}e^{-g}(K_- - 2fK_3 + f^2K_+)\right\}B(t) \\
&= \left\{(\dot{f} - \dot{g}f + \dot{h}e^{-g}f^2)K_+ + (\dot{g} - 2\dot{h}e^{-g}f)K_3 + \dot{h}e^{-g}K_-\right\}B(t)
\end{aligned}$$

where we have used relations

$$\begin{aligned}
&e^{fK_+}K_3e^{-fK_+} = K_3 - fK_+, \\
&e^{gK_3}K_-e^{-gK_3} = e^{-g}K_- \quad \text{and} \quad e^{fK_+}K_-e^{-fK_+} = K_- - 2fK_3 + f^2K_+.
\end{aligned}$$

The proof is easy. By comparing coefficients of $\dot{A}(t)$ and $\dot{B}(t)$ we have

$$\begin{cases} \dot{h}e^{-g} = \mu, \\ \dot{g} - 2\dot{h}e^{-g}f = -(\mu+\nu), \\ \dot{f} - \dot{g}f + \dot{h}e^{-g}f^2 = \nu \end{cases} \Longrightarrow \begin{cases} \dot{h}e^{-g} = \mu, \\ \dot{g} - 2\mu f = -(\mu+\nu), \\ \dot{f} - \dot{g}f + \mu f^2 = \nu \end{cases} \Longrightarrow \begin{cases} \dot{h}e^{-g} = \mu, \\ \dot{g} - 2\mu f = -(\mu+\nu), \\ \dot{f} + (\mu+\nu)f - \mu f^2 = \nu. \end{cases}$$

Note that the equation

$$\dot{f} + (\mu+\nu)f - \mu f^2 = \nu$$

is a (famous) Riccati equation. If we can solve the equation then we obtain solutions like

$$f \Longrightarrow g \Longrightarrow h.$$

Unfortunately, it is not easy. However there is an ansatz for the solution, G, F and E. That is,

$$f(t) = G(t), \quad g(t) = -2\log(F(t)), \quad h(t) = E(t)$$

in (3.15). To check these equations is left to readers (as a good exercise). From this

$$A(0) = B(0), \quad \dot{A}(0) = \dot{B}(0) \Longrightarrow A(t) = B(t) \quad \text{for all } t$$

and we finally obtain the disentangling formula

$$e^{t\{\nu K_+ + \mu K_- - (\mu+\nu)K_3\}} = e^{G(t)K_+} e^{-2\log(F(t))K_3} e^{E(t)K_-} \tag{3.16}$$

with (3.15).

Therefore (3.8) becomes

$$\widehat{\rho}(t) = e^{\frac{\mu-\nu}{2}t} \exp\left(G(t)a^\dagger \otimes a^\dagger\right) \exp\left(-\log(F(t))(N\otimes \mathbf{1} + \mathbf{1}\otimes N + \mathbf{1}\otimes\mathbf{1})\right) \exp\left(E(t)a\otimes a\right)\widehat{\rho}(0)$$

with (3.15). Some calculation by use of (2.22) gives

$$\begin{aligned}\widehat{\rho}(t) &= \frac{e^{\frac{\mu-\nu}{2}t}}{F(t)} \exp\left(G(t)a^\dagger \otimes a^T\right)\left\{\exp\left(-\log(F(t))N\right)\otimes \exp\left(-\log(F(t))N\right)^T\right\} \times \\ &\quad \exp\left(E(t)a\otimes (a^\dagger)^T\right)\widehat{\rho}(0)\end{aligned} \tag{3.17}$$

where we have used $N^T = N$ and $a^\dagger = a^T$. By coming back to matrix form by use of (3.6) like

$$\begin{aligned}\exp\left(E(t)a\otimes (a^\dagger)^T\right)\widehat{\rho}(0) &= \sum_{m=0}^{\infty}\frac{E(t)^m}{m!}\left(a\otimes (a^\dagger)^T\right)^m \widehat{\rho}(0) \\ &= \sum_{m=0}^{\infty}\frac{E(t)^m}{m!}\left(a^m\otimes ((a^\dagger)^m)^T\right)\widehat{\rho}(0) \longrightarrow \sum_{m=0}^{\infty}\frac{E(t)^m}{m!}a^m\rho(0)(a^\dagger)^m\end{aligned}$$

we finally obtain

$$\begin{aligned}\rho(t) &= \frac{e^{\frac{\mu-\nu}{2}t}}{F(t)} \times \\ &\sum_{n=0}^{\infty}\frac{G(t)^n}{n!}(a^\dagger)^n\left[\exp\left(-\log(F(t))N\right)\left\{\sum_{m=0}^{\infty}\frac{E(t)^m}{m!}a^m\rho(0)(a^\dagger)^m\right\}\exp\left(-\log(F(t))N\right)\right]a^n.\end{aligned} \tag{3.18}$$

This form is very beautiful but complicated !

3.3. General solution

Last, we treat the full equation (3.2)

$$\frac{\partial}{\partial t}\rho = -i\omega(a^\dagger a\rho - \rho a^\dagger a) - \frac{\mu}{2}\left(a^\dagger a\rho + \rho a^\dagger a - 2a\rho a^\dagger\right) - \frac{\nu}{2}\left(aa^\dagger\rho + \rho aa^\dagger - 2a^\dagger\rho a\right).$$

From the lesson in the preceding subsection it is easy to rewrite this as

$$\frac{\partial}{\partial t}\widehat{\rho} = \left\{-i\omega K_0 + \nu K_+ + \mu K_- - (\mu+\nu)K_3 + \frac{\mu-\nu}{2}\mathbf{1}\otimes\mathbf{1}\right\}\widehat{\rho} \tag{3.19}$$

in terms of $K_0 = N\otimes\mathbf{1} - \mathbf{1}\otimes N$ (note that $N^T = N$). Then it is easy to see

$$[K_0, K_+] = [K_0, K_3] = [K_0, K_-] = 0 \tag{3.20}$$

from (3.9), which is left to readers. That is, K_0 commutes with all $\{K_+, K_3, K_-\}$. Therefore

$$\begin{aligned}
\widehat{\rho}(t) &= e^{-i\omega t K_0} e^{t\{\nu K_+ + \mu K_- - (\mu+\nu)K_3 + \frac{\mu-\nu}{2}\mathbf{1}\otimes\mathbf{1}\}}\widehat{\rho}(0) \\
&= e^{\frac{\mu-\nu}{2}t}\exp(-i\omega t K_0)\exp(G(t)K_+)\exp(-2\log(F(t))K_3)\exp(E(t)K_-)\widehat{\rho}(0) \\
&= e^{\frac{\mu-\nu}{2}t}\exp(G(t)K_+)\exp(\{-i\omega t K_0 - 2\log(F(t))K_3\})\exp(E(t)K_-)\widehat{\rho}(0),
\end{aligned}$$

so that the general solution that we are looking for is just given by

$$\begin{aligned}
\rho(t) = \frac{e^{\frac{\mu-\nu}{2}t}}{F(t)}\sum_{n=0}^{\infty}\frac{G(t)^n}{n!}(a^\dagger)^n[\exp(\{-i\omega t - \log(F(t))\}N)\times \\
\left\{\sum_{m=0}^{\infty}\frac{E(t)^m}{m!}a^m\rho(0)(a^\dagger)^m\right\}\exp(\{i\omega t - \log(F(t))\}N)]a^n
\end{aligned} \tag{3.21}$$

by use of (3.17) and (3.18).

Particularly, if $\nu = 0$ then

$$\begin{aligned}
E(t) &= \frac{2\sinh\left(\frac{\mu}{2}t\right)}{\cosh\left(\frac{\mu}{2}t\right)+\sinh\left(\frac{\mu}{2}t\right)} = 1 - e^{-\mu t}, \\
F(t) &= \cosh\left(\frac{\mu}{2}t\right) + \sinh\left(\frac{\mu}{2}t\right) = e^{\frac{\mu}{2}t}, \\
G(t) &= 0
\end{aligned}$$

from (3.15), so that we have

$$\rho(t) = e^{-(\frac{\mu}{2}+i\omega)tN}\left\{\sum_{m=0}^{\infty}\frac{\left(1-e^{-\mu t}\right)^m}{m!}a^m\rho(0)(a^\dagger)^m\right\}e^{-(\frac{\mu}{2}-i\omega)tN} \tag{3.22}$$

from (3.21).

4. Quantum counterpart

In this section we explicitly calculate $\rho(t)$ for the initial value $\rho(0)$ given in the following.

4.1. Case of $\rho(0) = |0\rangle\langle 0|$

Noting $a|0\rangle = 0$ ($\Leftrightarrow 0 = \langle 0|a^\dagger$), this case is very easy and we have

$$\rho(t) = \frac{e^{\frac{\mu-\nu}{2}t}}{F(t)} \sum_{n=0}^{\infty} \frac{G(t)^n}{n!} (a^\dagger)^n |0\rangle\langle 0| a^n = \frac{e^{\frac{\mu-\nu}{2}t}}{F(t)} \sum_{n=0}^{\infty} G(t)^n |n\rangle\langle n| = \frac{e^{\frac{\mu-\nu}{2}t}}{F(t)} e^{\log G(t) N} \tag{4.1}$$

because the number operator $N\ (= a^\dagger a)$ is written as

$$N = \sum_{n=0}^{\infty} n|n\rangle\langle n| \implies N|n\rangle = n|n\rangle$$

, see for example (2.15). To check the last equality of (4.1) is left to readers. Moreover, $\rho(t)$ can be written as

$$\rho(t) = (1 - G(t)) e^{\log G(t) N} = e^{\log(1-G(t))} e^{\log G(t) N}, \tag{4.2}$$

see the next subsection.

4.2. Case of $\rho(0) = |\alpha\rangle\langle\alpha|$ ($\alpha \in \mathbf{C}$)

Remind that $|\alpha\rangle$ is a coherent state given by (2.16) ($a|\alpha\rangle = \alpha|\alpha\rangle \Leftrightarrow \langle\alpha|a^\dagger = \langle\alpha|\bar{\alpha}$). First of all let us write down the result :

$$\rho(t) = e^{|\alpha|^2 e^{-(\mu-\nu)t} \log G(t) + \log(1-G(t))} \exp\left\{ -\log G(t) \left(\alpha e^{-\left(\frac{\mu-\nu}{2}+i\omega\right)t} a^\dagger + \bar{\alpha} e^{-\left(\frac{\mu-\nu}{2}-i\omega\right)t} a - N \right) \right\} \tag{4.3}$$

with $G(t)$ in (3.15). Here we again meet a term like (2.3)

$$\alpha e^{-\left(\frac{\mu-\nu}{2}+i\omega\right)t} a^\dagger + \bar{\alpha} e^{-\left(\frac{\mu-\nu}{2}-i\omega\right)t} a$$

with $\lambda = \frac{\mu-\nu}{2}$.

Therefore, (4.3) is just our quantum counterpart of the classical damped harmonic oscillator.

The proof is divided into four parts.

[First Step] From (3.21) it is easy to see

$$\sum_{m=0}^{\infty} \frac{E(t)^m}{m!} a^m |\alpha\rangle\langle\alpha| (a^\dagger)^m = \sum_{m=0}^{\infty} \frac{E(t)^m}{m!} \alpha^m |\alpha\rangle\langle\alpha| \bar{\alpha}^m = \sum_{m=0}^{\infty} \frac{(E(t)|\alpha|^2)^m}{m!} |\alpha\rangle\langle\alpha| = e^{E(t)|\alpha|^2} |\alpha\rangle\langle\alpha|.$$

[Second Step] From (3.21) we must calculate the term

$$e^{\gamma N}|\alpha\rangle\langle\alpha|e^{\bar{\gamma}N} = e^{\gamma N}e^{\alpha a^\dagger - \bar{\alpha}a}|0\rangle\langle 0|e^{-(\alpha a^\dagger - \bar{\alpha}a)}e^{\bar{\gamma}N}$$

where $\gamma = -i\omega t - \log(F(t))$ (note $\bar{\gamma} \neq -\gamma$). It is easy to see

$$e^{\gamma N}e^{\alpha a^\dagger - \bar{\alpha}a}|0\rangle = e^{\gamma N}e^{\alpha a^\dagger - \bar{\alpha}a}e^{-\gamma N}e^{\gamma N}|0\rangle = e^{e^{\gamma N}(\alpha a^\dagger - \bar{\alpha}a)e^{-\gamma N}}|0\rangle = e^{\alpha e^{\gamma}a^\dagger - \bar{\alpha}e^{-\gamma}a}|0\rangle$$

where we have used

$$e^{\gamma N}a^\dagger e^{-\gamma N} = e^{\gamma}a^\dagger \quad \text{and} \quad e^{\gamma N}ae^{-\gamma N} = e^{-\gamma}a.$$

The proof is easy and left to readers. Therefore, by use of Baker–Campbell–Hausdorff formula (2.17) two times

$$\begin{aligned} e^{\alpha e^{\gamma}a^\dagger - \bar{\alpha}e^{-\gamma}a}|0\rangle &= e^{-\frac{|\alpha|^2}{2}}e^{\alpha e^{\gamma}a^\dagger}e^{-\bar{\alpha}e^{-\gamma}a}|0\rangle = e^{-\frac{|\alpha|^2}{2}}e^{\alpha e^{\gamma}a^\dagger}|0\rangle \\ &= e^{-\frac{|\alpha|^2}{2}}e^{\frac{|\alpha|^2}{2}e^{\gamma+\bar{\gamma}}}e^{\alpha e^{\gamma}a^\dagger - \bar{\alpha}e^{\bar{\gamma}}a}|0\rangle = e^{-\frac{|\alpha|^2}{2}(1-e^{\gamma+\bar{\gamma}})}|\alpha e^{\gamma}\rangle \end{aligned}$$

and we obtain

$$e^{\gamma N}|\alpha\rangle\langle\alpha|e^{\bar{\gamma}N} = e^{-|\alpha|^2(1-e^{\gamma+\bar{\gamma}})}|\alpha e^{\gamma}\rangle\langle\alpha e^{\gamma}|$$

with $\gamma = -i\omega t - \log(F(t))$.

[Third Step] Under two steps above the equation (3.21) becomes

$$\rho(t) = \frac{e^{\frac{\mu-\nu}{2}t}e^{|\alpha|^2(E(t)-1+e^{\gamma+\bar{\gamma}})}}{F(t)}\sum_{n=0}^{\infty}\frac{G(t)^n}{n!}(a^\dagger)^n|\alpha e^{\gamma}\rangle\langle\alpha e^{\gamma}|a^n.$$

For simplicity we set $z = \alpha e^{\gamma}$ and calculate the term

$$(\sharp) = \sum_{n=0}^{\infty}\frac{G(t)^n}{n!}(a^\dagger)^n|z\rangle\langle z|a^n.$$

Since $|z\rangle = e^{-|z|^2/2}e^{za^\dagger}|0\rangle$ we have

$$
\begin{aligned}
(\sharp) &= e^{-|z|^2} \sum_{n=0}^{\infty} \frac{G(t)^n}{n!} (a^\dagger)^n e^{za^\dagger} |0\rangle \langle 0| e^{\bar{z}a} a^n \\
&= e^{-|z|^2} e^{za^\dagger} \left\{ \sum_{n=0}^{\infty} \frac{G(t)^n}{n!} (a^\dagger)^n |0\rangle \langle 0| a^n \right\} e^{\bar{z}a} \\
&= e^{-|z|^2} e^{za^\dagger} \left\{ \sum_{n=0}^{\infty} G(t)^n |n\rangle \langle n| \right\} e^{\bar{z}a} \\
&= e^{-|z|^2} e^{za^\dagger} e^{\log G(t) N} e^{\bar{z}a}
\end{aligned}
$$

by (4.1). Namely, this form is a kind of disentangling formula, so we want to restore an entangling formula.

For the purpose we use the **disentangling formula**

$$
e^{\alpha a^\dagger + \beta a + \gamma N} = e^{\alpha\beta \frac{e^\gamma - (1+\gamma)}{\gamma^2}} e^{\alpha \frac{e^\gamma - 1}{\gamma} a^\dagger} e^{\gamma N} e^{\beta \frac{e^\gamma - 1}{\gamma} a} \tag{4.4}
$$

where α, β, γ are usual numbers. The proof is given in the fourth step. From this it is easy to see

$$
e^{ua^\dagger} e^{vN} e^{wa} = e^{-\frac{uw(e^v - (1+v))}{(e^v - 1)^2}} e^{\frac{uv}{e^v - 1} a^\dagger + \frac{vw}{e^v - 1} a + vN}. \tag{4.5}
$$

Therefore ($u \to z$, $v \to \log G(t)$, $w \to \bar{z}$)

$$
(\sharp) = e^{-|z|^2} e^{\frac{|z|^2 (1 + \log G(t) - G(t))}{(1 - G(t))^2}} e^{\frac{\log G(t)}{G(t) - 1} z a^\dagger + \frac{\log G(t)}{G(t) - 1} \bar{z} a + \log G(t) N},
$$

so by noting

$$
z = \alpha e^\gamma = \alpha \frac{e^{-i\omega t}}{F(t)} \quad \text{and} \quad |z|^2 = |\alpha|^2 e^{\gamma + \bar{\gamma}} = |\alpha|^2 \frac{1}{F(t)^2}
$$

we have

$$
\begin{aligned}
\rho(t) &= \frac{e^{\frac{\mu - \nu}{2} t}}{F(t)} e^{|\alpha|^2 (E(t) - 1)} e^{|\alpha|^2 \frac{1 + \log G(t) - G(t)}{F(t)^2 (1 - G(t))^2}} e^{\frac{\log G(t)}{F(t)(G(t) - 1)} \alpha e^{-i\omega t} a^\dagger + \frac{\log G(t)}{F(t)(G(t) - 1)} \bar{\alpha} e^{i\omega t} a + \log G(t) N} \\
&= \frac{e^{\frac{\mu - \nu}{2} t}}{F(t)} e^{|\alpha|^2 \left\{ E(t) - 1 + \frac{1 + \log G(t) - G(t)}{F(t)^2 (1 - G(t))^2} \right\}} e^{\frac{\log G(t)}{F(t)(G(t) - 1)} \alpha e^{-i\omega t} a^\dagger + \frac{\log G(t)}{F(t)(G(t) - 1)} \bar{\alpha} e^{i\omega t} a + \log G(t) N}.
\end{aligned}
$$

By the way, from (3.15)

$$G(t)-1=-\frac{e^{\frac{\mu-\nu}{2}t}}{F(t)},\quad \frac{1}{F(t)(G(t)-1)}=-e^{-\frac{\mu-\nu}{2}t},\quad E(t)-1=-\frac{e^{-\frac{\mu-\nu}{2}t}}{F(t)}$$

and

$$\begin{aligned}\frac{1-G(t)+\log G(t)}{F(t)^2(G(t)-1)^2}&=e^{-(\mu-\nu)t}\left\{\frac{e^{\frac{\mu-\nu}{2}t}}{F(t)}+\log G(t)\right\}\\&=\frac{e^{-\frac{\mu-\nu}{2}t}}{F(t)}+e^{-(\mu-\nu)t}\log G(t)\\&=-(E(t)-1)+e^{-(\mu-\nu)t}\log G(t)\end{aligned}$$

we finally obtain

$$\begin{aligned}\rho(t)&=(1-G(t))e^{|\alpha|^2e^{-(\mu-\nu)t}\log G(t)}e^{-\log G(t)\left\{\alpha e^{-i\omega t}e^{-\frac{\mu-\nu}{2}t}a^\dagger+\bar{\alpha}e^{i\omega t}e^{-\frac{\mu-\nu}{2}t}a-N\right\}}\\&=e^{|\alpha|^2e^{-(\mu-\nu)t}\log G(t)+\log(1-G(t))}e^{-\log G(t)\left\{\alpha e^{-\left(\frac{\mu-\nu}{2}+i\omega\right)t}a^\dagger+\bar{\alpha}e^{-\left(\frac{\mu-\nu}{2}-i\omega\right)t}a-N\right\}}.\end{aligned}$$

[Fourth Step] In last, let us give the proof to the disentangling formula (4.4) because it is not so popular as far as we know. From (4.4)

$$\begin{aligned}\alpha a^\dagger+\beta a+\gamma N&=\gamma a^\dagger a+\alpha a^\dagger+\beta a\\&=\gamma\left\{\left(a^\dagger+\frac{\beta}{\gamma}\right)\left(a+\frac{\alpha}{\gamma}\right)-\frac{\alpha\beta}{\gamma^2}\right\}\\&=\gamma\left(a^\dagger+\frac{\beta}{\gamma}\right)\left(a+\frac{\alpha}{\gamma}\right)-\frac{\alpha\beta}{\gamma}\end{aligned}$$

we have

$$\begin{aligned}e^{\alpha a^\dagger+\beta a+\gamma N}&=e^{-\frac{\alpha\beta}{\gamma}}e^{\gamma\left(a^\dagger+\frac{\beta}{\gamma}\right)\left(a+\frac{\alpha}{\gamma}\right)}\\&=e^{-\frac{\alpha\beta}{\gamma}}e^{\frac{\beta}{\gamma}a}e^{\gamma a^\dagger\left(a+\frac{\alpha}{\gamma}\right)}e^{-\frac{\beta}{\gamma}a}\\&=e^{-\frac{\alpha\beta}{\gamma}}e^{\frac{\beta}{\gamma}a}e^{-\frac{\alpha}{\gamma}a^\dagger}e^{\gamma a^\dagger a}e^{\frac{\alpha}{\gamma}a^\dagger}e^{-\frac{\beta}{\gamma}a}.\end{aligned}$$

Then, careful calculation gives the disentangling formula (4.4) ($N = a^{\dagger}a$)

$$
\begin{aligned}
e^{-\frac{\alpha\beta}{\gamma}} e^{\frac{\beta}{\gamma}a} e^{-\frac{\alpha}{\gamma}a^{\dagger}} e^{\gamma N} e^{\frac{\alpha}{\gamma}a^{\dagger}} e^{-\frac{\beta}{\gamma}a}
&= e^{-\frac{\alpha\beta}{\gamma}} e^{-\frac{\alpha\beta}{\gamma^2}} e^{-\frac{\alpha}{\gamma}a^{\dagger}} e^{\frac{\beta}{\gamma}a} e^{\gamma N} e^{\frac{\alpha}{\gamma}a^{\dagger}} e^{-\frac{\beta}{\gamma}a} \\
&= e^{-(\frac{\alpha\beta}{\gamma}+\frac{\alpha\beta}{\gamma^2})} e^{-\frac{\alpha}{\gamma}a^{\dagger}} e^{\frac{\beta}{\gamma}a} e^{\gamma N} e^{\frac{\alpha}{\gamma}a^{\dagger}} e^{-\frac{\beta}{\gamma}a} \\
&= e^{-(\frac{\alpha\beta}{\gamma}+\frac{\alpha\beta}{\gamma^2})} e^{-\frac{\alpha}{\gamma}a^{\dagger}} e^{\gamma N} e^{\frac{\beta}{\gamma}e^{\gamma}a} e^{\frac{\alpha}{\gamma}a^{\dagger}} e^{-\frac{\beta}{\gamma}a} \\
&= e^{-(\frac{\alpha\beta}{\gamma}+\frac{\alpha\beta}{\gamma^2})+\frac{\alpha\beta}{\gamma^2}e^{\gamma}} e^{-\frac{\alpha}{\gamma}a^{\dagger}} e^{\gamma N} e^{\frac{\alpha}{\gamma}a^{\dagger}} e^{\frac{\beta}{\gamma}e^{\gamma}a} e^{-\frac{\beta}{\gamma}a} \\
&= e^{\alpha\beta\frac{e^{\gamma}-1-\gamma}{\gamma^2}} e^{-\frac{\alpha}{\gamma}a^{\dagger}} e^{\frac{\alpha}{\gamma}e^{\gamma}a^{\dagger}} e^{\gamma N} e^{\beta\frac{e^{\gamma}-1}{\gamma}a} \\
&= e^{\alpha\beta\frac{e^{\gamma}-1-\gamma}{\gamma^2}} e^{\alpha\frac{e^{\gamma}-1}{\gamma}a^{\dagger}} e^{\gamma N} e^{\beta\frac{e^{\gamma}-1}{\gamma}a}
\end{aligned}
$$

by use of some commutation relations

$$
e^{sa}e^{ta^{\dagger}} = e^{st}e^{ta^{\dagger}}e^{sa}, \quad e^{sa}e^{tN} = e^{tN}e^{se^{t}a}, \quad e^{tN}e^{sa^{\dagger}} = e^{se^{t}a^{\dagger}}e^{tN}.
$$

The proof is simple. For example,

$$
e^{sa}e^{ta^{\dagger}} = e^{sa}e^{ta^{\dagger}}e^{-sa}e^{sa} = e^{te^{sa}a^{\dagger}e^{-sa}}e^{sa} = e^{t(a^{\dagger}+s)}e^{sa} = e^{st}e^{ta^{\dagger}}e^{sa}.
$$

The remainder is left to readers.

We finished the proof. The formula (4.3) is both compact and clear-cut and has not been known as far as we know. See [1] and [16] for some applications.

In last, let us present a challenging problem. A squeezed state $|\beta\rangle$ ($\beta \in \mathbf{C}$) is defined as

$$
|\beta\rangle = e^{\frac{1}{2}(\beta(a^{\dagger})^2 - \bar{\beta}a^2)}|0\rangle. \tag{4.6}
$$

See for example [4]. For the initial value $\rho(0) = |\beta\rangle\langle\beta|$ we want to calculate $\rho(t)$ in (3.21) like in the text. However, we cannot sum up it in a compact form like (4.3) at the present time, so we propose the problem,

Problem sum up $\rho(t)$ in a compact form.

5. Concluding remarks

In this chapter we treated the quantum damped harmonic oscillator, and studied mathematical structure of the model, and constructed general solution with any initial condition, and gave a quantum counterpart in the case of taking coherent state as an initial condition. It is in my opinion perfect.

However, readers should pay attention to the fact that this is not a goal but a starting point. Our real target is to construct general theory of **Quantum Mechanics with Dissipation**.

In the papers [7] and [8] (see also [13]) we studied a more realistic model entitled "Jaynes–Cummings Model with Dissipation" and constructed some approximate solutions under any initial condition. In the paper [5] we studied "Superluminal Group Velocity of Neutrinos" from the point of view of Quantum Mechanics with Dissipation.

Unfortunately, there is no space to introduce them. It is a good challenge for readers to read them carefully and attack the problems.

Acknowledgments

I would like to thank Ryu Sasaki and Tatsuo Suzuki for their helpful comments and suggestions.

Author details

Kazuyuki Fujii

* Address all correspondence to: fujii@yokohama-cu.ac.jp

International College of Arts and Sciences, Yokohama City University, Yokohama, Japan

References

[1] H. -P. Breuer and F. Petruccione : The theory of open quantum systems, Oxford University Press, New York, 2002.

[2] P. Dirac : The Principles of Quantum Mechanics, Fourth Edition, Oxford University Press, 1958.
This is a bible of Quantum Mechanics.

[3] R. Endo, K. Fujii and T. Suzuki : General Solution of the Quantum Damped Harmonic Oscillator, Int. J. Geom. Methods. Mod. Phys, 5 (2008), 653, arXiv : 0710.2724 [quant-ph].

[4] K. Fujii : Introduction to Coherent States and Quantum Information Theory, quant-ph/0112090.
This is a kind of lecture note based on my (several) talks.

[5] K. Fujii : Superluminal Group Velocity of Neutrinos : Review, Development and Problems, Int. J. Geom. Methods Mod. Phys, 10 (2013), 1250083, arXiv:1203.6425 [physics].

[6] K. Fujii and T. Suzuki : General Solution of the Quantum Damped Harmonic Oscillator II : Some Examples, Int. J. Geom. Methods Mod. Phys, 6 (2009), 225, arXiv : 0806.2169 [quant-ph].

[7] K. Fujii and T. Suzuki : An Approximate Solution of the Jaynes–Cummings Model with Dissipation, Int. J. Geom. Methods Mod. Phys, 8 (2011), 1799, arXiv : 1103.0329 [math-ph].

[8] K. Fujii and T. Suzuki : An Approximate Solution of the Jaynes–Cummings Model with Dissipation II : Another Approach, Int. J. Geom. Methods Mod. Phys, 9 (2012), 1250036, arXiv : 1108.2322 [math-ph].

[9] K. Fujii and et al ; Treasure Box of Mathematical Sciences (in Japanese), Yuseisha, Tokyo, 2010.
I expect that the book will be translated into English.

[10] V. Gorini, A. Kossakowski and E. C. G. Sudarshan ; Completely positive dynamical semigroups of N–level systems, J. Math. Phys, 17 (1976), 821.

[11] H. S. Green : Matrix Mechanics, P. Noordhoff Ltd, Groningen, 1965.
This is my favorite textbook of elementary Quantum Mechanics.

[12] K. Hornberger : Introduction to Decoherence Theory, in "Theoretical Foundations of Quantum Information", Lecture Notes in Physics, 768 (2009), 221-276, Springer, Berlin, quant-ph/061211.

[13] E. T. Jaynes and F. W. Cummings : Comparison of Quantum and Semiclassical Radiation Theories with Applications to the Beam Maser, Proc. IEEE, 51 (1963), 89.

[14] J. R. Kauder and Bo-S. Skagerstam : Coherent States–Applications in Physics and Mathematical Physics, World Scientific, Singapore, 1985.
This is a kind of dictionary of coherent states.

[15] G. Lindblad ; On the generator of quantum dynamical semigroups, Commun. Math. Phys, 48 (1976), 119.

[16] W. P. Schleich : Quantum Optics in Phase Space, WILEY–VCH, Berlin, 2001.
This book is strongly recommended although it is thick.

[17] C. Zachos : Crib Notes on Campbell-Baker-Hausdorff expansions, unpublished, 1999, see http://www.hep.anl.gov/czachos/index.html.
I expect that this (crib) note will be published in some journal.

Section 3

Path Integrals

Chapter 8

Generalized Path Integral Technique: Nanoparticles Incident on a Slit Grating, Matter Wave Interference

Valeriy I. Sbitnev

Additional information is available at the end of the chapter

1. Introduction

One of the crises of contemporary mathematics belongs in part to the subject of the infinite and infinitesimals [1]. It originates from the barest necessity to develop a rigorous language for description of observable physical phenomena. It was a time when foundations of integral and differential calculi were developed. A theoretical foundation for facilitation of understanding of classical mechanics is provided by the concepts of absolute time and space originally formulated by Sir Isaac Newton [2]. Space is distinct from body. And time passes uniformly without regard to whether anything happens in the world. For this reason Newton spoke of absolute space and absolute time as of a "container" for all possible objects and events. The space-time container is absolutely empty until prescribed metric and a reference frame are introduced. Infinitesimals are main tools of differential calculus [3, 4] within chosen reference frames.

Infinitesimal increment being a cornerstone of theoretical physics has one receptee default belief, that increment δV tending to zero contains a lot of events to be under consideration. Probability of detection of a particle within this infinitesimal volume $\rho(\vec{r})\delta V$ is adopted as a smooth differentiable function with respect to its argument. From experience we know that for reproducing the probability one needs to accumulate enormous amount of events occurring within this volume. On the other hand we know, that as δV tends to zero we lose information about amount of the events. What is more, the information becomes uncertain. It means the infinitesimal increment being applied in physics faces with a conflict of depth of understanding physical processes on such minuscule scales. This trouble is avoided in quantum mechanics by proclamation that infinitesimal increments are operators, whereas observables are averaged on an ensemble.

In light of classical views Newton maintained the theory that light was made up of tiny particles, corpuscles. They spread through space in accordance with law of classical mechanics. Christian Huygens (a contemporary of Newton), believed that light was made up of waves vibrating up and down perpendicularly to direction of the light propagation. It comes into contradiction with Newtonian idea about corpuscular nature of light. Huygens was a proclaimer of wave mechanics as opposite to classical mechanics [5].

We abstain from allusion to physical vacuum but expand Huygensian idea to its logical completion. Let us imagine that all Newtonian absolute space is not empty but is populated everywhere densely by Huygensian vibrators. The vibrators are silent at absence of wave propagating through. But as soon as a wave front reaches some surface all vibrators on this surface begin to radiate at a frequency resonant with that of incident wave. From here it follows, that in each point of the space there are vibrators with different frequencies ranging from zero frequency up to infinite. All are silent in absence of an external wave perturbance. Thus, the infinitesimal volume δV is populated by infinite amount of the vibrators with frequencies ranging from zero to infinity. They are virtual vibrators facilitating propagation of waves through space.

Let us return to our days. One believes that besides matter and physical fields there is nothing more in the universe. Elementary particles are only a building material of "eternal and indestructible" substance of the cosmos. However we should avow that all observed matter and physical fields, are not the basis of the world, but they are only a small part of the total quantum reality. Physical vacuum in this picture constitutes a basic part of this reality. In particular, modern conception of the physical vacuum covers Huygens's idea perfectly. All space is fully populated by virtual particle-antiparticle pairs situated on a ground level. Such a particle-antiparticle pair has zero mass, zero charge, and zero magnetic moment. Famous Dirac' sea (Dirac postulated that the entire universe is entirely filled by particles with negative energy) is boundless space of electron-positron pairs populated everywhere densely - each quantum state possessing a positive energy is accompanied by a corresponding state with negative energy. Electron has positive mass and positron has the same mass but negative; electron has negative charge and positron has the same positive charge; when electron and positron dance in pair theirs magnetic moments have opposite orientations, so magnetic moment of the pair is zero.

Let electron and positron of a virtual pair rotate about mass center of this pair. Rotation of the pairs happens on a Bohr orbit. Energy level of the first Bohr orbit, for example, $mv^2/2$, is about 14 eV. Here m is electron mass and v is its velocity (on the first Bohr orbit the velocity is about 2.188 10^6 m/s). Energy of the pair remains zero since positron has the same energy but with negative sign. Quantum fluctuations around this zero energy are as zero oscillations of electromagnetic field. Observe that, energy releasing of electron and positron from vacuum occurs at mc^2 = 1.022 MeV. So we see that there are about 7.5 10^4 Bohr orbits lying below this energy. That is, there is a vast scope for occupation of different Bohr orbits by the virtual electron-positron pairs.

A short outline given above is a basis for understanding of interference effects to be described below.

Ones suppose that random fluctuations of electron-positron pairs take place always. What is more, these fluctuations are induced by other pairs and by particles traveling through this random conglomerate. Edward Nelson has described mathematical models [6, 7] representing the above random fluctuations as Brownian motions of particles that are subjected by random impacts from particles populating aether (Nelson's title of a lower environment). The model is viewed as a Markov process

$$dx(t) = b(x(t),t)dt + dw(t) \tag{1}$$

loaded by a Wiener term w with diffusion coefficient equal to $\hbar/2m$, where m is a mass of the particle and $\hbar$ is the reduced Planck constant. In this perspective Nelson has considered two Markov processes complementary to each other. One is described by forward-difference operator in the time, here $b(x(t),t)$ is a velocity calculated forward. And other equation is described by backward-difference operator in the time with the velocity $b^{\dagger}(x(t),t)$ calculated backward. In general $b(x(t),t) \neq b^{\dagger}(x(t),t)$ [6, 7]. The two complementary processes, by means of transition to two new variables, real and imaginary, finally lead to emergence of the Schrödinger equation.

Nelson' vision that aether fluctuations look as Brownian movements of subparticles with $\hbar/2m$ being the diffusion coefficient of the movements, correlates with Feynman's ideas about quantum fluctuations of virtual particles in vacuum [8]. The Feynman path integral is akin to the Einstein-Smoluchowski integral equation [9]. The latter computes transition probability density. We shall deal with the modified Feynman path integral loaded by a temperature multiplied by the Boltzmann constant. At that, probability amplitude stays as a fundamental mathematical object at all stages of computations. As an example we shall consider emergence of interference patterns at scattering heavy particles on gratings. The particles are heavy in the sense that they adjoin to both realms, quantum and classical. They are nanoparticles. Such nanoparticles have masses about 100 amu and more, as, for example, fullerene molecules [10] shown in Fig. 1. It is remarkable that there are many experiments with such molecules showing interference patterns in the near field [11-18]. On the other hand these molecules are so large, that they behave themselves as classical particles at ordinary conditions.

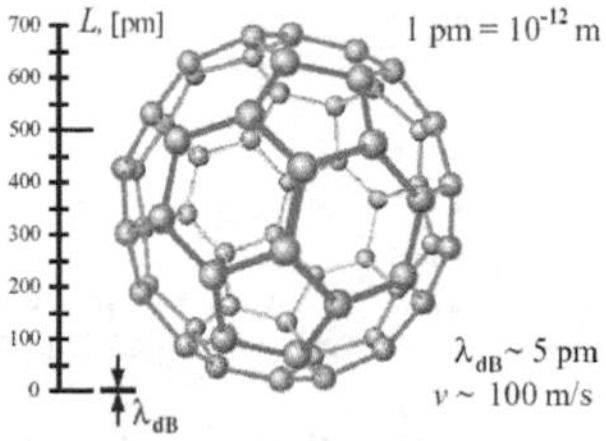

Figure 1. The fullerene molecule C_{60} consists of 60 carbon atoms. Its radius is about 700 pm. De Broglie wavelength of the molecule, λ_{dB}, is about 5 pm at a flight velocity v =100 m/s [18]. The molecules are prepared in a thermal emission gun which has temperature about 1000 K. It means that carbon atoms accomplish thermal fluctuations.

The article consists of five sections. Sec. 2 introduces a general conception of the path integral that describes transitions along paths both of classical and quantum particles. Here we fulfill expansions in the Taylor series of terms presented in the path integral. Depending on type of presented parameters we disclose either the Schrödinger equation or diffusion-drift equation containing extra term, osmotic diffusion. In the end of the section we compute passing nanoparticles through N-slit grating. Sec. 3 deals with interference patterns from the N-slit grating. Specifically, we study blurring of the Talbot carpet (an interference pattern emergent under special conditions imposed on the grating [19, 20]) arising under decoherence of incident on the grating nanoparticles. In Sec. 4 we find equations for computing the Bohmian trajectories. Also we compute variance of momenta along the trajectories. These computations lead to emergence of the uncertainty conditions. In concluding Sec. 5 we summarize results. For confirmation of existence of the Bohmian trajectories here we mention interference experiments with single silicon oil droplet [21].

2. Generalized path integral

Let many classical particles occupy a volume V and they move with different velocities in different directions. Let us imagine that there is a predominant orientation along which ensemble of the particles drifts. As a rule, one chooses a small volume δV in order to evaluate such a drift, Fig. 2. Learning of statistical mechanics begins with assumption that the volume should contain many particles.

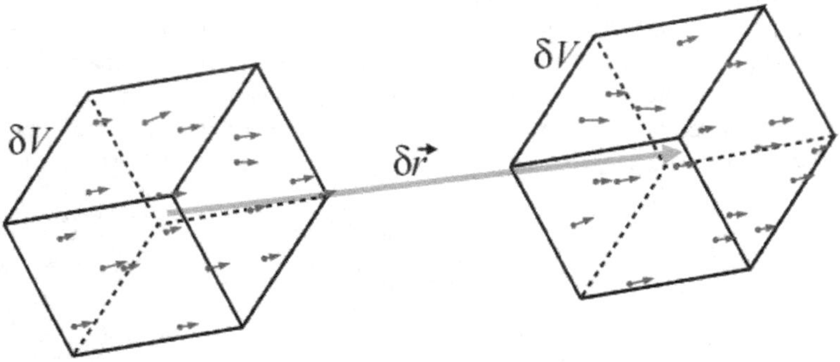

Figure 2. Infinitesimal volume δV contains many particles moving with different velocities having predominant orientation along blue arrow. The infinitesimal volume δV, as a mental image, is shifted along the same orientation.

The problem is to find transition probabilities that describe transition of the particle ensemble from one statistical state to another. These transient probabilities can be found through solution of the integral Einstein-Smoluchowski equation [9]. This equation in mathematical physics is known as Chapman-Kolmogorov [22-24] equation.. This equation looks as follows

$$p(\vec{r}_2,\vec{r}_1;t+\delta\tau)=\int_R p(\vec{r}_2,\vec{r};t+\delta\tau,t)p(\vec{r},\vec{r}_1;t)dV. \tag{2}$$

This equation describes a Markovian process without memory. That is, only previous state bears information for the next state. Integration here is fulfilled over all a working scene R, encompassing finite space volume. Infinitesimal volume δV in the integral tends to zero. It should be noted, however, that this volume should contain as many particles as possible for getting a satisfactory statistical pattern. One can see that this claim enters in conflict with the assumption $\delta V \to 0$.

Next we shall slightly modify approach to this problem. Essential difference from the classical probability theory is that instead of the probabilities we shall deal with probability amplitudes. The transition amplitudes can contain also imaginary terms. They bear information about phase shifts accumulated along paths. In that way, a transition from an initial state $\vec{q}_0$ to a final state $\vec{q}_1$ through all intermediate positions $\vec{q}_x$ given on a manifold R^3 (see Fig. 2) is represented by the following path integral

$$\psi(\vec{q}_1,\vec{q}_0;t+\delta t)=\int_R K(\vec{q}_1,\vec{q}_x;t+\delta t,t)\psi(\vec{q}_x,\vec{q}_0;t)D\,^3q_x. \tag{3}$$

Here function $\psi(...)$ is a probability amplitude. Probability density $p(...)$, in turn, is represented by square of modulo of the probability amplitude, namely, $p(...)=|\psi(...)|^2$. Integral kernel $K(\vec{q}_1, \vec{q}_x; t+\delta t, t)$ represents the transition amplitude from an intermediate state $\vec{q}_x$ to a final state $\vec{q}_1$. It is called propagator [8, 25, 26]. We suppose that the propagator has the following standard form

$$K(\vec{q}_1,\vec{q}_x;t+\delta t,t)=\frac{1}{A}\exp\left\{-\frac{L(\vec{q}_x,\dot{\vec{q}}_x)\delta t}{\Gamma}\right\}, \tag{4}$$

where denominator Γ under exponent is a complex-valued quantity, i.e., $\Gamma = \beta+\mathbf{i}\hbar$. The both parameters, β and the reduced Planck constant $\hbar$, have dimensionality of energy multiplied by time. From here it follows that $\beta = 2k_B T\delta t$. Here k_B is Boltzmann constant and T is temperature. So, we can write

$$\Gamma = 2k_B T\delta t + \mathbf{i}\hbar. \tag{5}$$

Factor 2 at the first term is conditioned by the fact that the kernel K relates to transitions of the probability amplitude ψ, not the probability p. Observe that a fullerene molecular beam for interference experiment prepared in a Knudsen cell at T=1070 K [27] spreads further in a vacuum. That is, fullerene molecules keep this temperature. From here it follows, that thermal fluctuations of carbon atoms from equilibrium occur at that temperature as the fullerene molecules propagate further within the vacuum chamber. One can see that at T=1000 K the term $2k_B T\delta t$ may be about $\hbar$ if δt is about 10^{-14} s.

Next let us imagine that particles pass through a path length one by one. That is, they do not collide with each other along the path length. The particles are complex objects, however. They are nanoparticles. Fullerene molecule, for example, contains 60 carbon atoms, Fig. 1. Conditionally we can think that the atoms are connected with each other by springs simulating elastic vibrations. In this view the Lagrangian $L\,(\vec{q}_x, \dot{\vec{q}}_x)$ can be written in the following form

$$L(\vec{q}_x,\dot{\vec{q}}_x) = \sum_{k=1}^{N}\left[\frac{m}{2}\frac{(\vec{q}_{1,k} - \vec{q}_{x,k}))^2}{\delta t^2} - U(\vec{q}_{x,k})\right]. \tag{6}$$

Here N is amount of atoms, constituent complex molecule, and m is mass of a single atom. By supposing that there are no quantum permutations between atoms we may expand the Lagrangian further

$$\begin{aligned} L(\vec{q}_x,\dot{\vec{q}}_x) &= \frac{m}{2}\sum_{k=1}^{N}\frac{(\vec{q}_{1,k} - \vec{q}_1 + \vec{q}_1 - \vec{q}_{x,k} + \vec{q}_x - \vec{q}_x)^2}{\delta t^2} - \sum_{k=1}^{N} U(\vec{q}_{x,k}) \\ &= \frac{m_N}{2}\frac{(\vec{q}_1 - \vec{q}_x)^2}{\delta t^2} - m_N\frac{(\vec{q}_1 - \vec{q}_x)}{\delta t}\frac{1}{N}\sum_{k=1}^{N}\frac{(\vec{\delta}_{1,k} - \vec{\delta}_{x,k})}{\delta t} + \frac{m}{2}\sum_{k=1}^{N}\frac{(\vec{\delta}_{1,k} - \vec{\delta}_{x,k})^2}{\delta t^2} - \sum_{k=1}^{N} U(\vec{q}_{x,k}). \end{aligned} \tag{7}$$

Here we admit that $\vec{q}_x$ and $\vec{q}_1$ are coordinates of center of mass of that complex molecule in intermediate and final positions. And $m_N = Nm$ is a mass of the molecule referred to its center of mass. Small deviations $\vec{\delta}_{1,k} = (\vec{q}_1 - \vec{q}_{1,k})$ and $\vec{\delta}_{x,k} = (\vec{q}_x - \vec{q}_{x,k})$ are due to oscillations of the kth atoms with respect to the center of mass.

Let us consider the second row in Eq. (7). First of all we note that the term $(\vec{q}_1 - \vec{q}_x)/\delta t$ represents a velocity of movement of the center of mass. In such a case the first term is a kinetic energy of the center of mass. The second term represents product of the center mass velocity on an averaged velocity of partial oscillations of atoms constituting this molecule. The averaged velocity we believe vanishes because of conservation of total momentum. The third term represents a thermal kinetic energy of partial oscillating atoms constituting this molecular object. This energy is small enough. But it is sufficient to exhibit itself in the Casimir effect. The last term is a total potential energy $U(\vec{q}_x)$in the point $\vec{q}_x$.

The path integral (3) contains functions depending only on coordinates $\vec{q}_0$, $\vec{q}_1$, $\vec{q}_x$relating to positions of the center of mass. Whereas the Lagrangian (6) gives description for behavior of each atom constituent the complex molecule. Here we shall suppose that oscillations of all atoms are noncoherent. And consequently they do not give contribution to interference effect on output. We believe that these oscillations provide a thermal noise. And next we shall replace this oscillating background by a corresponding thermal term. For this reason, we believe that along with the reduced Planck constant $\hbar$ the parameter $2k_B T\delta t$ in Eq. (5) can be different from zero as well.

2.1. Expansion of the path integral

The next step is to expand terms, ingoing into the integral (3), into Taylor series. The wave function written on the left is expanded up to the first term

$$\psi(\vec{q}_1,\vec{q}_0;t+\delta t)\approx\psi(\vec{q}_1,\vec{q}_0;t)+\frac{\partial\psi}{\partial t}\delta t. \tag{8}$$

As for the terms under the integral, here we preliminarily make some transformations. We define a small increment

$$\vec{\xi}=\vec{q}_1-\vec{q}_x\Rightarrow D^{\,3}q_x=-D^{\,3}\xi. \tag{9}$$

The Lagrangian (7) is rewritten, in such a case, in the following view

$$L(\vec{q}_x,\dot{\vec{q}}_x)=\frac{m_N}{2}\frac{\xi^2}{\delta t^2}-m_N\frac{\xi}{\delta t}\frac{1}{N}\sum_{k=1}^{N}\frac{(\vec{\delta}_{1,k}-\vec{\delta}_{x,k})}{\delta t}+\frac{m}{2}\sum_{k=1}^{N}\frac{(\vec{\delta}_{1,k}-\vec{\delta}_{x,k})^2}{\delta t^2}-U(\vec{q}_x) \tag{10}$$

Here $U(\vec{q}_x)$is sum of all potentials $U(\vec{q}_{x,k})$ given in the center of mass. Further we shall deal with the path integral (3) where the kernel K contains the Lagrangian given from Eq. (10). The under integral function $\psi(\vec{q}_x,\vec{q}_0;t)=\psi(\vec{q}_1-\vec{\xi},\vec{q}_0;t)$is subjected to expansion into the Taylor series up to the second terms of the expansion

$$\psi(\vec{q}_1-\vec{\xi},\vec{q}_0;t)\approx\psi(\vec{q}_1,\vec{q}_0;t)-\left(\nabla\psi,\vec{\xi}\right)+\nabla^2\psi\cdot\xi^2/2 \tag{11}$$

The potential energy $U(\vec{q}_x)=U(\vec{q}_1-\vec{\xi})$ is subjected to expansion into the Taylor series by the small parameter $\vec{\xi}$also. Here we restrict themselves by the first two terms of the expansion, $U(\vec{q}_1)-(\nabla U(\vec{q}_1),\vec{\xi})$.

Taking into account the expressions (8)-(11) and substituting theirs into Eq. (3) we get

$$\psi(\vec{q}_1,\vec{q}_0;t)+\frac{\partial\psi}{\partial t}\delta t=-\frac{1}{A}\int_{R^3}\exp\left\{-\frac{1}{\Gamma}\left(\left(\frac{m_N}{2}\frac{\xi^2}{\delta t}-\xi\underbrace{\frac{m_N}{N}\sum_{k=1}^{N}\frac{(\vec{\delta}_{1,k}-\vec{\delta}_{x,k})}{\delta t}}_{(a)}+\underbrace{\left[\frac{m}{2}\sum_{k=1}^{N}\frac{(\vec{\delta}_{1,k}-\vec{\delta}_{x,k})^2}{\delta t^2}\right]}_{(b)}\delta t\right.\right.\right. \tag{12}$$
$$\left.\left.\left.-\left(U(\vec{q}_1)-(\nabla U(\vec{q}_1),\vec{\xi})\right)\delta t\right)\right)\right\}\left(\psi(\vec{q}_1,\vec{q}_0;t)-\left(\nabla\psi,\vec{\xi}\right)+\nabla^2\psi\cdot\xi^2/2\right)D^3\xi.$$

First, we consider terms enveloped by braces (a) and (b): (a) here displacement $(\vec{\delta}_{1,k}-\vec{\delta}_{x,k})$ divided by δt is a velocity $\vec{v}_k$ of k*th* atom at its deviation from a steady position relative to the center of mass. Summation through all deviations of atoms divided by N gives averaged velocity, $\vec{v}$, of all atoms with respect to the center of mass. This averaged velocity, as we mentioned above, vanishes. The velocity can be nonzero only in a case when external forces push coherently all atoms. This case we do not consider here. (b) this term is a thermal kinetic energy, T_N, of the atoms oscillating about the center of mass. Observe that energy of thermal fluctuations, T_N, is proportional to k_BT. Because of its presence in the propagator intensity of an interference pattern diminishes in general. Further we shall add this term into the potential energy as some constant component.

In the light of the above observation we rewrite Eq. (12) as follows

$$\psi(\vec{q}_1,\vec{q}_0;t)+\frac{\partial\psi}{\partial t}\delta t=-\frac{1}{A}\int_{R^3}\exp\left\{-\frac{m_N}{2\Gamma}\frac{\xi^2}{\delta t}\right\}\left(1+\underbrace{m_N\vec{v}}_{(a)}\frac{\vec{\xi}}{\Gamma}+\underbrace{T_N}_{(b)}\frac{\delta t}{\Gamma}+\left(U(\vec{q}_1)-(\nabla U(\vec{q}_1),\vec{\xi})\right)\frac{\delta t}{\Gamma}\right) \tag{13}$$
$$\times\left(\psi(\vec{q}_1,\vec{q}_0;t)-\left(\nabla\psi,\vec{\xi}\right)+\nabla^2\psi\cdot\xi^2/2\right)D^3\xi.$$

Here we have expanded preliminarily exponents to the Taylor series up to the first term of the expansion. Exception relates to the term $\exp\{-m_N\xi^2/2\Gamma\delta t\}$ which remains in its original form. This exponent integrated over all space R^3 results in

$$-\frac{1}{A}\int_{R^3}\exp\left\{-\frac{m_N}{2\Gamma}\frac{\xi^2}{\delta t}\right\}D^3\xi=-\frac{1}{A}\left(\frac{2\pi\Gamma\delta t}{m_N}\right)^{3/2}=1\quad\Rightarrow\quad A=-\left(\frac{2\pi\Gamma\delta t}{m_N}\right)^{3/2}. \tag{14}$$

To derive outcomes of integration of terms containing factors $(\nabla\psi,\vec{\xi})$ and $\nabla^2\psi\cdot\xi^2/2$ we mention the following integrals [8]

$$\frac{1}{A}\int_{R^3}\exp\left\{-\frac{m_N}{2\Gamma}\frac{\xi^2}{\delta t}\right\}\xi D^3\xi=0\qquad\text{and}\qquad\frac{1}{A}\int_{R^3}\exp\left\{-\frac{m_N}{2\Gamma}\frac{\xi^2}{\delta t}\right\}\xi^2D^3\xi=\frac{\Gamma}{m_N}\delta t \tag{15}$$

In the light of this observation let us now solve integral (13) accurate to terms containing δt not higher the first order:

$$\psi(\vec{q}_1,\vec{q}_0;t)+\frac{\partial\psi}{\partial t}\delta t=\psi(\vec{q}_1,\vec{q}_0;t)+\frac{\Gamma}{2m_N}\nabla^2\psi\,\delta t+\frac{1}{\Gamma}\left(U(\vec{q}_1)-T_N\right)\psi\,\delta t. \tag{16}$$

We note that the term $(\vec{v},\nabla\psi)$ here is absent since we consider $\vec{v}=0$, as was mentioned above. By reducing from the both sides the function $\psi(\vec{q}_1,\vec{q}_0;t)$ we come to the following differential equation

$$\frac{\partial\psi(\vec{q}_1,\vec{q}_0;t)}{\partial t}=\frac{\Gamma}{2m_N}\nabla^2\psi(\vec{q}_1,\vec{q}_0;t)+\frac{1}{\Gamma}\left(U(\vec{q}_1)-T_N\right)\psi(\vec{q}_1,\vec{q}_0;t). \tag{17}$$

The parameter

$$\frac{\Gamma}{2m_N}=\frac{2k_BT\delta t}{2m_N}+\mathbf{i}\frac{\hbar}{2m_N} \tag{18}$$

is seen to be as a complex-valued diffusion coefficient consisting of real and imaginary parts.

2.1.1. *Temperature is zero*

Let $k_BT\delta t=0$. It means that $\Gamma=\mathrm{i}\hbar$. Also $T_N=0$. One can see that Eq. (17) is reduced to

$$\mathbf{i}\hbar\frac{\partial\psi(\vec{q}_1,\vec{q}_0;t)}{\partial t}=-\frac{\hbar^2}{2m_N}\Delta\psi(\vec{q}_1,\vec{q}_0;t)+U(\vec{q}_1)\psi(\vec{q}_1,\vec{q}_0;t). \tag{19}$$

It is the Schrödinger equation.

2.1.2. *Temperature is not zero*

Let $k_BT\delta t \gg \hbar$. We can suppose that $\Gamma=k_BT\delta t$. Eq. (17) takes a form

$$\frac{\partial\psi(\vec{q}_1,\vec{q}_0;t)}{\partial t}=D\Delta\psi(\vec{q}_1,\vec{q}_0;t)+\frac{\left(U(\vec{q}_1)-T_N\right)}{2m_ND}\psi(\vec{q}_1,\vec{q}_0;t). \tag{20}$$

Here

$$D = \frac{k_B T \delta t}{m_N} \tag{21}$$

is the diffusion coefficient. The coefficient has dimensionality of [length^2 time^{-1}]. It is a factor of proportionality representing amount of substance diffusing across a unit area in a unit time - concentration gradient in unit time.

We can see that Eq. (20) deals with the amplitude function ψ, not a concentration. However, a measurable function is $\rho=|\psi|^2$ - concentration having dimensionality of [(amount of substance) length^{-3}]. Let us multiply Eq. (20) from the left by 2ψ. First we note that the combination $2\psi\,\Delta\psi = 2\psi\,\nabla\,\psi^{-1}\psi\,\nabla\,\psi = 2(\psi\,\nabla\,\psi^{-1})(\psi\,\nabla\,\psi) + 2\nabla\,(\psi\,\nabla\,\psi)$ results in $-(1/2)(\nabla\ln(\rho)\cdot\nabla\,\rho) + \Delta\rho$. As a result we come to a diffusion-drift equation describing diffusion in a space loaded by a potential field $(U(\vec{q}_1) - T_N)$:

$$\frac{\partial\rho}{\partial t} + \frac{D}{2}\left(\nabla\ln(\rho), \nabla\rho\right) = D\Delta\rho + \frac{(U(\vec{q}_1) - T_N)}{2m_N D}\rho \tag{22}$$

Extra term $(D/2)\nabla\ln(\rho)$ in this diffusion-drift equation is a velocity of outflow of the particles from volume populated by much more number of particles than in adjacent volume. The term $-\ln(\rho)$ is entropy of a particle ensemble. From here it follows that $\nabla\ln(\rho)$ describes inflow of the particles to a region where the entropy is small. Observe that the velocity

$$\vec{u} = D\frac{\nabla\rho}{\rho} = D\nabla\ln(\rho) \tag{23}$$

is an osmotic velocity required of the particle to counteract osmotic effects [6]. Namely, imagine a suspension of many Brownian particles within a physical volume acted on by an external, virtual in general, force. This force is balanced by an osmotic pressure force of the suspension [6]:

$$\vec{K} = k_B T\frac{\nabla\rho}{\rho}. \tag{24}$$

From here it is seen, that the osmotic pressure force arises always when density difference exists and especially when the density tends to zero. And vice-versa, the force disappears in extra-dense media with spatially homogeneous distribution of particles. As states the second law of thermodynamics spontaneous processes happen with increasing entropy. The osmosis evolves spontaneously because it leads to increase of disorder, i.e., with increase of entropy. When the entropy gradient becomes zero the system achieves equilibrium, osmotic forces vanish.

Due to appearance of the term $(D/2)\nabla\ln(\rho)$ in Eq. (22) the diffusion equation becomes nonlinear. It is interesting to note, running ahead, that this osmotic term reveals many common with the quantum potential, which is show further. Observe that the both expressions contain the term $\nabla\rho/\rho = \nabla\ln(\rho)$ relating to gradient of the quantum entropy $S_Q = -\ln(\rho)/2$ [28].

Reduction to PDEs, Eq. (19) and Eq. (22), was done with aim to show that the both quantum and classical realms adjoin with each other much more closely, than it could seem with the first glance. Further we shall return to the integral path paradigm [8] and calculate patterns arising after passing particles through gratings. We shall combine quantum and classical realms by introducing the complex-valued parameter $\Gamma = 2k_B T\delta t + i\hbar$.

2.2. Paths through *N*-slit grating

Computation of a passing particle through a grating is based on the path integral technique [8]. We begin with writing the path integral that describes passing the particle through a slit made in an opaque screen that is situated perpendicularly to axis *z*, Fig. 3. For this reason we need to describe a movement of the particle from a source to the screen and its possible deflection at passing through the slit, see Fig. 4. At that we need to evaluate all possible deflections.

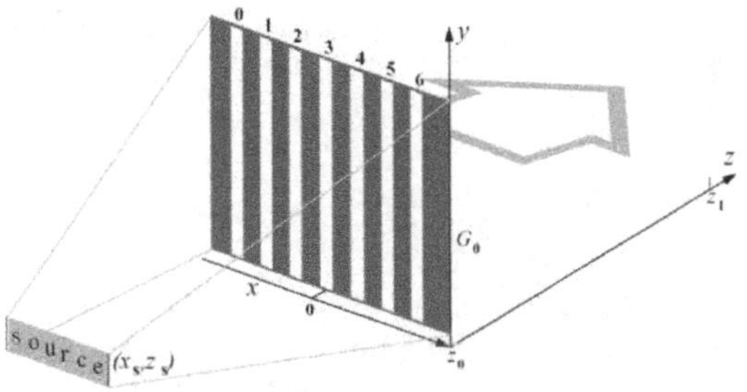

Figure 3. Interferometry from one grating G_0 situated transversely to a particle beam emitted from a distributed source.

We believe, that before the screen and after it, the particle (fullerene molecule, for example) moves as a free particle. Its Lagrangian, rewritten from Eq. (17), describes its deflection from a straight path in the following form

$$L = m_N \frac{\dot{x}^2}{2} - m_N(\dot{x}, \vec{v}) + T_N. \tag{25}$$

The first term relates to movement of the center of mass of the molecule. So that m_N is mass of the molecule and $\dot{x}$ is its transversal velocity, i.e., the velocity lies in transversal direction to the axis *z*. The second term is conditioned by collective fluctuations of atoms constituent this molecule. This term is nonzero when atoms have predominant fluctuations along axis x. For the sake of simplicity we admit that $\vec{v}$ is constant. The third term is a constant and comes from Eq. (17). It can be introduced into the normalization factor. For that reason we shall ignore this term in the following computations. A longitudinal momentum, p_z, is much greater than its transverse component [16, 17, 29] and we believe it is constant also. By translating a particle's position on a small distance $\delta x = (x_b - x_a) << 1$ for a small time $\delta t = (t_b - t_a) << 1$ we find that a weight factor of such a translation has the following form

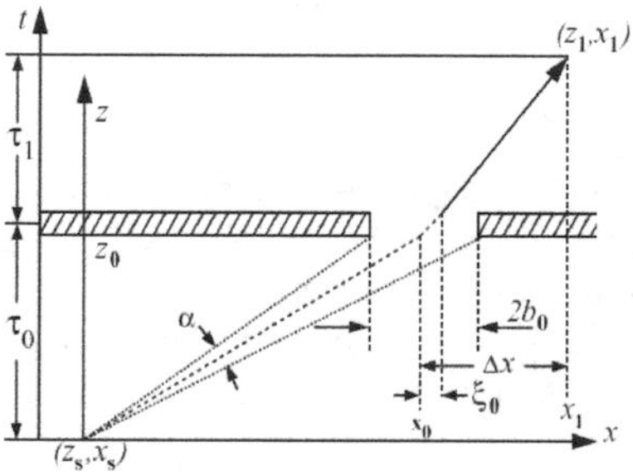

Figure 4. Passage of a particle along path $(z_s,x_s) \to (z_0,x_0) \to (z_1,x_1)$ through a screen containing one slit with a width equal to 2_{b0}. Divergence angle of particles incident on the slit, α, tends to zero as the source is removed to infinity.

$$\exp\left\{-\frac{L\delta t}{\Gamma}\right\} = \exp\left\{-\frac{m_N}{2\Gamma}\frac{(x_b - x_a)^2}{(t_b - t_a)} + \frac{m_N v}{\Gamma}(x_b - x_a)\right\}. \tag{26}$$

The particles flying to the grating slit along a ray α, Fig. 4, pass through the slit within a range from x_0-b_0 to x_0+b_0 with high probability. The path integral in that case reads

$$\psi(x_1,x_0,x_s) = \int_{-b_0}^{b_0} K(x_1,\tau_0+\tau_1;x_0+\xi_0,\tau_0)K(x_0+\xi_0,\tau_0;x_s,0)d\xi_0. \tag{27}$$

Integral kernel (propagator) for the particle freely flying is as follows [8, 26]

$$K(x_b,t_b;x_a,t_a) = \left[\frac{m_N(1-2v/v_{(a,b)})}{2\pi\Gamma(t_b - t_a)}\right]^{1/2} \exp\left\{-\frac{m_N(x_b - x_a)^2}{2\Gamma(t_b - t_a)}\left(1 - 2\frac{v}{v_{(a,b)}}\right)\right\} \tag{28}$$

Here $v_{(a,b)} = (x_b - x_a)/(t_b - t_b)$ is a velocity of the molecule on a segment from x_a to x_b. And v is an average velocity of collective deflection of atoms constituent the molecule. It was defined in Eq. (12). We shall believe that the ratio $v/v_{(a,b)}$ is small enough. We can define a new renormalized mass $\{m\}_N = m_N(1 - v/v_{(a,b)})$ and further we shall deal with this mass.

2.2.1. *Passing of a particle through slit*

By substituting the kernel (28) into the integral (27) we obtain the following detailed form

$$\psi(x_1,x_0,x_s) = \int_{-b_0}^{b_0} \left(\frac{\{m\}_N}{2\pi\Gamma\tau_1}\right)^{1/2} \exp\left\{-\frac{\{m\}_N(x_1-(x_0+\xi_0))^2}{2\Gamma\tau_1}\right\} \left(\frac{\{m\}_N}{2\pi\Gamma\tau_0}\right)^{1/2} \exp\left\{-\frac{\{m\}_N((x_0+\xi_0)-x_s)^2}{2\Gamma\tau_0}\right\} d\xi_0. \quad (29)$$

The integral is computed within a finite interval $[-b_0,+b_0]$. Observe, that the integrating can be broadened from $-\infty$ to $+\infty$. But in this case we need to load the integral by the step function equal to unit within the finite interval $[-b_0,+b_0]$ and it vanishes outside of the interval. The step function, that simulate a single slit, can be approximated by the following a set of the Gaussian functions [30]

$$G(\xi,b,\eta,K) = \frac{1}{\eta}\sqrt{\frac{2}{\pi}} \sum_{k=1}^{K} \exp\left\{-\frac{(K\xi - b(K-(2k-1)))^2}{2(b\eta)^2}\right\}. \quad (30)$$

Here parameter b is a half-width of the slit, real $\eta > 0$ is a tuning parameter, and K takes integer values. At $K = 1$ this form-factor degenerates to a single Gaussian function. And at $K \to \infty$ this function tends to an infinite collection of the Kronecker deltas which fill everywhere densely the step function. We rewrite Eq. (29) with inserting this form-factor

$$\psi(x_1,x_0,x_s) = \int_{-\infty}^{\infty} G(\xi_0,b_0) \frac{\{m\}_N}{2\pi\Gamma\sqrt{\tau_1\tau_0}} \exp\left\{-\frac{\{m\}_N}{2\Gamma}\left(\frac{(x_1-(x_0+\xi_0))^2}{\tau_1} + \frac{((x_0+\xi_0)-x_s)^2}{\tau_0}\right)\right\} d\xi_0. \quad (31)$$

We do not write parameters η and K in the Gaussian form-factor and for the sake of simplicity further we shall consider they equal to 1. That is, for simulating the slit we select a single Gaussian function.

2.2.2. *Definition of new working parameters*

First we replace the flight times τ_0 and τ_1 by flight distances (z_0-z_s) and (z_1-z_0), see Fig. 4. This replacement reads

$$\begin{cases} \tau_0 = (z_0 - z_s)/v_z \\ \tau_1 = (z_1 - z_0)/v_z \end{cases} \quad (32)$$

where v_z is a particle velocity along the axis z.

There is, however, one more parameter of time that is represented in definition of the coefficient $\Gamma = 2k_B T\delta t + \mathbf{i}\hbar$. It is a small time increment δt. The parameter δt first appears in the path integral (3) as the time increment along a path. In accordance with the uncertainty principle δt should be more or equal to the ratio of $\hbar$, Planck constant, to energy of occurring events. In a case of a flying particle through vacuum it can be minimal energy of vacuum fluctuations (it is about energy of the first Bohr orbit of electron-positron pair that is about 14 eV). Evaluation gives $\delta t \sim 2.8\ 10^{-16}$ s. From here it follows, that $2k_B T\delta t$ is less than $\hbar$ on about one order at $T = 1000$ K (almost temperature of fullerene evaporation from the Knudsen cell [27]).

Emergence of the term $2k_B T\delta t$ can be induced by existence of quantum drag [31] owing to different conditions for quantum fluctuations both inside of the fullerene molecule and outside what can induce weak Casimir forces. Because of the weak Casimir force the quantum drag does not lead to decoherence at least in the near zone. However further we shall see that a weak washing out of the Talbot interference pattern is due to existence of this term.

Let us divide the parameter Γ by $\{m\}_N v_z$

$$\frac{\Gamma}{\{m\}_N v_z} = \frac{2k_B T}{\{m\}_N v_z}\delta t + \mathbf{i}\frac{\hbar}{\{m\}_N v_z}. \tag{33}$$

Here $p_z = \{m\}_N v_z$ is a particle momentum along axis z. We can define the de Broglie wavelength $\lambda_{dB} = h/p_z$ where $h = 2\pi\hbar$ is the Planck constant. Let us also define a length $\delta_T = 4\pi k_B T\delta t/(\{m\}_N v_z)$. In this view we can rewrite Eq. (33) as follows

$$\frac{\Gamma}{\{m\}_N v_z} = \frac{\delta_T}{2\pi} + \mathbf{i}\frac{\lambda_{dB}}{2\pi} = \frac{1}{2\pi}\Lambda. \tag{34}$$

The length δ_T tends to zero as $T \to 0$. At $T = 1000$ K and at adopted $\delta t = 2.8\ 10^{-16}$ s we have $\delta_T \approx 0.4$ pm. On the other hand, the de Broglie wavelength, λ_{dB}, evaluated for the fullerene molecule moving with the velocity v_z =100 m/s is about 5 pm [18]. So, we can see that the length δ_T is less of the de Broglie wavelength on about one order and smaller. A signification of the length δ_T is that it determines decoherence of a particle beam. Decoherence of flying particles occurs the quickly, the larger δ_T. Observe that the length δ_T has a close relation with the coherence width - a main parameter in the generalized Gaussian Schell-model [32, 33].

3. Wave function behind the grating

Wave function from one slit after integration over ξ_0 from $-\infty$ to $+\infty$ has the following view [28]

$$\psi(x,z,x_0,x_s) = \sqrt{\frac{2}{\pi\Lambda\Sigma_0(z_0-z_s)}}\exp\left\{-\frac{\pi}{\Lambda}\left[\frac{(x-x_0)^2}{(z-z_0)}\left(1-\frac{\Xi_0^2}{\Sigma_0}\right)+\frac{(x_0-x_s)^2}{(z_0-z_s)}\right]\right\}. \tag{35}$$

Here argument of ψ-function contains apart x also z in order to emphasize that we carry out observation in the point (z,x), see Fig. 4. The factor $(2/\pi)^{1/2}$ comes from (30). Parameters Ξ_0 and Σ_0 read

$$\Xi_0 = 1-\frac{(x_0-x_s)(z-z_0)}{(z_0-z_s)(x-x_0)} \quad \text{and} \quad \Sigma_0 = \frac{(z-z_s)}{(z_0-z_s)}+\frac{\Lambda(z-z_0)}{2\pi b_0^2} \tag{36}$$

In order to simplify records here we omit subscript 1 at x and z - an observation point that is situated after the slit.

Let us consider that an opaque screen contains N_0 slits spaced through equal distance, d, from each other. Numeration of the slits is given as it is shown in Fig. 3, $n_0 - 0,1,2, \ldots, N_0 - 1$. Sum of all wave functions (35), each of which calculates outcome from an individual slit, gives a total effect in the point (z,x) where a detector is placed:

$$\left|\Psi_0(x,z,d,x_s,\Lambda)\right\rangle = \sum_{n_0=0}^{N_0-1}\psi\left(x,z,\left(n_0-\frac{N_0-1}{2}\right)d,x_x\right). \tag{37}$$

Probability density in the vicinity of the observation point (x,z) reads

$$p(x,z) = \left\langle\Psi_0(x,z,d,x_s,\Lambda)\middle|\Psi_0(x,z,d,x_s,\Lambda)\right\rangle. \tag{38}$$

Calculation of the wave function (37) is fulfilled for the grating containing $N_0 = 32$ slits. Distance between slits is $d = 10^5\lambda_{dB}$. So at $\lambda_{dB} = 5$ pm the distance is equal to 500 nm. Requirement $\lambda_{dB} \ll d$ and N_0 tending to infinity together with a condition that the particle beam is paraxial, that is, $x_s = 0$ and $z_s \to -\infty$, provides emergence in the near-field of an interference pattern, named Talbot carpet [19,20]. Here a spacing along interference patterns is measured in the Talbot length

$$z_T = 2\frac{d^2}{\lambda_{dB}}, \tag{39}$$

which is a convenient natural length at representation of interference patterns. Since we restrict themselves by finite N_0 we have a defective carpet, which progressively collapses as a spacing from the slit increases. Fig. 5 shows the Talbot carpet, being perfect in the vicinity of the grating slit; it is destroyed progressively with increasing z_T. As for the Talbot carpet we have the following observation. We see that in a cross-section $z = z_T/2$ image reproduces radiation of the slits but phase-shifted by half period between them. At $z = z_T$ radiation of the slits is reproduced again on the same positions where the slits are placed. And so forth.

Evaluation of sizes of the interference pattern is given by ratio of the Talbot length to a length of the slit grating. In our case the Talbot length is $z_T = 0.1$ m. And length of the slit grating is about $N_0 d = 1.6\ 10^{-5}$ m. From here we find that the ratio is 6250. It means that the interference pattern shown in Fig. 5 represents itself a very narrow strip.

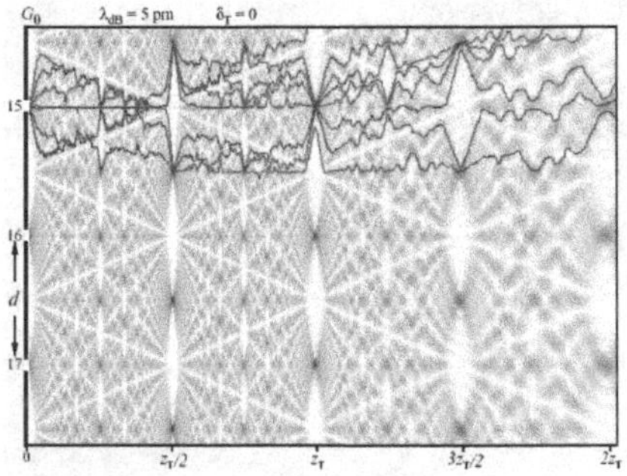

Figure 5. Interference pattern in the near field. It is shown only the central part of the grating containing $N_0 = 32$ slits, $\lambda_{dB} = 5$ pm, $d = 500$ nm, and $\delta_T = 0$. In the upper part of the figure a set of the Bohmian trajectories, looking like on zigzag curves, drawn in dark blue color is shown.

Zigzag curves, drawn in the upper part of Fig. 5 by dark blue color, show Bohmian trajectories that start from the slit No. 15. One can see that particles prefer to move between nodes having positive interference effect and avoid empty lacunas. However the above we noted, that the ratio of the Talbot length to the length of the grating is about 6250 >> 1. It means that really the Bohmian trajectories look almost as straight lines slightly divergent apart. Zigzag-like behavior of the trajectories is almost invisible. Such an almost feebly marked zigzag-like behavior may be induced by fluctuations of virtual particles escorting the real particle.

As soon as we add the term $k_B T\delta t$ different from zero ($T > 0$ K) we observe blurring the interference pattern. The blurring is the stronger, the larger $k_B T\delta t$. For comparison see Figs. 6 and 7. Here instead of $k_B T\delta t$ we write a more convenient parameter, the coherence length δ_T. This length characterizes a dispersed divergence from initially tuned the de Broglie wavelength. Such a disperse medium can be due to quantum drag on the vacuum fluctuations. Here the Bohmian trajectories are not shown, since because of the thermal term $k_B T\delta t > 0$ a Brownian like scattering of the trajectories arises. This scattering we shall discuss later on.

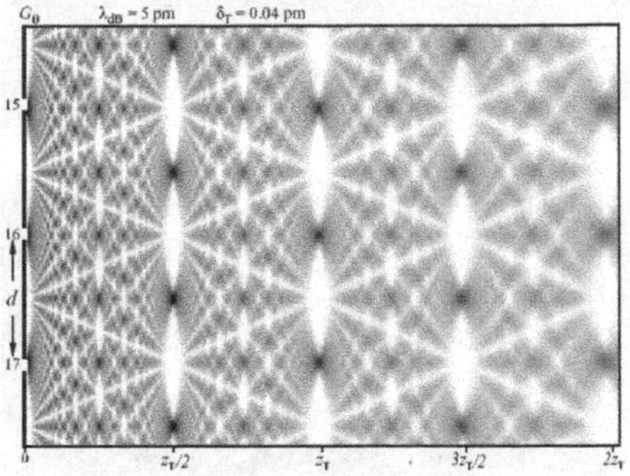

Figure 6. Blurred interference pattern in the near field. It is shown only the central part of the grating containing N_0 = 32 slits; d = 500 nm, δ_T = 0.04 pm << λ_{dB} = 5 pm.

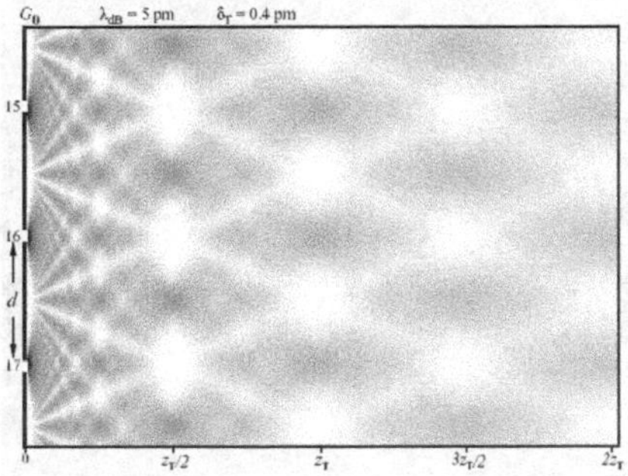

Figure 7. Blurred interference pattern in the near field. It is shown only the central part of the grating containing N_0 = 32 slits; d = 500 nm, δ_T = 0.4 pm < λ_{dB} = 5 pm.

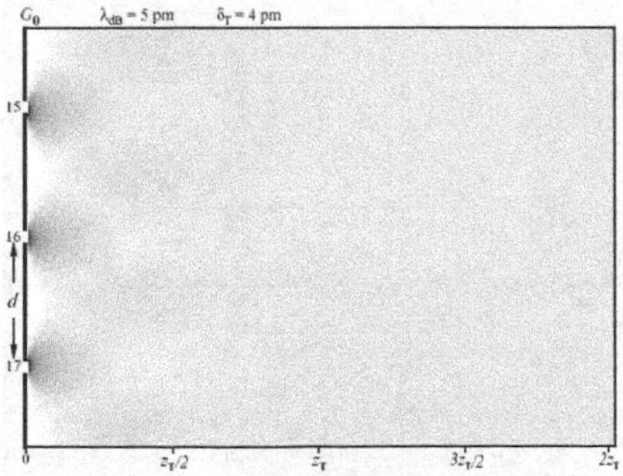

Figure 8. Destroyed interference pattern because of large δ_T = 4 pm ~ λ_{dB} = 5 pm. N_0 = 32 slits, d = 500 nm.

We may think that the technical vacuum can be not perfect. It causes additional scattering of particles on residual gases. Because of this additional scattering the interference pattern can be destroyed entirely, as shown in Fig. 8.

Now let us draw dependence of the probability density $p(x, z)$ as a function of x at fixed z. In other words, we calculate interference fringes in a cross-section of the interference patterns at $z = z_T/2$ for different values of the length δ_T. Such a cross-section is chosen because a self image of the slit grating appears phase-shifted by half period of the grating. For that reason we should see the interference fringes spaced between the slit sources of radiation.

Fig. 9 shows three characteristic patterns of the interference fringes. In Fig. 9(a) almost ideal interference fringes are shown obtained at $\delta_T < \lambda_{dB}$. Fig. 9(b) shows interference fringes obtained at $\delta_T \sim \lambda_{dB}$. It is instructive to compare these interference fringes with those that have been measured in experiments [14, 34]. And Fig. 9(c) shows disappearance of interference fringes because of strong scattering of the particles on residual gases in vacuum, $\delta_T > \lambda_{dB}$.

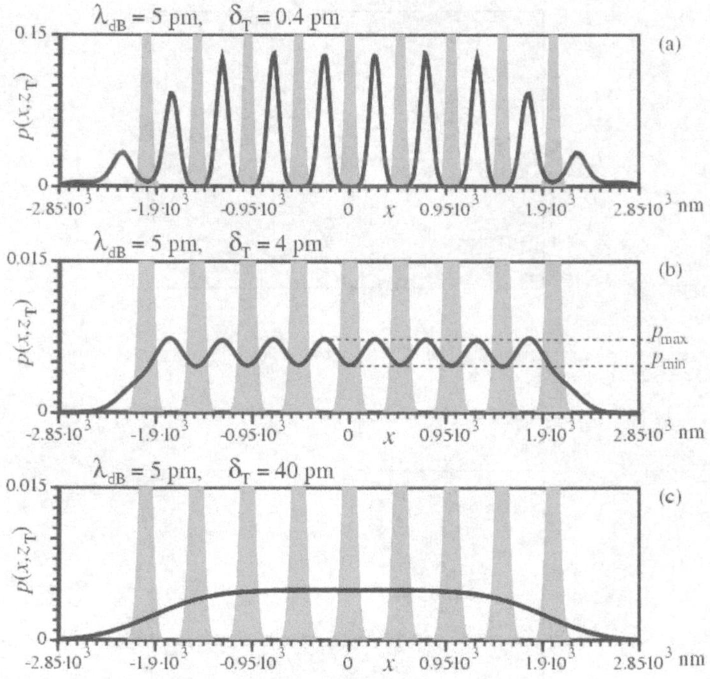

Figure 9. Interference fringes in cross-section of the density distribution pattern by the Talbot half-length, $z = z_T/2$, (the fringes are drawn in red): (a) $\delta_T = 0.4$ pm, almost coherent beam; (b) $\delta_T = 4$ pm, weak coherence; (c) $\delta_T = 40$ pm, entirely noncoherent beam. Cyan strips show luminosity of slits. The grating consists of $N_0 = 9$ slits. Collapse of the interference pattern on edges of the grating is due to its finite length. Therefore visibility of the interference fringes is evaluated only for 5 central slits.

Disappearance of interference fringes is numerically evaluated by calculating a characteristic called visibility [14, 34]. The fringe visibility [27] is represented as a ratio of difference between maximal and minimal intensities of the fringes to their sum:

$$V = \frac{P_{\max} - P_{\min}}{P_{\max} + P_{\min}}. \tag{40}$$

Evaluation of $P_{\max}$ and $P_{\min}$ is shown in Fig. 9(b). As follows from the figure, the evaluations are fulfilled in a central region of the grating. That is, edges of the grating have to be left far off from the measured zone. The visibility V as a function of the parameter δ_T is shown in Fig. 10. One can see that crossover from almost perfect interference fringes, $V = 1$, up to their absence, $V = 0$, begins near $\delta_T. \sim \lambda_{dB}$. Transition from almost coherent particle beam to incoherent is a cause of such a crossover [30].

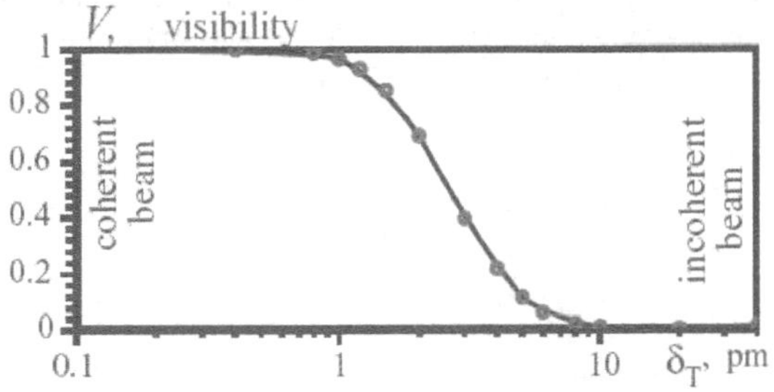

Figure 10. Visibility of interference fringes as a function of the parameter δ_T ranging from 0.1 to 40 pm. Wavelength of a matter wave is λ_{dB} = 5 pm.

4. Bohmian trajectories and variance of momenta and positions along paths

Here we repeat computations of David Bohm [35] which lead to the Hamiltoton-Jacobi equation loaded by the quantum potential and, as consequence, to finding Bohmian trajectories. But instead of the Schrödinger equation we choose Eq. (17) that contains complex-valued parameter $\Gamma=\beta+i\hbar$:

$$(\beta - \mathrm{i}\hbar)\frac{\partial\psi}{\partial t} = \frac{(\beta^2 + \hbar^2)}{2m_N}\nabla^2\psi + \frac{(\beta - \mathrm{i}\hbar)}{(\beta + \mathrm{i}\hbar)}V\psi. \tag{41}$$

Here $\beta = 2k_B T\delta t$ (in particular, the diffusion coefficient reads $D = \beta/2m_N$) and $V = U(\vec{q}_1) - T_N$.

Further we apply polar representation of the wave function, $\psi = R \exp\{iS/\hbar\}$. It leads to obtaining two equations for real and imaginary parts that deal with real-valued functions R and S. The function R is the amplitude of the wave function and $S/\hbar$ is its phase. After series of computations, aim of which is to put together real and imaginary terms, we obtain the following equations

$$\frac{\partial S}{\partial t} + \beta \frac{1}{R}\frac{\partial R}{\partial t} = -\underbrace{\frac{(\beta^2+\hbar^2)}{2m_N\hbar^2}(\nabla S)^2}_{(a)} + \underbrace{\frac{(\beta^2+\hbar^2)}{2m_N}\frac{\nabla^2 R}{R}}_{(b)} + \frac{(\beta^2-\hbar^2)}{(\beta^2+\hbar^2)}V, \tag{42}$$

$$\frac{\beta}{\hbar^2}\frac{\partial S}{\partial t} - \frac{1}{R}\frac{\partial R}{\partial t} = \frac{(\beta^2+\hbar^2)}{2m_N\hbar^2}\underbrace{(\nabla^2 S + 2\frac{1}{R}\nabla R \nabla S)}_{(c)} - \frac{2\beta}{(\beta^2+\hbar^2)}V. \tag{43}$$

Firstly, we can see that at $\beta = 0$ Eq. (42) reduces to the modified Hamilton-Jacobi equation due to loaded the quantum potential that is enveloped here by brace (b). And Eq. (43) reduces to the continuity equation. These equations read

$$\frac{\partial S}{\partial t} = -\underbrace{\frac{1}{2m_N}(\nabla S)^2}_{(a)} + \underbrace{\frac{\hbar^2}{2m_N}\frac{\nabla^2 R}{R}}_{(b)} + V, \tag{44}$$

$$-\frac{1}{R}\frac{\partial R}{\partial t} = \frac{1}{2m_N}\underbrace{(\nabla^2 S + 2\frac{1}{R}\nabla R \nabla S)}_{(c)}. \tag{45}$$

Terms enveloped by braces (a), (b), and (c) are the kinetic energy of the particle, the quantum potential, and the right part is a kernel of the continuity equation (45), respectively. In particular, the term $2\nabla R / R = \nabla \ln(R^2)$ relates to the osmotic velocity, see Eq. (23). Eqs. (44) and (45) are the same equations obtained by Bohm [35]. From historical viewpoint it should be noted that the same equations were published by Madelung[1] in 1926 [36].

Momentum of the particle reads

$$\vec{p} = m_N \vec{v} = \nabla S, \tag{46}$$

where $\vec{v}$ is its current velocity. The de Broglie equation relates the momentum p to the wavelength $\lambda_{dB} = h/p$, where $h = 2\pi\hbar$ is the Planck constant. Now, as soon as we found the current velocity

1 My attention to the Madelung' article was drawn by Prof. M. Berry.

$$\vec{v} = \frac{1}{m_N}\nabla S = \frac{\hbar}{m_N}\mathrm{Im}\left(\left|\psi\right\rangle^{-1}\nabla\left|\psi\right\rangle\right) \tag{47}$$

positions of the particle in each current time beginning from the grating' slits up to a detector is calculated by the following formula

$$\vec{r}(t+\delta t) = \vec{r}(t) + \vec{v}(t)\delta t. \tag{48}$$

Here t is a current time that starts from t=0 on a slit source and δt is an arbitrarily small increment of time. Some calculated trajectories of particles, the Bohmian trajectories, are shown in upper part of Fig. 5. It should be noted that the Bohmian trajectories follow from exact solutions of Eqs. (44)-(45). These equations give a rule for finding geodesic trajectories and secants of equal phases, $S/\hbar$, at given boundary conditions. The geodesic trajectories point to tendency of the particle migration along paths. And the secant surfaces describe a coherence of all the passing particles created on a single source.

In case of $\beta >> \hbar$ we have the following two equations

$$\frac{\partial R}{\partial t} = \underbrace{-\frac{\beta}{2m_N\hbar^2}R(\nabla S)^2}_{(a)} + \underbrace{\frac{\beta}{2m_N}\nabla^2 R}_{(b)}, \tag{49}$$

$$\frac{\partial S}{\partial t} = \frac{\beta}{2m_N}(\nabla^2 S + \underbrace{2\frac{1}{R}\nabla R\nabla S}_{(c)}), \tag{50}$$

Here we take into consideration that in the first equation we may replace $(\nabla S/\hbar)^2 = 4\pi^2/\lambda_{\mathrm{dB}}^2$. In the second equation we may replace the term ∇S by $m_N\vec{v}$ as follows from Eq. (46). We notice also, that $2R^{-1}\nabla R = 2\nabla\ln(R) = \nabla\ln(\rho)$. And $\beta/2m_N = D$ is the diffusion coefficient. Now we may rewrite Eqs. (49)-(50) as follows

$$\frac{\partial R}{\partial t} = \underbrace{D\nabla^2 R}_{(b)} - \underbrace{\frac{4\pi^2}{\lambda_{\mathrm{dB}}^2}DR}_{(a)}, \tag{51}$$

$$\frac{\partial S}{\partial t} = \underbrace{D\left(\nabla^2 S + m_N(\nabla\ln(\rho),\vec{v})\right)}_{(c)} = D\nabla^2 S + m_N(\vec{u},\vec{v}). \tag{52}$$

Here $\vec{u} = D\,\nabla \ln(\rho)$ is the osmotic velocity defined in Eq. (23). We got the two diffusion equations coupled with each other through sources. Namely, this coupling is provided due to the de Broglie wavelength and the osmotic velocity which can change with time. These diffusion equations cardinally differ from Eqs. (44)-(45). Because of diffusive nature of these supplementary parts blurring of interference patterns occurs. It leads to degeneration of the Bohmian trajectories to Brownian ones.

4.1. Dispersion of trajectories and the uncertainty principle

As for the Bohmian trajectories there is a problem concerning their possible existence. As follows from Eqs. (46) and (48) in each moment of time there are definite values of the momentum and the coordinate of a particle moving along the Bohmian trajectory. This statement enters in conflict with the uncertainty principle.

Here we try to retrace emergence of the uncertainty principle stemming from standard probability-theoretical computations of expectation value and variance of a particle momentum. We adopt a wave function in the polar representation

$$|\Psi\rangle = R\exp\{\mathbf{i}S/\hbar\}, \tag{53}$$

where $R = (\rho)^{1/2}$ is the amplitude of the wave function ($\rho = R^2 = \langle\Psi|\Psi\rangle$ is the probability density) and $S/\hbar$ is its phase. Momentum operator $\hat{p} = -i\hbar\nabla$ and corresponding velocity operator $\hat{v} = -i(\hbar/m)\nabla$ are kinetic operators in quantum mechanics. Here m is mass of the particle. Expectation value of the velocity operator reads

$$\vec{V}_g = \frac{1}{\langle\Psi|\Psi\rangle}\langle\Psi|-\mathbf{i}\frac{\hbar}{m}\nabla|\Psi\rangle = \frac{1}{m}\left(\nabla S + \mathbf{i}\hbar\nabla S_Q\right). \tag{54}$$

The velocity V_g is seen to be complex-valued. Here $S_Q = -\ln(R) = -\ln(\rho)/2$ is the quantum entropy [28] and ($\hbar/2m$) is the quantum diffusion coefficient [6, 7]. Therefore its imaginary part is a quantum osmotic velocity

$$\vec{u}_Q = -(\hbar/m)\nabla S_Q = (\hbar/2m)\nabla\ln(\rho). \tag{55}$$

It is instructive to compare this velocity with the classical osmotic velocity given in Eq. (23). As can see the osmotic velocity stems from gradient of entropy that evaluates degree of order and disorder on a quantum level, likely of vacuum fluctuations.

Real part of Eq. (54) gives the current velocity $\vec{v}$ defined by Eq. (47). It should be noted that because of existence of imaginary unit in definition of the momentum operator, real part of Eq. (54) is taken as the current velocity. Whereas imaginary unit is absent in Eq. (47). Therefore at computing the current velocity by Eq. (47) we take imaginary part.

Let us now calculate variance of the velocity V_g. This computation reads

$$\mathrm{Var}(V_g) = \frac{1}{\langle\Psi|\Psi\rangle}\left\langle\Psi\left|\left(-\mathbf{i}\frac{\hbar}{m}\nabla - \vec{V}_g\right)^2\right|\Psi\right\rangle = \frac{1}{\langle\Psi|\Psi\rangle}\left\langle\Psi\left|\mathbf{i}\frac{\hbar}{m}\nabla\vec{V}_g + \underbrace{\mathbf{i}\frac{\hbar}{m}\vec{V}_g\nabla + V_g^2}_{(d)} - \frac{\hbar^2}{m^2}\Delta\right|\Psi\right\rangle. \quad (56)$$

Terms over bracket (d) kill each other as follows from Eq. (54). It is reasonable in the perspective to multiply Var(V_g) by $m/2$

$$\frac{m}{2}\mathrm{Var}(V_g) = -\frac{1}{\langle\Psi|\Psi\rangle}\left\langle\Psi\left|\frac{\hbar^2}{2m}\Delta\right|\Psi\right\rangle + \mathbf{i}\frac{\hbar}{2}\nabla\vec{V}_g. \quad (57)$$

So, this expression has a dimensionality of energy. The first term to be computed represents the following result

$$-\frac{1}{\langle\Psi|\Psi\rangle}\left\langle\Psi\left|\frac{\hbar^2}{2m}\Delta\right|\Psi\right\rangle = \underbrace{\frac{1}{2m}(\nabla S)^2}_{(a)} - \underbrace{\frac{\hbar^2}{2m}\frac{\nabla^2 R}{R}}_{(b)} - \mathbf{i}\underbrace{\frac{\hbar}{2m}\left(\Delta S + 2\frac{(\nabla S, \nabla R)}{R}\right)}_{(c)}. \quad (58)$$

Here the term enveloped by bracket (a) is a kinetic energy of the particle, the term enveloped by bracket (b) with negative sign added is the quantum potential Q, and the term enveloped by bracket (c) comes from the continuity equation. See for comparison Eqs. (44) and (45). We rewrite the quantum potential as follows

$$Q = -\frac{\hbar^2}{2m}\frac{\nabla^2 R}{R} = -\frac{\hbar^2}{2m}(\nabla S_Q)^2 + \frac{\hbar^2}{2m}\nabla^2 S_Q. \quad (59)$$

As for the second term in Eq. (57) we have $i(\hbar/2)\nabla\vec{V}_g = i(\hbar/2m)\Delta S - (\hbar^2/2m)\Delta S_Q$. It is follows from computation by Eq. (54). As a result, the expression (57) takes the following view

$$\frac{m}{2}\mathrm{Var}(V_g) = \frac{1}{2m}(\nabla S)^2 - \frac{\hbar^2}{2m}(\nabla S_Q)^2 - \mathbf{i}\frac{\hbar}{m}\frac{(\nabla S, \nabla R)}{R}. \quad (60)$$

One can see that the variance consists of real and imaginary parts. Observe that the right side is represented through square of gradient of the complexified action [28], namely $(\nabla(S + i\hbar S_Q))^2/2m$. We shall not consider here the imaginary part. Instead we shall consider the real part of this expression. It reads

$$\frac{m}{2}\mathrm{Re}(\mathrm{Var}(V_g))=\frac{1}{2m}(\nabla S)^2-\hbar\left(\frac{\hbar}{2m}(\nabla S_Q)^2\right). \quad (61)$$

The first term in this expression represents kinetic energy, E, of the particle. The second term, stemming from the quantum potential, contains under braces a term having dimensionality of inverse time, that is, of frequency

$$\omega_Q=\frac{\hbar}{2m}(\nabla S_Q)^2 \quad (62)$$

This frequency multiplied by $\hbar$ represents an energy binding a particle with vacuum fluctuations. This energy, as follows from Eq. (61), is equal to the particle mass multiplied by squared the osmotic velocity (55) and divided by 2. It is an osmotic kinetic energy. In the light of the above said we rewrite Eq. (61) in the following way

$$\frac{m}{2}\mathrm{Re}(\mathrm{Var}(V_g))=E-\hbar\omega_Q \ge 0. \quad (63)$$

Let we have two Bohmian trajectories. Along one trajectory we have E_1 - $\hbar\omega_{Q,1}$, and along other trajectory we have a perturbed value E_2 - $\hbar\omega_{Q,2}$. Subtracting one from other we have

$$\delta E-\hbar\delta\omega_Q \ge 0. \quad (64)$$

One can suppose that emergence of the second trajectory was conditioned by a perturbation of the particle moving along the first trajectory. If it is so, then emergence of the second trajectory stems from an operation of measurement of some parameters of the particle. One can think that duration of the measurement is about $\delta t = 1/\delta\omega_Q$. From here we find

$$\boxed{\delta E\delta t \ge \hbar} \quad (65)$$

Now let us return to Eq. (48) and rewrite it in the following view

$$\delta\vec{r}(t)=\vec{v}_1(t)\delta t \ge \vec{v}_1(t)\hbar/\delta E. \quad (66)$$

The initial Bohmian trajectory is marked here by subscript 1. Observe that $\delta E=m(v_2^2-v_1^2)/2\approx m\vec{v}_1\delta\vec{v}$. Here we have calculated $v_2^2=(\vec{v}_1+\delta\vec{v})^2\approx v_1^2+2\vec{v}_1\delta\vec{v}$. Substituting computations of δE into Eq. (66) we obtain finally

$$\boxed{\delta\vec{p}\delta\vec{r} \geq \hbar} \tag{67}$$

Here we take into account $\delta\vec{p}=m\delta\vec{v}$.

5. Concluding remarks

Each nanoparticle incident on a slit grating passes only through a single slit. Its path runs along a Bohmian trajectory which is represented as an optimal path for the nanoparticle migrating from a source to a detector. Unfortunately, the Bohmian trajectory can not be observable since a serious obstacle for the observation comes from the uncertainty principle. In other words, an attempt to measure any attribute of the nanoparticle, be it position or orientation, i.e., the particle momentum, leads to destroying information relating to future history of the nanoparticle. What is more, any collision of the nanoparticle with a foreign particle destroys the Bohmian trajectory which could give a real contribution to the interference pattern. It relates closely with quality of vacuum. In the case of a bad vacuum such collisions will occur frequently. They lead to destruction of the Bohmian trajectories. Actually, they degenerate to Brownian trajectories.

Excellent article [21] of Couder & Fort with droplets gives, however, a clear hint of what happens when the nanoparticle passes through a single grating slit. In the light of this hint we may admit that the particle "bouncing at moving through vacuum" generates a wave at each bounce. So, a holistic quantum mechanical object is the particle + wave. Here the wave to be generated by the particle plays a role of the pilot-wave first formulated by Lui de Broglie and later developed by Bohm [37]. It is interesting to note in this context, that the pilot-waves have many common with Huygens waves [5].

A particle passing through vacuum generates waves with wavelength that is inversely proportional to its momentum (it follows from the de Broglie formula, $\lambda = h/p$, where h is Planck's constant). One can guess that a role of the vacuum in the experiment of Couder & Fort takes upon itself a silicon oil surface with subcritical Faraday ripples activated on it [21]. Observe that pattern of the ripples is changed in the vicinity of extraneous bodies immersed in the oil which simulate grating slits. Interference of the ripples with waves generated by the bouncing droplets provides optimal paths for the droplets traveling through the slits and further. As a result we may observe an interference pattern emergent depending on amount of slits in the grating and distance between them.

Now we may suppose that the subcritical Faraday ripples on the silicon oil surface simulate vacuum fluctuations. Consequently, the vacuum fluctuations change their own pattern near the slit grating depending on amount of slits and distance between them. We may imagine that the particle passing through vacuum (bouncing through, Fig. 11) initiates waves which interfere with the vacuum fluctuations. As a result of such an interference the particle moves along an optimal path - along the Bohmian trajectory. Mathematically the bounce is imitated by an exponential term $\exp\{iS/\hbar\}$, where the angle $S/\hbar$ parametrizes the group of rotation given on a circle of unit radius. So, the path along which the particle moves is scaled by this unitary group, U(1), due to the exponential mapping of the phase $S/\hbar$ on the circle.

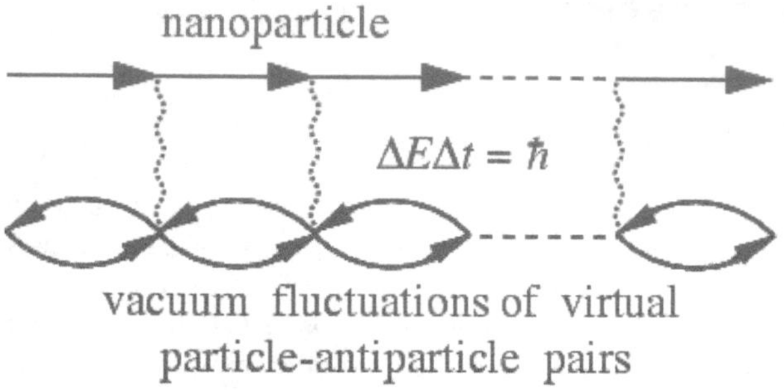

Figure 11. Bouncing a nanoparticle at moving through vacuum. Vertical dotted sinusoidal curves depict exchange by energy ΔE with vacuum virtual particle-antiparticle pairs over period of about $\Delta t = \hbar/\Delta E$.

In conclusion it would be like to remember remarkable reflection of Paul Dirac. In 1933 Paul Dirac drew attention to a special role of the action S in quantum mechanics [38] - it can exhibit itself in expressions through $\exp\{iS/\hbar\}$. In 1945 he emphasized once again, that the classical and quantum mechanics have many general points of crossing [39]. In particular, he had written in this article: "We can use the formal probability to set up a quantum picture rather close to the classical picture in which the coordinates q of a dynamical system have definite values at any time. We take a number of times t_1, t_2, t_3, ... following closely one after another and set up the formal probability for the q's at each of these times lying within specified small ranges, this being permissible since the q's at any time all commutate. We then get a formal probability for the trajectory of the system in quantum mechanics lying within certain limits. This enables us to speak of some trajectories being improbable and others being likely."

Acknowledgement

Author thanks Miss Pipa (administrator of Quantum Portal) for preparing programs that permitted to calculate and prepare Figures 5 to 8. Author thanks also O. A. Bykovsky for taking my attention to a single-particle interference observed for macroscopic objects by Couder and Fort and V. Lozovskiy for some remarks relating to the article.

Author details

Valeriy I. Sbitnev

St.-Petersburg Nuclear Physics Institute, NRC Kurchatov Institute, Gatchina, Russia

References

[1] Dauben JW. Abraham Robinson: The creation of nonstandard analysis. A personal and mathematical odyssey. Princeton, NJ: Princeton University Press; 1995.

[2] Stanford Encyclopeia of Philosophy: Newton's Views on Space, Time, and Motion. http://plato.stanford.edu/entries/newton-stm/ (accessed 12 August 2004).

[3] Henson CW. Foundations of nonstandard analysis. In: Arkeryd L.O., Cutland N. J., & Henson C. W., (eds.) Nonstandard Analysis, Theory and application. The Netherlands: Kluwer Acad. Publ.; 1997,p. 1-51.

[4] Nelson E. Radically Elementary Probability Theory. Princeton, New Jersey: Princeton Univ. Press; 1987.

[5] Huygens C. Treatise on Light. Gutenberg eBook, No. 14725. http://www.gutenberg.org/files/14725/14725-h/14725-h.h (accessed 18 January 2005)

[6] Nelson E. Dynamical theories of Brownian motion. Princeton, New Jersey: Princeton Univ. Press; 2001.

[7] Nelson E. Quantum fluctuations. Princeton Series in Physics. Princeon, New Jersey: Princeton Univ. Press; 2001.

[8] Feynman RP., Hibbs A. Quantum mechanics and path integrals. N. Y.: McGraw Hill; 1965.

[9] Einstein A., von Smoluchowski M. Brownian motion. Moscow-Leningrad: ONTI (in Russian); 1936.

[10] Sidorov LN., Troyanov SI. At the dawn of a new chemistry of fullerenes. PRIRODA 2011; 1153(9) 22-30.

[11] Arndt M., Hackermüller L., Reiger E. Interferometry with large molecules: Exploration of coherence, decoherence and novel beam methods. Brazilian Journal of Physics 2005; 35(2A) 216-223.

[12] Brezger B., Arndt M., Zeilinger A. Concepts for near-field interferometers with large molecules. J. Opt. B: Quantum Semiclass. Opt. 2003; 5(2) S82-S89.

[13] Brezger B., Hackermüller L., Uttenthaler S., Petschinka J., Arndt M., Zeilinger A. Matter-Wave Interferometer for Large Molecules. Phys. Rev. Lett. 2002; 88 100404.

[14] Gerlich S., Eibenberger S., Tomandl M., Nimmrichter S., Hornberger K., Fagan P. J., Tuxen J., Mayor M., Arndt, M. Quantum interference of large organic molecules. Nature Communications 2011; (2) 263. http://www.nature.com/ncomms/journal/v2/n4/full/ncomms1263.html (accessed 5 April 2011)

[15] Hornberger K., Hackermüller L., Arndt M. Inuence of molecular temperature on the coherence of fullerenes in a near-field interferometer. Phys. Rev. A. 2005; 71, 023601.

[16] Hornberger K., Sipe JP., Arndt M. Theory of decoherence in a matter wave Talbot-Lau interferometer. Phys. Rev. A. 2004; 70, 053608.

[17] Nimmrichter S., Hornberger K. Theory of near-fieldmatter wave interference beyond the eikonal approximation. Phys. Rev. A. 2008; 78, 023612.

[18] Juffmann,T., Truppe S., Geyer P., Major AG., Deachapunya S., Ulbricht H., Arndt, M. Wave and particle in molecular interference lithography. Phys. Rev. Lett. 2009; 103, 263601.

[19] Berry M., Klein S. Integer, fractional and fractal Talbot effects, Journal of Modern Optics, 1996; 43(10) 2139-2164.

[20] Berry M., Marzoli L., Schleich W. Quantum carpets, carpets of light. Physics World, 2001; (6), 39-44.

[21] Couder Y., Fort E. Single-Particle Diffraction and Interference at a Macroscopic Scale. Phys. Rev. Lett. 2006; 97 (6), 154101(4).

[22] Stratonovich RL. Topics in the theory of random noise. N.Y.: Gordon and Breach; 1963.

[23] Gnedenko BV. Theory of Probability. N.Y.: Gordon & Breach; 1997.

[24] Ventzel AD. The course of stochastic processes theory. Moskow: Nauka (in Russian); 1975.

[25] Grosche C. An introduction into the Feynman path integral. http://arXiv.org/abs/hep-th/9302097 (accessed 20 February 1993).

[26] MacKenzie R. Path integral methods and applications. http://arXiv.org/abs/quant-ph/0004090 (accessed 20 April 2000).

[27] Juffmann T., Milic A., Müllneritsch M., Asenbaum P., Tsukernik A., Tüxen J., Mayor M., Cheshnovsky O., Arndt, M. Real-time single-molecule imaging of quantum interference. Nature Nanotechnology - Letter. 2012; 7, 297–300. http://www.nature.com/nnano/journal/v7/n5/full/nnano.2012.34.html?WT.ec_id=NNANO-201205 (accessed 25 March 2012)

[28] Sbitnev VI. Bohmian trajectories and the path integral paradigm - complexified Lagrangian mechanics. In: Pahlavani MR. (ed.) Theoretical Concepts of Quantum Mechanics. Rijeka: InTech; 2011 p313-340. http://www.intechopen.com/books/theoretical-concepts-of-quantum-mechanics/the-path-integral-paradigm-and-bohmian-trajectories-from-the-lagrangian-mechanics-to-complexified-on (accessed 2004-2012)

[29] Nairz O., Arndt M., Zeilinger A. Quantum interference experiments with large molecules. Am. J. Phys. 2003; 71 (4), 319-325.

[30] Sbitnev VI. Matter waves in the Talbot-Lau interferometry, http://arxiv.org/abs/1005.0890 (accessed 17 September 2010).

[31] Volokitin AI. Quantum drag and graphene. PRIRODA, 2011; 1153(9), 13-21.

[32] Mandel L., Wolf E. Optical coherence and quantum optics. Cambridge: Cambridge University Press; 1995.

[33] McMorran B.,Cronin AD. Model for partial coherence and wavefront curvature in grating interferometers. Phys. Rev. A. 2008; 78 (1), 013601(10).

[34] Gerlich S., Hackermüller L., Hornberger K., Stibor A., Ulbricht H., Gring M., Goldfarb F., Savas T., Müri M., Mayor M., Arndt, M. A Kapitza-Dirac-Talbot-Lau interferometer for highly polarizable molecules. Nature Physics. 2007; 3, 711-715.

[35] Bohm D. A suggested interpretation of the quantum theory in terms of "hidden variables", I & II. Physical Review, 1952; 85, 166-179 & 180-193.

[36] Madelung E. Quantentheorie in Hydrodynamischer form. Zts. f. Phys. 1926; 40, 322-326.

[37] Stanford Encyclopeia of Philosophy: Bohmian Mechanics. http://plato.stanford.edu/entries/qm-bohm/ (accessed 26 October 2001).

[38] Dirac PAM. The Lagrangian in Quantum Mechanics. Physikalische Zeitschrift der Sowjetunion. 1933; 3, 64-72.

[39] Dirac PAM. On the analogy between classical and quantum mechanics. Rev. Mod. Phys. 1945; 17 (2 & 3), 195-199.

Chapter 9

The Schwinger Action Principle and Its Applications to Quantum Mechanics

Paul Bracken

Additional information is available at the end of the chapter

1. Introduction

In physics it is generally of interest to understand the dynamics of a system. The way the dynamics is to be specified and studied invariably depends on the scale of the system, that is whether it is macroscopic or microscopic. The formal machinery with which the world is explained and understood depends at which of these two levels an experiment is conducted [1]. At the classical level, the dynamics of the system can be understood in terms of such things as trajectories in space or space-time.

In fact, classical mechanics can be formulated in terms of a principle of stationary action to obtain the Euler-Lagrange equations. To carry this out, an action functional has to be defined. It is written as S and given by

$$S[\mathbf{q}(t)] = \int_{t_1}^{t_2} L(q^i(t), \dot{q}^i(t), t)\, dt. \tag{1}$$

The action depends on the Lagrangian, written L in (1). It is to be emphasized that the action is a functional, which can be thought of as a function defined on a space of functions. For any given trajectory or path in space, the action works out to be a number, so S maps paths to real numbers.

One way to obtain equations of motion is by means of Hamilton's principle. Hamilton's principle states that the actual motion of a particle with Lagrangian L is such that the action functional is stationary. This means the action functional achieves a minimum or maximum value. To apply and use this principle, Stationary action must result in the Euler-Lagrange equations of motion. Conversely, if the Euler-Lagrange equations are imposed, the action functional should be stationary. As is well known, the Euler-Lagrange equations provide a system of second-order differential equations for the path. This in turn leads to other approaches to the same end.

As an illustration, the momentum canonically conjugate to the coordinate q^i is defined by

$$p_i = \frac{\partial L}{\partial \dot{q}^i}. \tag{2}$$

The dependence on the velocity components $\dot{q}^i$ can be eliminated in favor of the canonical momentum. This means that (2) must be solved for the $\dot{q}^i$ in terms of the q^i and p_i, and the inverse function theorem states this is possible if and only if $(\partial p_i / \partial \dot{q}^j) \neq 0$. Given a non-singular system all dependence on $\dot{q}^i$ can be eliminated by means of a Legendre transformation

$$H(\mathbf{q}, \mathbf{p}, t) = p_i \dot{q}^i - L(\mathbf{q}, \dot{\mathbf{q}}, t). \tag{3}$$

The Hamiltonian equations are obtained by considering the derivatives of the Hamiltonian with respect to q^i and p_i. The action in terms of H is written

$$S[\mathbf{q}, \mathbf{p}] = \int_{t_1}^{t_2} dt\, [p_i \dot{q}^i - H(\mathbf{q}, \mathbf{p}, t)], \tag{4}$$

such that the action now depends on both $\mathbf{q}$ and $\mathbf{p}$. Using this the principle of stationary action can then be modified so that Hamilton's equations result.

In passing to the quantum domain, the concept of path or trajectory is of less importance, largely because it has no meaning. In quantum physics it can no longer be assumed that the interaction between system and measuring device can be made arbitrarily small and that there are no restrictions on what measurements can be made on the system either in terms of type or in accuracy. Both these assumptions tend to break down at the scales of interest here, and one is much more interested in states and observables, which replace the classical idea of a trajectory with well defined properties [2,3].

Now let us follow Dirac and consider possible measurements on a system as observables. Suppose A_i denotes any observable with a_i as a possible outcome of any measurement of this observable. As much information as possible can be extracted with regard to a quantum mechanical system by measuring some set of observables $\{A_i\}_{i=1}^n$ without restriction. Thus, the observables should be mutually compatible because the measurement of any observable in the set does not affect the measurement of any of the other observables. The most information that can be assembled about a system is the collection of numbers $\{a_i\}_{i=1}^n$, which are possible values for the set of mutually compatible observables, and this set specifies the state of the system.

2. Schwinger's action principle

In Dirac's view of quantum mechanics, the state of a system is associated with a vector in a complex vector space V. The knowledge of the values for a complete set of mutually compatible observables gives the most information about a state. It can then be assumed that $\{|a\rangle\}$, where $|a\rangle = |a_1, \cdots, a_n\rangle$, the set of all possible states, forms a basis for V. Associated with any vector space V is the dual space V^* whose elements are referred to as bras in Dirac's terminology. A basis for V^* is denoted by $\{\langle a|\}$ and is dual to $\{|a\rangle\}$. The quantities satisfy

$$\langle a'|a''\rangle = \delta(a', a''), \tag{5}$$

where $\delta(a', a'')$ is the Kronecker delta if a' is a discrete set, and the Dirac delta if it is continuous. The choice of a complete set of mutually compatible observables is not unique. Suppose $\{|b\rangle\}$ also provides a basis for V relevant to another set of mutually compatible observables $B_1, B_2, \cdots$. Since $\{|a\rangle\}$ and $\{|b\rangle\}$ are both bases for V, this means that one set of basis vectors can be expressed in terms of the other set

$$|b\rangle = \sum_a |a\rangle\langle a|b\rangle, \tag{6}$$

and the $\langle a|b\rangle$ coefficients are some set of complex numbers, so that $\langle b|a\rangle^* = \langle a|b\rangle$. If there is a third basis for V provided by $\{|c\rangle\}$, then these complex numbers are related by means of

$$\langle a|c\rangle = \sum_b \langle a|b\rangle\langle b|c\rangle. \tag{7}$$

Schwinger's action principle is based on the types of transformation properties of the transformation functions which can be constructed from this basis set [4].

Suppose the transformation function is subjected to, as Schwinger asserted, any conceivable infinitesimal variation. Then, by performing an arbitrary variation of (7), it follows that

$$\delta\langle a|c\rangle = \sum_b [(\delta\langle a|b\rangle)\langle b|c\rangle + \langle a|b\rangle(\delta\langle b|c\rangle)], \tag{8}$$

and moreover,

$$\delta\langle a|b\rangle = \delta\langle b|a\rangle^*. \tag{9}$$

Now a new operator can be defined which evaluates this actual variation when it is placed between the relevant state vectors. Define this operator to be δW_{ab}, so that it has the following action between states

$$\delta\langle a|b\rangle = \frac{i}{\hbar}\langle a|\delta W_{ab}|b\rangle. \tag{10}$$

Including the factor of $\hbar$ gives the operator units of action. Using (10) in (8) produces,

$$\langle a|\delta W_{ac}|c\rangle = \sum_b [\langle a|\delta W_{ab}|b\rangle\langle b|c\rangle + \langle a|b\rangle\langle b|\delta W_{bc}|c\rangle] = \langle a|\delta W_{ab} + \delta W_{bc}|c\rangle, \tag{11}$$

using the completeness relation (7). Now it follows from (11) that

$$\delta W_{ac} = \delta W_{ab} + \delta W_{bc}. \tag{12}$$

In the case in which the a and b descriptions are identified then using $\delta\langle a|a'\rangle = 0$, there results

$$\delta W_{aa} = 0. \tag{13}$$

Identifying the a and c pictures in (11) gives,

$$\delta W_{ba} = -\delta W_{ab}. \tag{14}$$

The complex conjugate of (10) with a and b descriptions reversed implies

$$-\frac{i}{\hbar}\langle b|\delta W_{ba}|a\rangle^* = \frac{i}{\hbar}\langle a|\delta W_{ab}|b\rangle. \tag{15}$$

This has the equivalent form

$$-\langle a|\delta W_{ba}^{\dagger}|b\rangle = \langle a|\delta W_{ab}|b\rangle. \tag{16}$$

Using (14), this yields the property

$$\delta W_{ba}^{\dagger} = \delta W_{ab}. \tag{17}$$

The basic properties of the transformation function and the definition (10) have produced all of these additional properties [5,6].

In the Heisenberg picture, the basis kets become time dependent. The transformation function relates states which are eigenstates of different complete sets of commuting observables at different times. Instead of using different letters, different subscripts $1, 2$ can be used to denote different complete sets of commuting observables. In this event, (10) takes the form

$$\delta\langle a_2', t_2|a_1', t_1\rangle = \frac{i}{\hbar}\langle a_2', t_2|\delta W_{21}|a_1', t_1\rangle. \tag{18}$$

The assumption at the heart of this approach is that the operator δW_{21} in (18) is obtained from the variation of a single operator W_{21}. This is referred to as the action operator.

To adapt the results of the previous notation to the case with subscripts, we should have

$$W_{31} = W_{32} + W_{21}, \qquad W_{11} = 0, \qquad W_{21} = -W_{12} = W_{21}^{\dagger}. \tag{19}$$

At this point, a correspondence between the Schwinger action principle and the classical principle of stationary action can be made. Suppose the members of a complete set of commuting observables A_1 which have eigenvectors $|a_1, t\rangle$ in the Heisenberg picture are deformed in some fashion at time t_1. For example, take the alteration in the observables to correspond to a unitary transformation $A \to U^{\dagger}AU$ such that $U^{\dagger} = U^{-1}$. To remain eigenstates of the transformed operator, it must be that states transform as $|a\rangle \rightarrow U^{\dagger}|a\rangle$. Thinking of the transformation as being infinitesimal in nature, the operator U can be written $U = I + \frac{i}{\hbar}G$, where G is Hermitean. It is then possible to define a variation

$$\delta|a_1, t_1\rangle = -\frac{i}{\hbar}G_1|a_1, t_1\rangle. \tag{20}$$

Here operator G is a Hermitian operator and depends only on the observables A_1 at the time t_1. Similarly, if observables A_2 are altered at t_2, it is the case that

$$\delta\langle a_2, t_2| = \frac{i}{\hbar}\langle a_2, t_2|G_2, \tag{21}$$

and the operator G_2 depends only on observables A_2 at time t_2. If both sets A_1, A_2 are altered infinitesimally, then the change in the transformation function is given by

$$\delta\langle a_2, t_2|a_1, t_1\rangle = \frac{i}{\hbar}\langle a_2, t_2|G_2 - G_1|a_1, t_1\rangle. \tag{22}$$

Comparing this with (18), it is concluded that

$$\delta W_{21} = G_2 - G_1. \tag{23}$$

If the time evolution from state $|a_1, t_1\rangle$ to $|a_2, t_2\rangle$ can be thought of as occurring continuously in time, then W_{21} can be expressed as

$$W_{21} = \int_{t_1}^{t_2} L(t)\, dt, \tag{24}$$

where $L(t)$ is called the Lagrange operator. As a consequence of (23), it follows that if the dynamical variables which enter $L(t)$ are altered during an arbitrary infinitesimal change between t_1 and t_2, then it must be that

$$\delta W_{21} = 0. \tag{25}$$

The operator equations of motion are implied in this result. The usual form for the Lagrange operator is

$$L(t) = \frac{1}{2}(p_i \dot{x}^i + \dot{x}^i p_i) - H(\mathbf{x}, \mathbf{p}, t). \tag{26}$$

The first term has been symmetrized to give a Hermitian $L(t)$, as the operators $\dot{x}^i$ and p_i do not as usual commute. It is assumed the Hamiltonian H is a Hermitian operator. The action operator (24) is used to calculate the variation δW_{21}. In order to vary the endpoints t_1, t_2, we follow Schwinger exactly and change the variable of integration from t to τ such that $t = t(\tau)$. This allows for the variation of the functional dependence of t to depend on τ with the variable of integration τ held fixed. Then W_{21} takes the form,

$$W_{21} = \int_{\tau_1}^{\tau_2} d\tau \, [\frac{1}{2}(P_i \frac{dy^i}{d\tau} + \frac{dy^i}{d\tau} P_i) - \tilde{H}(\mathbf{y}, \mathbf{P}, \tau) \, \frac{dt}{d\tau}], \tag{27}$$

where in (27),

$$\tilde{H}(\mathbf{y}, \mathbf{P}, \tau) = H(\mathbf{x}, \mathbf{p}, t). \tag{28}$$

Thus $y^i(\tau) = x^i(t)$ and $P_i(\tau) = p_i(t)$ when the transformation $t = t(\tau)$ is implemented. Evaluating the infinitesimal variation of (27), it is found that

$$\delta W_{21} = \int_{\tau_1}^{\tau_2} d\tau [\frac{1}{2} \delta P_i \frac{dy^i}{d\tau} + \frac{1}{2} P_i \delta(\frac{dy^i}{d\tau}) + \frac{1}{2} \delta(\frac{dy^i}{d\tau}) P_i + \frac{1}{2} \frac{dy^i}{d\tau} \delta P_i - \delta \tilde{H} \frac{dt}{d\tau} - \tilde{H} \delta(\frac{dt}{d\tau})]. \tag{29}$$

Moving the operator δ through the derivative, this becomes

$$\delta W_{21} = \int_{t_1}^{t_2} d\tau \, \{\frac{1}{2}(\delta P_i \frac{dy^i}{d\tau} + \frac{dy^i}{d\tau} \delta P_i - \frac{dP_i}{d\tau} \delta y^i - \delta y^i \frac{dP_i}{d\tau}) - \delta \tilde{H} \frac{dt}{d\tau} + \frac{d\tilde{H}}{d\tau} \delta t$$

$$+ \frac{d}{d\tau} [\frac{1}{2}(P_i \delta x^i + \delta x^i P_i) - \tilde{H} \delta t]\}. \tag{30}$$

No assumptions with regard to the commutation properties of the variations with the dynamical variables have been made yet. It may be assumed that the variations are multiples of the identity operator, which commutes with everything. After returning to the variable t in the integral in δW_{21}, the result is

$$\delta W_{21} = \int_{t_1}^{t_2} dt (\delta p_i \dot{x}^i - \dot{p}_i \delta x^i + \frac{dH}{dt} \delta t - \delta H) + G_2 - G_1. \tag{31}$$

Here G_1 and G_2 denote the quantity

$$G = p_i \, \delta x^i - H \, \delta t, \tag{32}$$

when it is evaluated at the two endpoints $t = t_1$ and $t = t_2$. If we define,

$$\delta H = \delta x^i \frac{\partial H}{\partial x^i} + \delta p_i \frac{\partial H}{\partial p_i} + \delta t \frac{\partial H}{\partial t}, \tag{33}$$

then δW_{21} can be expressed in the form,

$$\delta W_{21} = \int_{t_1}^{t_2} dt \, \{\delta p_i(\dot{x}^i - \frac{\partial H}{\partial p_i}) - \delta x^i(\dot{p}_i + \frac{\partial H}{\partial x^i}) + (\frac{dH}{dt} - \frac{\partial H}{\partial t})\delta t\} + G_2 - G_1. \quad (34)$$

Taking the variations with endpoints fixed, it follows that $G_1 = G_2 = 0$. Consequently, the operator equations of motion which follow from equating δW_{21} in (34) to zero are then

$$\dot{x}^i = \frac{\partial H}{\partial p_i}, \qquad \dot{p}_i = -\frac{\partial H}{\partial x^i}, \qquad \frac{dH}{dt} = \frac{\partial H}{\partial t}. \quad (35)$$

The results produced in this way are exactly of the form of the classical Hamilton equations of motion, and the derivatives in the first two equations of (35) are with respect to operators.

3. Commutation relations

Let B represent any observable and consider the matrix element $\langle a|B|a'\rangle$. If the variables A are subjected to a unitary transformation $A \to \bar{A} = UAU^\dagger$, where U is a unitary operator, then the eigenstates $|a\rangle$ are transformed into $|\bar{a}\rangle = U|a\rangle$ having the eigenvalue a. Define the operator $\bar{B}$ to be

$$\bar{B} = UBU^\dagger. \quad (36)$$

These operators have the property that

$$\langle \bar{a}|\bar{B}|\bar{a}'\rangle = \langle a|B|a'\rangle. \quad (37)$$

Let U now be an infinitesimal unitary transformation which can be expressed in the form

$$U = I - \frac{i}{\hbar}G_a. \quad (38)$$

In (38), G_a is a Hermitian quantity and can depend on observables A. Consequently, if δ_{G_a} is the change produced by the canonical transformation whose generator is G_a, then

$$\delta_{G_a}|a\rangle = |\bar{a}\rangle - |a\rangle = -\frac{i}{\hbar}G_a|a\rangle, \quad (39)$$

and $\bar{B}$ is given by

$$\bar{B} = (I - \frac{i}{\hbar}G_a)B(I + \frac{i}{\hbar}G_a) = B - \frac{i}{\hbar}[G_a, B]. \quad (40)$$

The change in the matrix element $\langle a|B|a'\rangle$ can then be considered to be entirely due to the change in the state vector, with B held fixed. To first order in the operator G_a, there results

$$\delta_{G_a}\langle a|B|a'\rangle = \langle \bar{a}|B|\bar{a}'\rangle - \langle a|B|a'\rangle = \langle a|\frac{i}{\hbar}[G_a, B]|a'\rangle. \tag{41}$$

Alternatively, this can be approached by taking the change in $\langle a|B|a'\rangle$ to be due to a change in the operator B, but with the states held fixed. Define then the change in B as $\delta_{G_a}B$ such that

$$\delta_{G_a}\langle a|B|a'\rangle = \langle a|\delta_{G_a}B|a'\rangle. \tag{42}$$

If the result obtained for the change in $\langle a|B|a'\rangle$ is not to depend on which of these approaches is taken, by comparing (41) and (42) it is concluded that

$$\delta_{G_a}B = \frac{i}{\hbar}[G_a, B]. \tag{43}$$

As an example, let the change in the operators A at times t_1 and t_2 be due to a change in time $t \to t + \delta t$ with the x^i fixed at times t_1 and t_2 so that the generator for this tranformation is given by $G(t) = -H(t)\delta t$ and (43) in this case is

$$\delta_{G_a}B = \frac{i}{\hbar}[B, H]\delta t, \tag{44}$$

where B and H are evaluated at the same time. It is only the change in B resulting from a change in operators which is considered now. Specifically, if $B = B[A(t), t]$, then $\bar{B} = B[A(t + \delta t), t]$ and so

$$\delta B = -\delta t(\frac{dB}{dt} - \frac{\delta B}{\delta t}). \tag{45}$$

Using (45) in (44), a result in agreement with the Heisenberg picture equation of motion results,

$$\frac{dB}{dt} = \frac{\partial B}{\partial t} + \frac{i}{\hbar}[H, B]. \tag{46}$$

The Heisenberg equation of motion has resulted by this process. If B is taken to be the operator $B = x^i$, then (46) becomes,

$$\dot{x}^i = \frac{i}{\hbar}[H, x^i]. \tag{47}$$

If $B = p_i$ is used in (46), there results

$$\dot{p}_i = \frac{i}{\hbar}[H, p_i]. \tag{48}$$

The general equations of motion have been produced [7].

The canonical commutation relations are also a consequence of the action principle. To see this, first fix δt at the times t_1 and t_2 but permit δx^i to vary. Then (32) gives the generator of this transformation

$$G = p_i \delta x^i. \tag{49}$$

Then for any operator B, equation (43) gives

$$\delta B = \frac{i}{\hbar} \delta x^i \, [p_i, B]. \tag{50}$$

Putting $B = x^i$ and $B = p_j$ respectively in (50) leads to the following pair of commutation relations

$$[x^i, p_j] = i\hbar \, \delta^i_j, \qquad [p_i, p_j] = 0. \tag{51}$$

If B has a dependence on x, then $\delta B = B[x(t)] - B[x(t) - \delta x(t)]$ implies

$$[p_i, B] = -\frac{i}{\hbar} \frac{\partial}{\partial x^i} B. \tag{52}$$

In order to obtain the commutator $[x^i, x^j]$, the freedom of altering the Lagrangian operator by the addition of a total time derivative can be used. Suppose that

$$F = \frac{1}{2}(p_i x^i + x^i p_i), \tag{53}$$

which satisfies $F^\dagger = F$, and consequently

$$\delta F = \frac{1}{2}(\delta p_i x^i + p_i \delta x^i + \delta x^i p_i + x^i \delta p_i) = \delta p_i x^i + p_i \delta x^i. \tag{54}$$

Then $\bar{G}$ is calculated to be

$$\bar{G} = G - \delta F = -\delta p_i x^i - H \, \delta t. \tag{55}$$

Taking $\delta t = 0$ so that G_p can be defined as $G_p = -\delta p_i \, x^i$, it follows that

$$\delta B = -\frac{i}{\hbar} \delta p_i \, [x^i, B], \tag{56}$$

for any operator B. Taking $B = p_j$, the bracket $[x^i, p_j] = i\hbar \delta^i_j$ is obtained from (56), and when $B = x^j$, using the independence of δx^i and δp_i there results the bracket

$$[x^i, x^j] = 0. \tag{57}$$

It has been shown that the set of canonical commutation relations can be obtained from this action principle.

Consider a matrix element $\langle a|F(A,B)|b\rangle$ where A and B each represent a complete set of mutually compatible observables and $F(A,B)$ is some function, and it is not assumed the observables from the two different sets commute with each other. If the commutator $[B,A]$ is known, it is always possible to order the operators in $F(A,B)$ so all A terms are to the left of B, which allows the matrix element to be evaluated. Let

$$\mathcal{F}(A,B) = F(A,B) \tag{58}$$

denote the operator where the commutation relation for $[A,B]$ has been used to move all occurrences of A in $F(A,B)$ to the left of all B, and (58) is said to be well-ordered

$$\langle a|F(A,B)|b\rangle = \mathcal{F}(a,b)\langle a|b\rangle. \tag{59}$$

The matrix element of $F(A,B)$ is directly related to the transformation function $\langle a|b\rangle$.

The idea of well-ordering operators can be used in the action principle. Define

$$\delta\mathcal{W}_{21} = \delta W_{21}. \tag{60}$$

be the well-ordered form of δW_{21}. Then

$$\delta\langle a_2, t_2|a_1, t_1\rangle = \frac{i}{\hbar}\delta\mathcal{W}_{21}\langle a_2, t_2|a_1, t_1\rangle, \tag{61}$$

where $\delta\mathcal{W}_{21}$ denotes the replacement of all operators with their eigenvalues. Equation (61) can be integrated to yield

$$\langle a_2, t_2|a_1, t_1\rangle = \exp(\frac{i}{\hbar}\mathcal{W}_{21}). \tag{62}$$

Using (23) and (32) in the action principle,

$$\delta\langle a_2, t_2|a_1, t_1\rangle = \frac{i}{\hbar}\delta x^i(t_2)\langle a_2, t_2|p(t_2)|a_1, t_1\rangle - \frac{i}{\hbar}\delta t_2\langle a_2, t_2|H(t_2)|a_1, t_1\rangle$$

$$-\frac{i}{\hbar}\delta x^i(t_1)\langle a_2, t_2|p_i(t_1)|a_1, t_1\rangle + \frac{i}{\hbar}\delta t_1\langle a_2, t_2|H(t_1)|a_1, t_1\rangle. \tag{63}$$

Since $\delta \mathcal{W}_{21}$ is well-ordered, it is possible to write

$$\delta \langle a_2, t_2 | a_1, t_1 \rangle = \frac{i}{\hbar} \langle a_2, t_2 | \delta \mathcal{W}_{21} | a_1, t_1 \rangle. \tag{64}$$

Comparing this with (63), it must be that

$$\frac{\partial \mathcal{W}_{21}}{\partial x^i(t_2)} = p_i(t_2), \quad \frac{\partial \mathcal{W}_{21}}{\partial t_2} = -H(t_2), \quad \frac{\partial \mathcal{W}_{21}}{\partial x^i(t_1)} = -p_i(t_1), \quad \frac{\partial \mathcal{W}_{21}}{\partial t_1} = H(t_1). \tag{65}$$

If we consider a matrix element $\langle x_2, t_2 | x_1, t_1 \rangle$, which could be regarded as the propagator, we then have a form of (62),

$$\langle \mathbf{x}_2, t_2 | \mathbf{x}_1, t_1 \rangle = \exp(\frac{i}{\hbar} \mathcal{W}_{21}). \tag{66}$$

The arbitrary integration constant is determined by requiring that $\langle \mathbf{x}_2, t_2 | \mathbf{x}_1, t_1 \rangle = \delta(\mathbf{x}_2, \mathbf{x}_1)$. Any other transformation function can be recovered from the propagator equation

$$\langle a_2, t_2 | a_1, t_1 \rangle = \int d^n x_2 \int d^n x_1 \, \langle a_2, t_2 | \mathbf{x}_2, t_2 \rangle \langle \mathbf{x}_2, t_2 | \mathbf{x}_1, t_1 \rangle \langle \mathbf{x}_1 t_1 | a_1, t_1 \rangle, \tag{67}$$

which follows from the completeness relation. Another way of formulating transition amplitudes will be seen when the path integral approach is formulated.

4. Action principle adapted to case of quantum fields

The Schwinger action principle, much like the Feynman path integral, concentrates on the transition amplitude between two quantum states. The action principle will be formulated here for a local field theory. Classically, a local field is a function which depends only on a single spacetime point, rather than on an extended region of spacetime. The theory can be quantized by replacing the classical fields $\phi^i(x)$ with field operators $\varphi^i(x)$.

Let Σ denote a spacelike hypersurface. This means that any two points on Σ have a spacelike separation and consequently must be causally disconnected. As a consequence of this, the values of the field at different points of the surface Σ must be independent. If x_1 and x_2 are two spacetime points of Σ, then

$$[\varphi^i(x_1), \varphi^j(x_2)] = 0. \tag{68}$$

This follows since it must be the case that a measurement at x_1 must not influence one at x_2. A fundamental assumption of local field theory is that a complete set of commuting observables can be constructed based on the fields and their derivatives on the surface Σ. Let τ denote such a complete set of commuting observables on Σ such that τ' represents the eigenvalues of the observables. A quantum state is then denoted by

$$|\tau', \Sigma\rangle. \tag{69}$$

Since causality properties are important in field theory, the surface Σ is written explicitly in the ket. The Heisenberg picture has been adopted here. The states are then time-independent with the time dependence located in the operators. This is necessary for manifest covariance, since the time and space arguments of the field are not treated differently.

Suppose that Σ_1 and Σ_2 are two spacelike hypersurfaces such that all points of Σ_2 are to the future of Σ_1. Let τ_1 be a complete set of commuting observables defined on Σ_1, and τ_2 a complete set of observables defined on Σ_2, such that these observables have the same eigenvalue spectrum. Then τ_1 and τ_2 should be related by a unitary transformation

$$\tau_2 = U_{12}\tau_1 U_{12}^{-1}. \tag{70}$$

The eigenstates are related by

$$|\tau_2', \Sigma_2\rangle = U_{12}|\tau_1', \Sigma_1\rangle. \tag{71}$$

In (71), U_{12} is a unitary operator giving the evolution of the state in the spacetime between the two spacelike hypersurfaces. The transition amplitude is defined by

$$\langle\tau_2', \Sigma_2|\tau_1', \Sigma_1\rangle = \langle\tau_1', \Sigma_1|U_{12}^{-1}|\tau_1', \Sigma_1\rangle. \tag{72}$$

The unitary operator U_{12} depends on a number of details of the quantum system, namely, the choice made for the commuting observables τ, and the spacelike hypersurfaces Σ_1 and Σ_2. A change in any of these quantities will induce a change in the transformation function according to

$$\delta\langle\tau_2', \Sigma_2|\tau_1', \Sigma_1\rangle = \langle\tau_1', \Sigma_1|\delta U_{12}^{-1}|\tau_1', \Sigma_1\rangle. \tag{73}$$

The unitary operator U_{12} can be expressed in the form

$$U_{12} = \exp(-\frac{i}{\hbar}S_{12}), \tag{74}$$

in which $S_{12}^{\dagger} = S_{12}$ is a Hermitian operator. Moreover, beginning with $U_{12}U_{12}^{-1} = I$ and using (74), it is found that

$$\delta U_{12}^{-1} = -U_{12}^{-1}\delta U_{12}U_{12}^{-1} = \frac{i}{\hbar}U_{12}^{-1}\delta S_{12}. \tag{75}$$

The change in the transformation function can be written in terms of the operator in (74) as

$$\delta\langle\tau_2', \Sigma_2|\tau_1', \Sigma_1\rangle = \frac{i}{\hbar}\langle\tau_2', \Sigma_2|\delta S_{12}|\tau_1', \Sigma_1\rangle. \tag{76}$$

Equation (76) can be regarded as a definition of δS_{12}. In order that δS_{12} be consistent with the basic requirement that

$$\langle\tau_2', \Sigma_2|\tau_1', \Sigma_1\rangle^* = \langle\tau_1', \Sigma_1|\tau_2', \Sigma_2\rangle,$$

it must be that δS_{12} is Hermitean. If Σ_3 is a spacelike hypersurface, all of whose points lie to the future of those on Σ_2, the basic law for composition of probability amplitudes is

$$\langle \tau_3', \Sigma_3 | \tau_1', \Sigma_1 \rangle = \sum_{\tau_2'} \langle \tau_3', \Sigma_3 | \tau_2', \Sigma_2 \rangle \langle \tau_2', \Sigma_2 | \tau_1', \Sigma_1 \rangle. \tag{77}$$

Varying both sides of the expression in (77),

$$\langle \tau_3', \Sigma_3 | \delta S_{13} | \tau_1', \Sigma_1 \rangle = \sum_{\tau_2'} \{ \langle \tau_3', \Sigma_3 | \delta S_{23} | \tau_2', \Sigma_2 \rangle \langle \tau_2', \Sigma_2 | \tau_1', \Sigma_1 \rangle + \langle \tau_3', \Sigma_3 | \tau_2', \Sigma_2 \rangle \langle \tau_2', \Sigma_2 | \delta S_{12} | \tau_1', \Sigma_1 \rangle$$

$$= \langle \tau_3', \Sigma_3 | \delta S_{23} + \delta S_{12} | \tau_1', \Sigma_1 \rangle. \tag{78}$$

Comparing both ends of the result in (78), it follows that

$$\delta S_{13} = \delta S_{23} + \delta S_{12}. \tag{79}$$

In the limit $\Sigma_2 \to \Sigma_1$, it must be that

$$\delta S_{12} = 0.$$

In the limit in which $\Sigma_3 \to \Sigma_1$, it follows that

$$\delta S_{21} = -\delta S_{12}.$$

If the operators in τ_1 and τ_2 undergo infinitesimal, unitary transformations on the hypersurfaces Σ_1 and Σ_2, respectively, and only on these two hypersurfaces, then the change in the transformation function has the form

$$\delta \langle \tau_2', \Sigma_2 | \tau_1', \Sigma_1 \rangle = \frac{i}{\hbar} \langle \tau_2', \Sigma_2 | F_2 - F_1 | \tau_1', \Sigma_1 \rangle. \tag{80}$$

Here F_1 and F_2 are Hermitean operators constructed from a knowledge of the fields and their derivatives on Σ_1 and Σ_2. The result (80) is of the form (76) provided that

$$\delta S_{12} = F_2 - F_1. \tag{81}$$

A generator F of this type on a spacelike hypersurface Σ should be expressible in the form,

$$F = \int_{\Sigma} d\sigma_x n^{\mu} F_{\mu}(x), \tag{82}$$

and $d\sigma_x$ is the area element on Σ, n^{μ} is the outward unit normal to Σ, and $F_{\mu}(x)$ may be put together based on a knowledge of the fields on the surface Σ.

The points of Σ are all spacelike separated, hence independent and so the result follows by adding up all of these independent contributions. Applying (82) to Σ_1 and Σ_2 assuming that $F_\mu(x)$ is defined throughout the spacetime region bounded by these two surfaces, δS_{12} can be expressed as

$$\delta_{12}S = \int_{\Sigma_2} d\sigma_x\, n^\mu F_\mu(x) - \int_{\Sigma_1} d\sigma_x\, n^\mu F_\mu(x) = \int_{\Omega_{12}} dv_x\, \nabla^\mu F_\mu(x). \tag{83}$$

In (83), Ω_{12} is the spacetime region bounded by Σ_1 and Σ_2, and dv_x is the invariant volume. This assumes that the operators are changes only on Σ_1 and Σ_2. However, suppose the operators are changed in the spacetime region between Σ_1 and Σ_2. Assume once more that δS_{12} can be expressed as a volume integral as,

$$\delta S_{12} = \int_{\Omega_{12}} dv_x\, \delta\mathcal{L}(x) \tag{84}$$

for some $\delta\mathcal{L}(x)$. Combining these two types of variation yields

$$\delta S_{12} = F_2 - F_1 + \int_{\Omega_{12}} dv_x\, \delta\mathcal{L}(x) = \int_{\Omega_{12}} dv_x[\delta\mathcal{L}(x) + \nabla^\mu F_\mu(x)]. \tag{85}$$

It is an important result of (85) that altering $\delta\mathcal{L}$ by the addition of the divergence of a vector field will result in a unitary transformation of the states on Σ_1 and Σ_2.

To summarize, the fundamental assumption of the Schwinger action principle is that δS_{12} may be obtained from a variation of

$$S_{12} = \int_{\Omega_{12}} dv_x\, \mathcal{L}(x), \tag{86}$$

where $\mathcal{L}(x)$ is a Lagrangian density. The density depends on the fields and their derivatives at a single spacetime point. Since δS_{12} is required to be Hermitian, S_{12} must be Hermitian and similarly, so must the Lagrangian density.

5. Correspondence with Feynman path integrals

Suppose a classical theory described by the action $S[\varphi]$ is altered by coupling the field to an external source $J_i = J_I(x)$. By external it is meant that it has no dependence on the field φ^i, and i stands for (I, x). For example, $F_{,i}[\varphi]\sigma^i$ is an abbreviation for

$$F_{,i}[\varphi]\sigma^i = \int d^n x'\, \frac{\delta F[\varphi(x)]}{\delta\varphi^I(x')}\, \sigma^i(x').$$

The idea of introducing external sources originates with Schwinger. As he states, causality and space-time uniformity are the creative principles of source theory. Uniformity in space-time also has a complementary momentum-energy implication, illustrated by the source idea [5,8]. Not only for the special balance of energy and momentum involved in the emission or absorption of a single particle is the source defined and meaningful. Given a sufficient excess of energy over momentum, or an excess of mass, several particles can be emitted or absorbed. For example, consider the emission of a pair of charged particles by an extended photon source. This process is represented as the conversion of a virtual photon into a pair of real particles. In ordinary scattering, particle-particle scattering, the particles persist while exchanging a space-like virtual photon. Another is an annihilation of the particle-antiparticle pair, producing a time-like virtual photon, which quickly decays back into particles.

Let us choose a simple scalar theory

$$S_J[\varphi] = S[\varphi] + \int J_I(x)\varphi^I(x)\, d^n x, \tag{87}$$

and try to establish a connection between the action principle introduced here and a path integral picture [9]. The transition amplitude $\langle \tau_2', \Sigma_2 | \tau_1', \Sigma_1 \rangle$ discussed in the previous section may be regarded as a functional of the source term $J_I(x)$. This will be denoted explicitly by writing $\langle \tau_2', \Sigma_2 | \tau_1', \Sigma_1 \rangle_J$. If the Schwinger action principle is applied to this modified theory

$$\delta \langle \tau_2', \Sigma_2 | \tau_1', \Sigma_1 \rangle_J = \frac{i}{\hbar} \langle \tau_2', \Sigma_2 | \delta S_J | \tau_1', \Sigma_1 \rangle_J. \tag{88}$$

The classical field $\varphi^I(x)$ has been replaced by the operator $\phi^I(x)$. Moreover, in addition to its previous meaning, δ now includes a possible change in the source. By considering the variation to be with respect to the dynamical variables which are held fixed on Σ_1 and Σ_2, the operator field equations are obtained as

$$\frac{\delta S[\phi]}{\delta \phi^I(x)} + J_I(x) = 0. \tag{89}$$

Assume that the variation in (88) is one in which the dynamical variables are held fixed and only the source is altered. Since the source enters in the simple way given in (87),

$$\delta \langle \tau_2', \Sigma_2 | \tau_1', \Sigma_1 \rangle = \frac{i}{\hbar} \int \delta J_I(x) \langle \tau_2', \Sigma_2 | \phi^I(x) | \tau_1', \Sigma_1 \rangle_J \, d^n x. \tag{90}$$

The result in (90) may be rewritten in the equivariant form,

$$\frac{\delta \langle \tau_2', \Sigma_2 | \tau_1', \Sigma_1 \rangle}{\delta J_i(x)} = \frac{i}{\hbar} \langle \tau_2', \Sigma_2 | \phi^i(x) | \tau_1', \Sigma_1 \rangle [J]. \tag{91}$$

To simplify this, it can be written in the alternate form,

$$\frac{\delta \langle 2|1 \rangle}{\delta J_i} = \frac{i}{\hbar} \langle 2 | \phi^i | 1 \rangle [J]. \tag{92}$$

An abbreviated notation for the initial and final state has been introduced. This result can be varied with respect to the source which gives

$$\delta\frac{\delta\langle 2|1\rangle[J]}{\delta J_i} = \frac{i}{\hbar}\delta\langle 2|\phi^i|1\rangle[J]. \tag{93}$$

To evaluate (93), a spatial hypersurface Σ' is introduced which resides to the future of Σ_1 and the past of Σ_2 and contains the spacetime point on which the $\phi^i(x)$ depend. Any source variation can be represented as the sum of a variation which vanishes to the future of Σ', but is non-zero to the past, and one which vanishes to the past of Σ' but is nonzero to the future. Consider the case where δJ_i vanishes to the future. In this event, any amplitude of the form $\langle 2|\phi^i|\tau'\rangle[J]$, where $|\tau'\rangle$ represents a state on Σ', can not be affected by the variation of the source since δJ_i will vanish to the future of Σ'. By using the completeness relation

$$\langle 2|\phi^i|1\rangle[J] = \sum_{\tau'}\langle 2|\phi^i|\tau'\rangle\langle\tau'|1\rangle[J],$$

it follows that the right-hand side of (93) may be reexpressed with the use of

$$\delta\langle 2|\phi^i|1\rangle[J] = \sum_{\tau'}\langle 2|\phi^i|\tau'\rangle\delta\langle\tau'|1\rangle[J]. \tag{94}$$

The Schwinger action principle then implies that

$$\delta\langle\tau'|1\rangle[J] = \frac{i}{\hbar}\delta J_k\langle\tau'|\phi^k|1\rangle[J]. \tag{95}$$

Substituting (95) into (94) leads to the conclusion that

$$\delta\langle 2|\phi^i|1\rangle[J] = \frac{i}{\hbar}\delta J_j\sum_{\tau'}\langle 2|\phi^i|\tau'\rangle\langle\tau'|\phi^j|1\rangle[J] = \frac{i}{\hbar}\delta J_j\,\langle 2|\phi^i\phi^j|1\rangle[J]. \tag{96}$$

Since it can be said that δJ_j vanishes to the future of Σ', which contains the spacetime point of ϕ^i, the spacetime point of ϕ^j must lie to the past of the former.

Consider the case in which δJ_j vanishes to the past of Σ'. A similar argument yields the same conclusion as (96), but with ϕ^j to the left of ϕ^i. Combining this set of results produces the following conclusion

$$\frac{\delta\langle 2|\phi^i|1\rangle}{\delta J_j} = \frac{i}{\hbar}\langle 2|T(\phi^i\phi^j)|1\rangle[J]. \tag{97}$$

In (97), T is the chronological, or time, ordering operator, which orders any product of fields in the sequence of increasing time, with those furthest to the past to the very right.

Differentiating (91) and using the result (97), we get

$$\frac{\delta^2\langle 2|1\rangle[J]}{\delta J_i \delta J_j} = (\frac{i}{\hbar})\langle 2|T(\phi^i\phi^j)|1\rangle[J]. \tag{98}$$

This can be generalized, and omitting details,

$$\frac{\delta^n\langle 2|1\rangle[J]}{\delta J_{i_1}\cdots\delta J_{i_n}} = (\frac{i}{\hbar})^n\langle 2|T(\phi^{i_1}\cdots\phi^{i_n})|1\rangle[J]. \tag{99}$$

The amplitude $\langle 2|1\rangle[J]$ may be defined by a Taylor expansion about $J_i = 0$, so using the previous result

$$\langle 2|1\rangle[J] = \sum_{n=0}^{\infty}\frac{1}{n!}(\frac{i}{\hbar})^n J_{i_1}\cdots J_{i_n}\langle 2|T(\phi^{i_1}\cdots\phi^{i_n})|1\rangle[J=0]. \tag{100}$$

The series may be formally summed to yield

$$\langle 2|1\rangle[J] = \langle 2|T(\exp(\frac{i}{\hbar}J_i\phi^i))|1\rangle[J=0] \tag{101}$$

and J_i is set to zero everywhere on the right-hand side except in the exponential.

The action $S[\phi]$ can be expanded in a Taylor series about $\phi^i = 0$, indicating differentiation with a comma,

$$S[\phi] = \sum_{n=0}^{\infty}\frac{1}{n!}S_{,i_1\cdots i_n}[\phi=0]\phi^{i_1}\cdots\phi^{i_n}. \tag{102}$$

Similarly, the derivative of S with respect to the field has the expansion [7],

$$S_{,i}[\phi] = \sum_{n=0}^{\infty}\frac{1}{n!}S_{,ii_1\cdots i_n}[\phi=0]\phi^{i_1}\cdots\phi^{i_n}. \tag{103}$$

If ϕ^i is replaced by $\frac{\hbar}{i}\frac{\delta}{\delta J_i}$ in this expression and then operate on $\langle 2|1\rangle[J]$ with $S_{,i}[\frac{\hbar}{i}\frac{\delta}{\delta J_i}]$, and use (102), the following differential equation arises

$$S_{,i}[\frac{\hbar}{i}\frac{\delta}{\delta J_i}]\langle 2|1\rangle[J] = \langle 2|T(S_{,i}[\phi]\exp(\frac{i}{\hbar}J_i\phi^i))|1\rangle[J=0]. \tag{104}$$

The operator equation of motion (89) implies

$$S_{,i}[\frac{\hbar}{i}\frac{\delta}{\delta J_i}]\langle 2|1\rangle[J] = -J_i\langle 2|1\rangle[J]. \tag{105}$$

This results in a differential equation for the transition amplitude. In order to solve equation (105), the functional analogue of a Fourier transform may be used

$$\langle 2|1\rangle = \int (\prod_I d\varphi^I(x))F[\varphi]\exp(\frac{i}{\hbar}\int J_I(x')\varphi^I(x')\,d^nx'). \tag{106}$$

The integration in (106) extends over all fields which correspond to the choice of states described by $|1\rangle$ and $|2\rangle$. The functional $F[\varphi]$ which is to be thought of as the Fourier transform of the transformation function, is to be determined by requiring (106) satisfy (105)

$$0 = \int (\prod_{I,x} d\varphi^I(x))\{S_{,i}[\varphi] + J_i\}F[\varphi]\exp(\frac{i}{\hbar}\int J_I(x')\varphi^I(x')\,d^nx')$$

$$= \int (\prod_I d\varphi^I(x))\{S_{,i}[\varphi]F[\varphi] + \frac{\hbar}{i}F[\varphi]\frac{\delta}{\delta\varphi^i}\}\exp(\frac{i}{\hbar}\int J_I(x')\varphi^I(x')\,d^nx').$$

Upon carrying out an integration by parts on the second term here

$$0 = \int (\prod_i d\varphi^i)\{S_{,i}[\varphi]F[\varphi] - \frac{\hbar}{i}F_{,i}[\varphi]\}\exp(\frac{i}{\hbar}\int J_I(x')\varphi^I(x')\,d^nx')$$

$$+\frac{\hbar}{i}F[\varphi]\exp(\frac{i}{\hbar}\int J_I(x')\varphi^I(x')\,d^nx')|_{\varphi_1}^{\varphi_2}. \tag{107}$$

Assuming the surface term at the end vanishes, it follows from (107) that

$$F[\varphi] = \mathcal{N}\exp(\frac{i}{\hbar}S[\varphi]), \tag{108}$$

where $\mathcal{N}$ is any field-independent constant. The condition for the surface term to vanish is that the action $S[\varphi]$ be the same on both surfaces Σ_1 and Σ_2. This condition is usually fulfilled in field theory by assuming that the fields are in the vacuum state on the initial and final hypersurface.

The transformation function can then be summarized as

$$\langle 2|1\rangle[J] = \mathcal{N}\int (\prod_i d\varphi^i)\exp(\frac{i}{\hbar}\{S[\varphi] + J_i\varphi^i\}). \tag{109}$$

This is one form of the Feynman path integral, or functional integral, which represents the transformation function. This technique turns out to be very effective with further modifications applied to the quantization of gauge theories. These theories have been particularly successful in understanding the weak and strong interactions [10].

6. QED - A physical example and summary

A given elementary interaction implies a system of coupled field equations. Thus for the photon and the charged spin 1/2 particle as in quantum electrodynamics [5],

$$[\gamma(-i\partial - eqA(x)) + m]\psi(x) = \eta^A(x),$$

$$-\partial^2 A^\mu(x) + \partial^\mu \partial A(x) = J^\mu(x) + \frac{1}{2}\psi(x)\gamma^0\gamma^\mu eq\psi(x) - \int dx' f^\mu(x-x')\psi(x')\gamma^0 ieq\eta^A(x'). \tag{110}$$

Since this is a nonlinear system, the construction of the fields in terms of the sources will be given by doubly infinite power series. The succesive terms of this series $W_{n\nu}$ with n paricles and ν photon sources represent increasingly complicated physical processes. One of the simplest terms in the interaction skeleton will be discussed below to the point of obtaining experimental consequences.

There are two asymmetrical ways to eliminate the fields. First, introduce the formal solution of the particle field equation

$$\psi^A(x) = \int dx' G_+^A(x,x')\eta^A(x),$$

and $G_+^A(x,x')$ is the Green function

$$[\gamma(-i\partial - eqA(x)) + m]G_+^A(x,x') = \delta(x-x').$$

This gives the partial action

$$W = \int dx[J^\mu A_\mu - \frac{1}{4}F^{\mu\nu}F_{\mu\nu}] + \frac{1}{2}\int dxdx'\,\eta^A(x)\gamma^0 G_+^A(x,x')\eta^A(x').$$

The stationarity requirement on variations of A_μ recovers the Maxwell equation above. If we eliminate the vector potential

$$A_\mu^f(x) = \int dx' D_+(x-x')[J^\mu(x) + j^\mu_{cons}] + \partial_\mu\lambda(x).$$

$$j^\mu_{cons}(x) = j^\mu(x) - \int dx'\, f^\mu(x-x')\partial'_\nu(x'),$$

and the gauge condition determines $\lambda(x)$.

Finally, the first few successive $W_{2\nu}$ are written out, noting each particle source is multiplied by a propagation function $G_+(x,x')$ to form the field ψ,

$$W_{21} = \frac{1}{2}\int d^4x\psi(x)\gamma^0 eq\gamma A(x)\psi(2),$$

$$W_{22} = \frac{1}{2} \int dx dx' \psi(x) \gamma^0 eq\gamma A(x) G_+(x - x') eq\gamma A(x') \psi(x').$$

As a brief introduction to how this formalism can lead to important physical results, let us look at a specific term like W_{21}, the interaction energy of an electron with a static electromagnetic field A_μ^{ext}

$$E = \int d^3x \, j_\mu A^\mu_{ext} = e \int d^3x \bar{\psi}_{p'} (\gamma_\mu + \Gamma^R_\mu(p', p) + \frac{i}{4\pi} \Pi^R_{\mu\nu} i D^{\nu\sigma} \gamma_\sigma) \psi_p A^\mu_{ext}. \tag{111}$$

These terms include the bare electron-photon term, the electron-photon correction terms and then the photon vacuum-polarization correction term, R means a renormalized quantity and γ_μ denote Dirac matrices. The self-energy correction is left out, because for free particles, it contributes only to charge and mass renormalization. The polarization tensor $\Pi_{\mu\nu}(q^2)$ is given by

$$\Pi_{\mu\nu}(q^2) = (q^2 g_{\mu\nu} - q_\mu q_\nu) \Pi(q^2), \tag{112}$$

where $\Pi(q^2)$ is the polarization function. A simple result is obtained in the limit of low momentum transfer, $q^2 \to 0$, which is also of special physical significance and the case of interest here. The renormalized polarization function is

$$\Pi^R(q^2) = -\frac{e^2}{\pi} \frac{q^2}{m^2} (\frac{1}{15} + \frac{1}{140} \frac{q^2}{m^2} + \cdots). \tag{113}$$

The regularized vertex function is

$$\Gamma^R_\mu(p', p) = \gamma_\mu F_1(q^2) + \frac{i}{2m} \sigma_{\mu\nu} q^\nu F_2(q^2). \tag{114}$$

The functions $F_1(q^2)$ and $F_2(q^2)$ are called form factors. The electron gets an apparent internal structure by the interaction with the virtual radiation field which alters it from a pure Dirac particle. In the limit, $q^2 \to 0$, these functions can be calculated to be

$$F_1(q^2) = \frac{\alpha}{3\pi} \frac{q^2}{m^2} (\ln(\frac{m}{\mu}) - \frac{3}{8}), \qquad F_2(q^2) = \frac{\alpha}{2\pi}. \tag{115}$$

Substituting all of these factors and $D_F^{\mu\nu}(q^2) = -4\pi g^{\mu\nu}/q^2$ into (111) yields for small values of q^2,

$$E = e \int d^3x \, \bar{\psi}_{p'} \{\gamma_\mu [1 + \frac{\alpha}{3\pi} \frac{q^2}{m^2} (\ln(\frac{m}{\mu}) - \frac{3}{8} - \frac{1}{5}) + \frac{\alpha}{2\pi} \frac{i}{2m} \sigma_{\mu\nu} q^\nu \} \psi_p A^\mu_{ext}. \tag{116}$$

Note μ appears in (115) as an elementary attempt to regularize a photon propagator in one of the terms and does not interfere further with the application at this level and α is the fine structure constant. The Gordon decomposition allows this to be written

$$E = e\int d^3x\bar{\psi}_{p'}\{\frac{1}{2m}(p+p')_\mu[1+\frac{\alpha}{3\pi}\frac{q^2}{m^2}(\ln(\frac{m}{\mu})-\frac{3}{8}-\frac{1}{5})]+(1+\frac{\alpha}{2\pi})\frac{i}{2m}\sigma_{\mu\nu}q^\nu\}\psi_p A^\mu_{ext}. \tag{117}$$

The momentum factors can be transformed into gradients in configuration space, thus $q_\mu \to i\partial_\mu$ acts on the photon field and $p'_\mu = -i\overleftarrow{\partial}_\mu$ $p_\mu = i\partial_\mu$ act on the spinor field to the left and right respectively. Then (117) becomes

$$E = e\int d^3x\{\frac{i}{2m}\bar{\psi}_{p'}(x)(\partial_\mu - \overleftarrow{\partial})\psi_p[1-\frac{\alpha}{3\pi}\frac{1}{m^2}(\ln(\frac{m}{\mu})-\frac{3}{8}-\frac{1}{5})]A^\mu_{ext}$$

$$-(1+\frac{\alpha}{2\pi})\frac{1}{2m}\bar{\psi}_P(x)\sigma_{\mu\nu}\psi_P(x)\partial^\nu A^\mu_{ext}\}. \tag{118}$$

The first term contains the convection current of the electron which interacts with the potential. In the special case of a purely magnetic field the second part can be identified as the dipole energy. By introducing the electromagnetic field strength tensor $F^{\mu\nu} = \partial^\mu A^\nu - \partial^\nu A^\mu$ and using the antisymmetry of $\sigma_{\mu\nu} = \frac{i}{2}[\gamma_\mu, \gamma_\nu]$, the second part is

$$W_{mag} = e(1+\frac{\alpha}{2\pi})\frac{1}{4m}\int d^3x\bar{\psi}(x)\sigma_{\mu\nu}\psi(x)F^{\mu\nu}. \tag{119}$$

When $F^{\mu\nu}$ represents a pure magnetic field, $F^{12} = -B^3$, $\sigma_{12} = \Sigma_3$ with cyclic permutations and the interaction energy becomes

$$W_{mag} = -\frac{e}{4m}(1+\frac{\alpha}{2\pi})2\int d^3x\,\bar{\psi}(x)\vec{\Sigma}\psi(x)\cdot\vec{B} = -\langle\vec{\mu}\rangle\cdot\vec{B}. \tag{120}$$

The magnetic moment is given by

$$\langle\vec{\mu}\rangle = \frac{e\hbar}{2mc}(1+\frac{\alpha}{2\pi})2\langle\vec{S}\rangle = g\mu_B\langle\vec{S}\rangle. \tag{121}$$

The magnetic moment is thus proportional to the spin expectation value of the electron. The proportionality factor is the so called g-factor

$$g = 2(1+\frac{\alpha}{2\pi}) = 2(1+0.00116141). \tag{122}$$

The first point to note is that the value of the g-factor obtained including quantum mechanics differs from the classical value of 2. The result in (122) was first calculated by Schwinger and it has been measured to remarkable accuracy many times. A modern experimental value for the g-factor is

$$g_{exp} = 2(1+0.001159652193), \tag{123}$$

and only the last digit is uncertain.

Of course Schwinger's calculation has been carried out to much further accuracy and is the subject of continuing work. At around order α^4, further corrections must be included, such as such effects as virtual hadron creation. The pure-QED contributions are represented by coefficients C_i as a power series in powers of α/π, which acts as a natural expansion parameter for the calculation

$$g_{theo} = 2[1 + C_1(\frac{\alpha}{\pi}) + C_2(\frac{\alpha}{\pi})^2 + C_3(\frac{\alpha}{\pi})^3 + \cdots]. \tag{124}$$

Not all of the assumptions made in classical physics apply in quantum physics. In particular, the assumption that it is possible, at least in principle to perform a measurement on a given system in a way in which the interaction between the measured and measuring device can be made as small as desired. In the absence of concepts which follow from observation, principles such as the action principle discussed here are extremely important in providing a direction in which to proceed to formulate a picture of reality which is valid at the microscopic level, given that many assumptions at the normal level of perception no longer apply. These ideas such as the action principle touched on here have led to wide ranging conclusions about the quantum world and resulted in a way to produce useful tools for calculation and results such as transition amplitudes, interaction energies and the result concerning the g factor given here.

Author details

Paul Bracken

* Address all correspondence to: bracken@panam.edu

Department of Mathematics, University of Texas, Edinburg, TX, USA

References

[1] A. L. Fetter and J. D. Walecka, Theoretical Mechanics of particles and Continua, Dover, Mineola, NY, 2003.

[2] J. Schwinger, Quantum Mechanics, Springer Verlag, Berlin Heidelberg, 2001.

[3] J. Schwinger, Selected Papers on Quantum Electrodynamics, Dover, NY, 1958.

[4] J. Schwinger, The theory of quantized fields. I, Phys. Rev. 82, 914-927 (1951).

[5] J. Schwinger, Particles Sources and Fields, Vols I-III, Addison-Wesley, USA, 1981.

[6] D. Toms, The Schwinger Action Principle and Effective Action, Cambridge University Press, Cambridge, 2007.

[7] S. S. Schweber, An Introduction to Relativistic Quantum Field Theory, Row, Peterson and Co, Evanston, 1961.

[8] W. K. Burton, Equivalence of the Lagrangian formulations of quantum field theory due to Feynman and Schwinger, Nuovo Cimento, 1, 355-357 (1955).

[9] P. Bracken, Quantum Mechanics in Terms of an Action Principle, Can. J. Phys. 75, 261-271 (1991).

[10] K. Fujikawa, Path integral measure for gauge theories with fermions, Phys. Rev. D 21, 2848-2858 (1980).

Chapter 10

Quantum Intentionality and Determination of Realities in the Space-Time Through Path Integrals and Their Integral Transforms

Francisco Bulnes

Additional information is available at the end of the chapter

1. Introduction

In the universe three fundamental realities exist inside our perception, which share messages and quantum processes: the physical, energy and mental reality. These realities happen at all times and they are around us like part of our existence spending one to other one across *organised transformations* which realise a linking field - energy-matter across the concept of conscience of a field on the interpretation of the matter and space to create a reality non-temporal that only depends on the nature of the field, for example, the gravitational field is a reality in the space - time that generates a curved space for the presence of masses. At macroscopic level and according to the Einsteinian models the time is a flexible band that acts in form parallel to the space. Nevertheless, studying the field at microscopic level dominated by particles that produce gravity, the time is an intrinsic part of the space (*there is no distinction between one and other*), since the particles contain a rotation concept (*called spin*) that is intrinsic to the same particles that produce gravity from quantum level [1].Then the gravitational field between such particles is an always present reality and therefore non-temporal. The time at quantum level is the distance between cause and effect, but the effect (*gravitational spin*) is contained in the proper particle that is their cause on having been interrelated with other particles and vice versa the effect contains the cause since the particle changed their direction [1].

Then the action of any field that is wished transforms their surrounding reality which must spill through the component particles of the space - time, their nature and to transmit it in organised form, which is legal, because the field is invariant under movements of the proper space, and in every particle there sublies a part of the field through their *spinor*.

Three fundamental realities perceived by our anthropometric development of the universe; field - energy- matter between three different but indistinguishable realities are realised at macroscopic level: one is the *material reality* which is determined by their atomic linkage between material particles (*atoms constituted by protons, neutrons and electrons*), an energy reality, called also quantum reality, since the information in this reality area exchanges the matter happen through sub-particles (*bosons, fermions, gluons, etc*) and finally a virtual reality that sublies like *fundamental field* and that is an origin of the changes of spin of the sub-particles and their support doing that they transform these into others and that they transform everything around him (*Higgs field*). The integration of these three realities will be called by us a *hyper-reality* by us. The hyper-reality contains to the *quantum reality* and to the reality perceived by our senses (material reality).

Consider $\mathbb{R}^d \times I_t$, like the space - time where happens the transitions of energy states into space - time, and let u, v, elements of this space, the integral of all the continuous possible paths to particle $x(s)$, that transit from energy state in u, to an energy state in v, in $\mathbb{R}^d \times I_t$, is

$$\mathrm{I}(L, x(t)) = \int_{C^{u,v}[0,t]} \exp\left\{\frac{i}{h}\Im(x)\right\} dx, \quad (1)$$

where h, is the constant of Max Planck, and the action $\Im$, is the one realised by their *Lagrangian L*.

Since we have mentioned, the action of a field is realised being a cause and effect, for which it must be a cause and effect in each of the component particles, "waking up" the particle conscience to particle being transmitted this way without any exception. This action must infiltrate to the field itself that it sublies in the space and that it is shaped by the proper particles that compose it creating a certain co-action that is major than their algebraic sum [2].The configuration space $C_{n,m} = \{\gamma_t \in \Omega(\Gamma) \mid \gamma t \rightarrow \Gamma \rightarrow \Gamma/\gamma\}$ [3], is the model created by the due action to each corresponding trajectory to the different splits it. Is clear here we must have in mind all the paths in the space-time M, that contribute to interference amplitude in this space, remaining the path of major statistical weight. The intention takes implicitly a space $C_{n,m}$. Any transformation that wants to realise of a space, has as constant the same energy that comes from the permanent field of the matter and which is determined by the quantum field of the particles $x(s)$, constituents of the space and matter. If we want to define a conscience in the above mentioned field, that is to say, an action that involves an intention is necessary to establish it inside the argument of the action. Likewise, if $x(s) \in \Omega$, and $\Im(x(s))$, there is their action due to a field of particles X, and there is spilled an intention defined by (1) the length and breadth of the space M, such that satisfies the property of synergy [2], for all the possible trajectories that they fill Ω, we have that

$$\Im_{\mathrm{TOTAL}} \geq \int_{E^-}^{E^+} \left\{ \sum_j \int_{\gamma_t} \Im_j(x(s)) d(x(s)) \right\} d\mu, \quad (2)$$

where the entire action (2) is an intentional action (*for the whole infinity of paths* γ_t, that defines Γ, and that are trajectories of the space - time $\Omega \subset \mathbb{R}^3 \times I_t$), if and only if $O_c(x, \dot{x}(s)) = L(\phi(x), \partial\phi(x))d\phi$, where then

$$\Im_{TOTAL} = \int_{\Gamma} \left\{ \Im \left[\int_{\Omega(\Gamma)} O_c(\phi(x))d(\phi(x)) \right] \right\} \mu_x = (E^+ - E^-) \int_{-\infty}^{+\infty} \left\{ \int_{\Omega(\Gamma)} \Im(x(s))dx(s) \right\} \mu_s, \qquad (3)$$

where the energy factor $E^+ - E^-$, represents the energy needed by the always present force to realise the action and O_c, is the conscience operator which defines the value or record of the field X (direction), on every particle of the space $\Omega(\Gamma)$, which along their set of trajectories Γ, realizes the action of permanent field $\Im_{O_c}$, it being fulfilled that

$$\Im_{O_c}(\phi(x)) = \int_{X(M)} O_c(\phi(x))d\phi(x), \qquad (4)$$

where the operator $\Im_{O_c}$, invests an energy quasi-infinite, encapsulated in a microscopic region of the space (*quantum space* $\mathcal{M}$), and with applications and influence in an unlimited space of the sub-particles (*boson space*). Likewise a photon of certain class $\phi(x)$, will be generated by the quantum field (*if it manages to change its field spin*) and will be moved for the intention on a trajectory Γ, by the path integral

$$\mathrm{I}(\Im_{O_c}(\phi(x))) = \int_{\Gamma} \left[\int_{X(\Omega(\Gamma))} O_c(\phi(x))d(\phi(x)) \right] \mu_{\Gamma}, \qquad (5)$$

Interesting applications of the formula (3) to nano-sciences will happen at the end of the present chapter. Also it will be demonstrated that (3) is a quantum integral transform of bundles or distortions of energy in the space - time if it involves a *special kernel*. The bundle stops existing if there is applied certain intention (*path integral transform*). The operator O_c, involves a *connection* of the tangent bundle of the space of trajectories $\Omega(\Gamma)$.

The operator O_c, include a connection of the tangent bundle of the space of trajectories $\Omega(\Gamma)$. The integral (5) will determine on certain hypotheses the interdependence between the material, quantum and virtual realities in $\mathbb{M}$.

Def. **2. 1** (*intentional action of the* X). Let X, be a field acting on the particles $x_1(s)$, $x_2(s)$, …, $\in \mathbb{M}$, and let $\Im$, be their action on the above mentioned particles under an operator who recognizes the "target" in $\mathbb{M}$, (*conscience operator*). We say that $\Im$, is an conscientious intentional action (*or simply intention*) of the field X, if and only if:

a. $\Im$, is the determination of the field X, to realise or execute, (their force $F(x)$),

b. $\Im$, recognizes well their target, it is known what the field X wants to do (their direction $\Leftrightarrow$ she follows a configuration patron)

Consider a particle system p_1, p_2,... in a space - time $\mathbb{M} \cong \mathbb{R}^4$. Let $x(t) \in \Omega(\Gamma) \subset \mathbb{R}^3 \times I_t$, be a trajectory which predetermines a position $x \in \mathbb{R}^3$, for all time $t \in I_t$. A field X, that infiltrates its action to the whole space of points predetermined by all the trajectories $x_1(t)$, $x_2(t)$, $x_3(t)$, ..., $\in \Omega(\Gamma)$, is the field that predetermines the points $\phi(x_i(t))$, which are fields whose determination is given by the action of the field X, and evaluated in the position of every particle. Every point have a defined force by the action $\Im$, of X, along the geodesic γ_t, and determined direction by their tangent bundle given for $T\mathbf{\ast}^1(\Omega(\Gamma))$, that is the cotangent space $T^*(\Omega(\Gamma))$ [4], which give the images of the states under Lagrangian, that is to say, the field provides of direction to every point ϕ_i, because their tangent bundle has a subjacent spinor bundle $\mathbb{S}$ [5], where the field X, comes given as $X = \sum_i \phi^i \frac{\partial}{\partial \phi^i}\Big|_{(x^i, \phi^i)}$, $\forall\ \phi_1, \phi_2, \phi_3, \ldots \in \mathbf{\ast}^1$, on every particle $p_i = x_i(t)$ (i = 1, 2, ...). Then *to direct an intention* is the map or connection:

$$O_c : T\Omega(\Gamma) \rightarrow T\mathbf{\ast}^1(\Omega(\Gamma)) \ (\cong T^*(\Omega(\Gamma)), \tag{6}$$

with rule of correspondence

$$(x^i, \partial_t x^i)I \rightarrow (\phi^i, \partial_m \phi^i) \tag{7}$$

which produces one to us *ith-* state of field energy ϕ [6], where the action $\Im$, of the field X, infiltrates and transmits from particle to particle in the whole space $\Omega(\Gamma)$, using a configuration given by their Lagrangian *L* (*conscience operator*), along all the trajectories of $\Omega(\Gamma)$. Then of a sum of trajectories $\int D_F(x(t))$, one has the sum $\int d(\phi(x))$, on all the possible field configurations $C_{n,m}$. Extending these intentions to whole space $\Omega(\Gamma) \subset \mathbb{M}$, on all the elections of possible paths whose statistical weight corresponds to the determined one by the intention of the field, and realising the integration in paths for an infinity of particles - fields in $T\Omega(\Gamma)$, it is had that

$$\mathrm{I}(\phi^i(x)) = \int_{T\Omega(\Gamma)} \omega(\phi(x)) = \lim_{\substack{N\to\infty \\ \delta s \to 0}} \frac{1}{B} \int_{-\infty}^{+\infty} \frac{d\phi^1}{B} \cdots \int_{-\infty}^{+\infty} \frac{d\phi^n}{B} \cdots = \prod_{i=1}^{\infty} \int_{-\infty}^{+\infty} e^{i\Im[\phi^i, \partial_\mu \phi^i]} d\phi^i(x(s)), \tag{8}$$

where $B = \left[\frac{m}{2\pi\hbar i \delta s}\right]^{1/2}$, is the amplitude of their propagator and in the second integral of (8), we have expressed the Feynman integral using the form of volume $\omega(\phi(x))$, of the space of all the paths that add in $T\Omega(\Gamma)$, to obtain the real path of the particle (*where we have chosen quantized trajectories,* that is to say, $\int d(\phi(x))$). Remember that the sum of all these paths is the interference amplitude between paths that is established under an action whose Lagrangian is $\omega(\phi(x)) = \Im_{\xi}(x) d\phi(x)$, where, if $\Omega(\mathbb{M})$, is a complex with $\mathbb{M}$, the space-time, and $C(\mathbb{M})$, is a complex or configuration space on $\mathbb{M}$, (*interfered paths in the experiment given by multiple split* [7]), endowed with a pairing

$$\int : C(\mathbb{M}) \times \Omega^*(\mathbb{M}) \rightarrow \mathbb{R}, \tag{9}$$

where $\Omega^*(\mathrm{M})$, is some dual complex (*"forms on configuration spaces"*), i.e. such that "Stokes theorem" holds:

$$\int_{\Omega\times C} \omega = <\Im, d\omega>, \tag{10}$$

then the integrals given by (8) can be written (to m-border points and n-inner points (see **figure 1a**))) as:

$$\begin{aligned}\int_{T\Omega(\Gamma)} \Im(\phi(x))d\phi &= \int_{\Omega(\Gamma_{t^1})\times\ldots\times\Omega(\Gamma_{t^m})\times\ldots} \Im_q d\phi_1^{m_1}\ldots d\phi_n^{m_n}\ldots \\ &= \int_{\Omega(\Gamma_{t^1})}(\int_{\Omega(\Gamma_{t^2})\ldots}(\int_{\Omega(\Gamma_{t^m})}\Im d\phi_1^{m_n})\ldots d\phi_n^{m_1})\ldots,\end{aligned} \tag{11}$$

This is an *infiltration* in the space-time by the direct action $\Im$ [2, 3], that happens in the space $\Omega\times C$, to each component of the space $\Omega(\Gamma)$, through the expressed Lagrangian in this case by ω, de (10). In (11), the integration of the space realises with the infiltration of the time, integrating only energy state elements of the field.

The design of some possible spintronic devices that show the functioning of this process of transformation in the space M, will be included in this chapter.

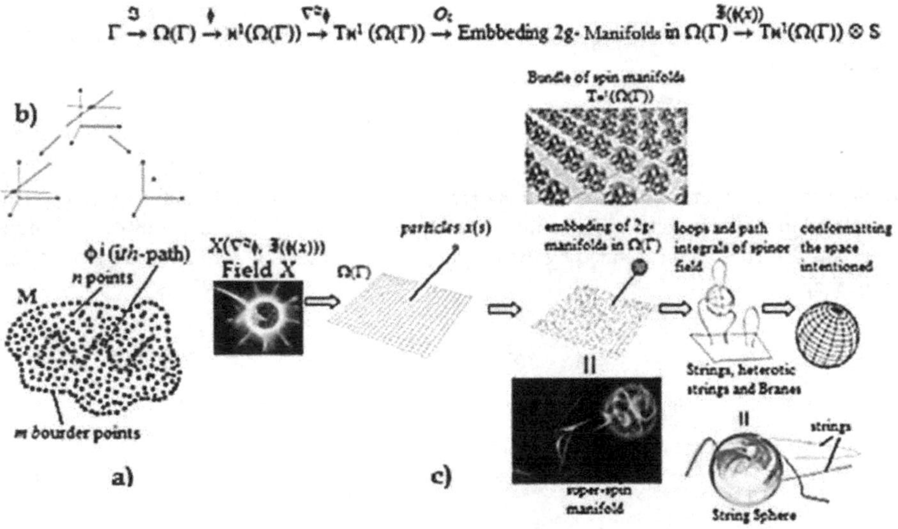

Figure 1. In a) The configuration space $C_{n,m}$, is the model created by the due action to each corresponding trajectory to the different splits. b) Example of a doble fibration to explain the relation between two realities of a space of particles: the bundle of lines L, and the ordinary space $\mathbb{R}^3$. c) Way in as a quantum field X, which acts on a space - time to change its reality, that is to say, to spill their intention.

2. Conscience operators and configuration spaces

We consider $M \cong R^3 \times I_t$, the space-time of certain particles $x(s)$, in movement, and let L, be an operator that explains certain law of movement that governs the movement of the set of particles in M, in such a way that the energy conservation law is applied for the total action of each one of their particles. The movement of all the particles of the space M, is given geometrically for their tangent vector bundle TM. Then the action due to L, on M, is defined like [8]:

$$\Im_L : TM \rightarrow R, \tag{12}$$

with rule of correspondence

$$\Im(x(s)) = FluxL(x(s))x(s), \tag{13}$$

and whose energy due to the movement is

$$E = \Im - L, \tag{14}$$

But this energy is due from their Lagrangian $L \in C^{\infty}(TM, R)$, defined like [9]

$$L(x(s),\dot{x}(s),s) = T(x(s),\dot{x}(s),s) - V(x(s),\dot{x}(s),s), \tag{15}$$

If we want to calculate the action defined in (7) and (8), along a given path $\Gamma = x(s)$, we have that the action is

$$\Im_\Gamma = \int_\Gamma L(x(s),x(s),s)ds, \tag{16}$$

If this action involves an intention (that is to say, it is an intentional action) then the action is translated in all the possible field configurations, considering all the variations of the action along the fiber derivative defined by the Lagrangian L. Of this way, the conscience operator is the map

$$O_c : TM \rightarrow T^*M, \tag{17}$$

with corresponding rule

$$O_c(v)w = \frac{d}{dt}L(v + tw)\Big|_{t=0}, \tag{18}$$

That is, $O_c(v)w$, is the derivative of L, along the fiber in direction w. In the case of $v = x'(s)$, and $q = x(s)$, $\forall\ q \in M$, $L(q, v) = E - V = ½<v, v> - V(q)$, we see that $O_c(v)w = <v, w>$, so we recover the usual map $s^b : TM \rightarrow T^*M$, (with ^{b} Euclidean in R^3) associated with the bilinear form $<, >$. Is here where the spin structure subjacent appears in the momentum of the particle $x(s)$.

As we can see, T^*M, carries a canonical symplectic form, which we call ω. Using O_c, we obtain a closed two-form ωL, on TM, by setting

$$\omega_L = \left(O_c\right)^* \omega, \tag{19}$$

Considering the local coordinates $(\phi^i, \partial_\mu \phi^i)$, to ω_L, modeling the space-time M, through $\mathcal{H}$-spaces, we have that (19) is

$$\omega_L = \frac{\partial^2 L}{\partial \phi^i \partial \partial_\mu \phi^j} d\phi^i \wedge d\phi^j + \frac{\partial^2 L}{\partial \phi^i \partial \partial_\mu \phi^j} d\phi^i \wedge d\partial_\mu \phi^j, \tag{20}$$

Likewise, the variation of the action from the operator $O_c = d\Im(\phi) = L(\phi, \partial_\mu \phi) d\phi$, is translated in the differential

$$d\Im(\phi)h = \int_\Gamma \left(\frac{\partial L}{\partial \phi} - \frac{d}{dt} \frac{\partial L}{\partial \dot{\phi}} \right) (\phi(s), \dot{\phi}(s)) h(s) ds, \tag{21}$$

where $h(s)$: $\Gamma \rightarrow TM$, and is such that $\tau_M \circ h = \Gamma$ and $h(x_1) = h(x_2) = 0$, to extreme points of Γ, $x(s_1) = p$ y $x(s_2) = q$. The total differential (21) is the symplectic form ω_L, that constructs the application of the field intention expanding $2n$-coordinates in (20). The space $\mathfrak{X}^1(\Omega(\Gamma))$, is the space of differentiable vector fields on $\Omega(\Gamma)$, and $\Omega(\Gamma)$, is the manifold of trajectories (space-time of curves) that satisfies the variation principle given by the Lagrange equation that expresses the force $F(x(s)^j)$, ($j = 1, 2, \ldots, n$) generated by a field that generates one "conscience" of order given by their Lagrangian (to see the figure 2).

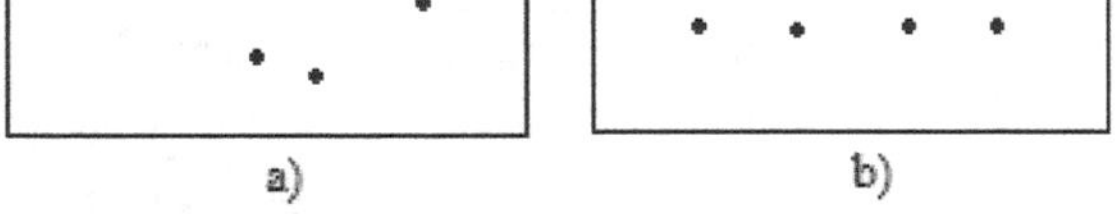

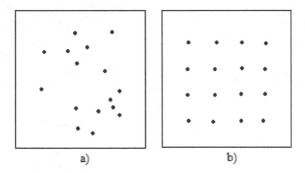

Figure 2. a) The particles act in free form in the space - time without the action of a quantum field that spills a force that generates an order conscience. b) A force $F(x(s)^j)$, is spilled, generated by a field that generates a "conscience" of order given by their Lagrangian. For it, there is not to forget the principle of conservation energy re-interpreted in the Lagrange equations and given for this force like $\frac{d}{dt}\left(\frac{\partial T}{\partial \dot{x}^j}\right) - \frac{\partial T}{\partial x^j} = F^j(x(s))$, (also acquaintances as "*living forces*") transmitting their momentum in every *ith*-particle of the space E, creating a infiltred region by path integrals of trajectories $\Omega(\Gamma)$, where the actions have effect. Here T, is their kinetic energy.

How does it influence the above mentioned intention in the space - time? what is the handling of the force $F^j(x(s))$? What is the quantum mechanism that makes possible the transformation of a body or space dictated by this intention?

It is necessary to have two aspects clear: the influence grade on the space, and a property that the field itself "wakes up" in the space or body to be transformed though the quantum information $\phi(x)$, their particles. Consider the integral (8) and their Green function for n, states $\phi(x_j)$ $(j = 1, 2, \ldots. n)$:

$$G^{(n)}(x_1, x_2, \ldots, x_n) = \frac{\prod_{i=1}^{n} \int_{-\infty}^{+\infty} e^{i\Im[\phi^i, \partial_\mu \phi^i]} d\phi^i(x(s))}{\int_{-\infty}^{+\infty} e^{i\Im[\phi^i, \partial_\mu \phi^i]} d\phi^i(x(s))}, \tag{22}$$

These Green's functions can most straightforwardly be evaluated by use of generating functional where we are using an external force $F^j(x(s))$, given by the intention

$$W[F^j(x(s))] = N \int_{-\infty}^{+\infty} \exp\left\{ i\Im(\phi^j, \partial_\mu \phi^j) + i \int_{\Omega(\Gamma)} F^j(x(s))\phi^j(x) \right\} d(x(s)), \tag{23}$$

This operator is the operator of execution $\mathfrak{exe}_{\Im(\phi)}$, which establishes in general form (5) that has been studied and applied in other developed research (see [2, 10, 11] as an example).

Then the influence realised on the space $\Omega(\Gamma) \subset$ M, that there bears the functional one (23) that involves the force of the intention given by the field (observe that the second addend of the argument of *exp*, is the action which is realised from the exterior on the space $\Omega(\Gamma)$) can go according to the functional derivative:

$$G^{(n)}(x_1, x_2, \ldots, x_n) = (-i)^n \frac{1}{W[0]} \frac{\delta^n}{\delta F^j(x_1) \cdots \delta F^j(x_n)} W[F^j(x(s))]\Big|_{F^j=0}, \tag{24}$$

where $\frac{\delta^n}{\delta F^j(x_1) \cdots \delta F^j(x_n)}$, describe the functional differentiation of nth-order, defined by the formula [6]

$$\frac{\delta^n (F^j(x))}{\delta F^j(x_1) \cdots \delta F^j(x_n)} = \frac{\delta^n}{\delta F^j(x_1) \cdots \delta F^j(x_n)} \int_{\Omega(\Gamma) \subset E^n} \delta\,(x - x_1) F^j(x_1) \delta(x - x_2) F^j(x_2) \cdots$$
$$\cdots \delta(x - x_n) F^j(x_n) dx_1 \cdots dx_n = \delta(x - x_1) \cdots \delta(x - x_n), \tag{25}$$

where these derivatives express impulses (force) of every particle placed in the positions x_1, ..., x_n. In case of receiving the influence of the field X, these impulses will be directed by the derivative of their Lagrangian density $\mathcal{L}^{(0)}$, that is a consequence of the differential (21), (whereas, by the application of their conscience operator O_c) know

$$\partial_\mu \frac{\delta L^{(0)}}{\delta \partial_\mu \phi(x)} - \frac{\delta L^{(0)}}{\delta \phi(x)} = 0, \tag{26}$$

But the equation (26) is the quantum wave equation (bearer of the information (*configuration and momentum* of the intention)) due to O_c, to the time *s*. Then the generating functional takes the form (23), considering the property of the operator O_c, given through the operator $\mathcal{O}(x - x_j)$[1]:

$$\begin{aligned} W[F^j(x(s))] &= N\int_{-\infty}^{+\infty} \exp\left\{-\frac{i}{2}\int_{\Omega(\Gamma)} \phi(x) o(x-x_j)\phi(x_j) - i\int_{\Omega(\Gamma)} F^j(x(s)) o^{-1}(x-x_j) F^j(x_j(s))\right\} d(\phi(x)) \\ &= N\int_{-\infty}^{+\infty} \exp\left\{-\frac{i}{2}\int_{\Omega(\Gamma)} \phi'(x) o(x-x_j)\phi'(x_j) - i\int_{\Omega(\Gamma)} F^j(x(s)) o^{-1}(x-x_j) F^j(x_j(s))\right\} d(\phi' x)), \end{aligned} \tag{27}$$

Where we have used $[d\phi(x)] = [d\phi'(x)]$.

The intention infiltrated by the conscience given for O_c, establishes that the differential of the action $d\Im(\phi)h$, given by (21) (using the energy (*amplitude*) that their propagator contributes D_F) can be visualised inside the configuration space through their boarder points ("targets" of the intention of the field X, and that happen in ∂M [12]),being also the interior points of the space M, *int*M, are *the proper sources* of the field (particles of the space M, that generate the field *X*). Then the intention of the field *X*, is the total action

$$\Im_T = \Im_{\partial M}(\Im_{int M}), \tag{28}$$

where this is a composition of the actions $\Im_{int M}$, and $\Im_{\partial M}$. These actions have codimensions strata *k*, and *n* – *k*, respectively, when we want to form the space M, using path integrals [].

To extract the intrinsic properties of integrals over configurating spaces, we will follow the proof of the formality theorem [13], and record the relevant facts in our homological-physical interpretation: admissible graphs are "cobordisms" $\omega(\gamma) \rightarrow [m]$, when Un is thought as a state-sum model [14]. The graphs are also interpreted as "extensions" $\gamma \rightarrow \Gamma \rightarrow \gamma'$, when considering the associated Hopf algebra structure. The implementations of these tools were done in the [3]. Remember that using the Stokes theorem (10) a Lagrangian on the class $\mathcal{G}$, of Feynman graphs is a *k*-linear map $\omega : H \rightarrow \Omega^\bullet(M)$, associating to any Feynman graph Γ, a

[1] The operator $\mathcal{O}(x - x_j) = (\Box_x + m^2 - i\varepsilon)\delta^n(x - x_j)$, and such that to their inverse $\mathcal{O}^{-1}(x - x_j)$, the functional property is had:

$$\int_{\Omega(\Gamma)} o^{-1}(x-x_j) o(x_j - y) d^n x_j = \delta^n(x-y),$$

closed volume form on $\Im(\Gamma)$, vanishing on the boundaries, i.e. for any subgraph $\gamma \to \Gamma$ (viewed as a sub-object) meeting the boundary of $\Gamma : [s] \to [t]$ (viewed as a cobordism), $\omega(\gamma) = 0$. Then an action given by $\Im_{int_M}$, is defined through their interior as:

Def. **2. 1.** An action on $\mathcal{G}$ ("$\Im_{int}$"), is a character $W : H \to \mathbb{R}$, which is a cocycle in the associated D$\mathcal{G}$-coalgebra (T(H^*), D), where $\mathcal{G}$, is a class of Feynman graph.

Let $C_{n,m}$, be the configuration space of n, interior points and m, boundary points in the manifold M, with boundary ∂M (that is to say. [13], upper half-plane H). Its elements will be thought as (geometric) "representations of cobordisms" (enabling degrees of freedom with constraints). Then the action in (28) takes the form

$$\left\{\{\omega(\gamma)=0\}\xrightarrow{[n]}[m]\right\}\xrightarrow{x(s)}\left\{\{\omega(\Gamma)=0\}\xrightarrow{[m]}\partial M\right\}, \quad (29)$$

Let H, be the Hopf algebra (*associative algebra used to the quantised action in the space-time*), of a class of Feynman graphs $\mathcal{G}$ [12]. If Γ, is such a graph, then configurations are attached to their vertices, while momentum are attached to edges in the two dual representations (Feynman rules in position and momentum spaces). This duality is represented by a pairing between a "configuration functor" (typically C_Γ, (configuration space of subgraphs and strings [15], and a "*Lagrangian*" (e.g. ω, determined by its value on an edge, i.e. by a propagator D_F). Together with the pairing (typically integration) representing the action, they are thought as part of the Feynman model of the state space of a quantum system. The differential (21) considering the D$\mathcal{G}$-structure [12], in the class of Feynman graphs $\mathcal{G}$, can be defined as one graph homology differential:

$$d\Gamma = \sum_{e \in E_\Gamma} \pm \Gamma / \gamma_e, \quad (30)$$

where the sum is over the edges of Γ, γ_e is the one-edge graph, and Γ/γ_e, is the quotient (forget about the signs for now).

We can give a major generalisation of this graphical homological version of the differential establishing the graduated derivation that comes from considering H = T($\mathfrak{g}$), the tensor algebra with reduced co-product

$$\Delta\Gamma = \sum_{\gamma \to \Gamma \to \gamma'} \gamma \otimes \gamma', \quad (31)$$

Consider the following basic properties of the operators O_c. Let $\mathcal{O}(x - x')$, defined in the footnote 1, and $\phi(x) \in \mathcal{H}$, where the space is the set of points $\mathcal{H} = \{\phi(x) \in [m] \mid [m] \subset T^*M\}^2$ [8].

[2] The corresponding cotangent space to vector fields is:
$T^*\mathfrak{X}^1(M) = \{(\phi, \partial_\mu\phi) \in \mathcal{H} \times T\mathfrak{X}^1(M) \mid \partial_\mu\phi = \nabla_\mu\xi, \forall\, \xi \in \mathfrak{X}^1(M)\}$.
Here $[m] = T^*C_{n,m}$.

Points of phase space are called states of the particle system acting in the cotangent space of M. Thus, to give the state of a system, one must specify their *configuration and momentum.*

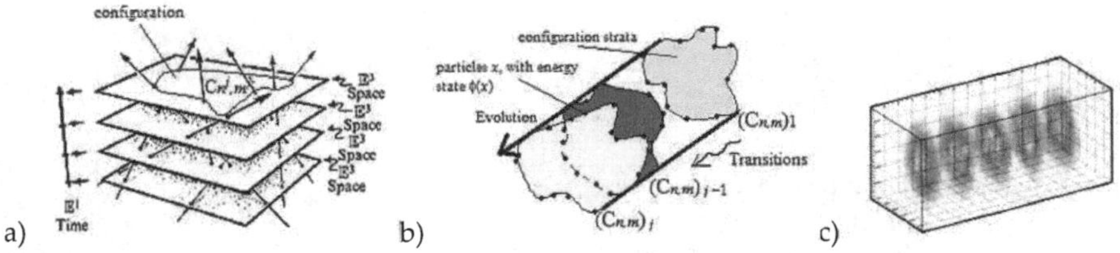

Figure 3. a) In every plane there is a particle configuration for a given time. b) The evolution of the particles along everything t, happens for a succession of configurations through which the particles system spends different strata codimension one. The causal structure of the space - time is invariant for every particle along the transformation process. d) Strata Evolution strata of the configuration space $C_{n,m}$, in the space-time E^4. The translation obeys to evolution of Lagrangian system given by $L(\phi_1, \ldots, \phi_n)$.

Example 1. Let $\pi : T^*M \rightarrow M$, be and $\gamma : \mathbb{R} \rightarrow TC_{n,m}$, then $\pi \circ \gamma : \mathbb{R} \rightarrow M$, describes the curve in the configuration space which describes the sequence of configurations through which the particles system passes to different strata of co-dimension one (see figure 2). Every strata correspond to a phase space of *m*, particles that are moved by curve γ and directed from their energy states $d\phi(x)$, by π, to *n*, particles $\phi(x)$.

This defines our intentional conscience. Then are true the following properties:

i. $O(x - x')\phi(x) = \delta(x - x')\phi(x)$, $\forall\, x, x' \in M$,
ii. $Oc(x(s))\phi(x'(s)) = O(x - x')$, $\forall\, x, x' \in M$, and $s \leq t$,
iii. $\int Oc(\phi(s))d\phi = \Im Oc$; $d\Im Oc(\phi(s))/d\phi = Oc(x(s))$, in the unlimited space H,
iv. $Oc = \delta(s - s')$, if and only if $\delta x(s)/\delta x(s') = \delta(s - s')$, $\forall\, s \leq t$, then $F(x(s)) = x(s)$,
v. $O^{-1}(x - x')Oc(x(s)) = -\Delta F(x - x')\delta(x - x')$, $\forall\, x, x' \in M$, and $s \leq t$,
vi. $\int H\, Oc(\phi(s))d\phi = \int \Omega O(x - x')x(s)d(x(s))$.

On the one hand, $\int_{\mathcal{H}} O_c(\phi(s))\phi(x)dx = \int_{\mathcal{H}} O_c(\phi(s))d\phi = \Sigma_j \int_{\Gamma} \mathcal{O}(x - x')x(s)ds$. Also

$$\int_{\Omega(\Gamma)} O_c(\phi(s))x(s)ds = \int_{H} o(x - x')\phi(x)dx,$$

then *i*, is satisfied. To the property *ii.* is necessary consider $\mathcal{O}(x - x') = (\Box_x + m^2 - i\varepsilon)\phi(x - x')$ [6]. But the operator O_c, is the defined as

$$O_c(x(s))\phi(x') = \left(\frac{d\Im_{O_c}}{d\phi} \right) \phi(x'),$$

Considering the operator $\mathcal{O}(x - x')$, we have

$$o(x - x')d\phi = d\Im_{O_c}\phi(x'),$$

Integrating both members on unlimited space $\mathcal{H}\times\Omega(\Gamma)$, (applying the principle of Stokes integration given by (10)) we have that integral identity is valid for whole space. Then is verifying *ii.* The property *iii.*, is directly consequence of (19), (20) and (21), considering the Stokes theorem given in (10), therefore $O_c(x(s))$, is such that $\omega_L = (O_c)^*\omega$, considering $\omega = d\phi$. Then

$$O_c = \int_{\Omega(\Gamma)\times H} \omega_L = \int_{\Omega(\Gamma)} (O_c)^*\omega, \tag{32}$$

which is a integral of type (10). Indeed,

$$\Im_{O_c} = \int_{H\times\Omega(\Gamma)} \omega_L = \int_H O_c(x(s))^*\omega = \int_H O_c(x(s))d\phi = \int_H d\Im_{O_c}(\phi), \tag{33}$$

The derivative in the last integral from (33) is the total differential given by (21) from where we have the derivative formula in the context of the unlimited space $\mathcal{H}$.

The property *iv.*, require demonstrate two implication where both implications are reciprocates. If $O_c(x(s')) = \delta(s - s')$, then all intention on trajectory defined Γ, its had that

$$\int_\Gamma O_c(x(s))x(s)ds = \int_\Gamma \delta(s - s')x(s)ds, \tag{34}$$

But for the differential (21) and the second member of the integral (34) we have

$$\delta\Im_{O_c}(x(s')) = \delta(\int_\Gamma \delta(s - s')x(s')ds'),$$

since $\frac{\delta}{\delta x(s')}(\int_\Gamma \delta(s - s')x(s')ds') = \delta(s - s')$, then $\delta(\int_\Gamma \delta(s - s')x(s')ds' = \delta x(s)$, and for other side

$$\delta(\int_\Gamma \delta(s - s')x(s')ds') = \delta(s - s')\delta x(s'),$$

from where $\frac{\delta x(s)}{\delta x(s')} = \delta(s - s'), s \le t$ [3]. But this implies directly $F(x(s)) = x(s)$. This property tell us that we can have influence on the space M, considering only a curve any of the space where the influence of the field exists like the force $\delta(s - s')$, since the space is infiltrated by

---

[3]In the general sense the functional derivative $\frac{\delta\phi_a(y)}{\delta\phi_b(x)} = \delta_{ba}\delta^n(y - x)$, implies

$$\delta\phi_b(y) = \sum_a \int \delta^n(y - x)\delta\phi_a(x)\delta_{ba}dx,$$

but does not imply

$$\delta\phi_b(y) = \delta_{ba}\delta^n(y - x)\delta\phi_a(x).$$

the force F, and this produces the permanent state of energy generated by every component of the space. On the other side, if $\frac{\delta x(s)}{\delta x(s')} = \delta(s - s'), s \leq t$ which is equivalent to $\frac{\delta}{\delta x(s')}(\int_{\Gamma} \delta(s - s')x(s')ds') = \delta(s - s'),$. But integrating (21) we have

$$\int_{\Gamma} \delta\Im_{O_c}(x(s')) = \int_{\Gamma} \delta(s - s')x(s')ds',$$

which for before implication is $\delta(s - s')\delta x(s')$. But

$$\int_{\Gamma} \{O_c(x(s'))x(s') + \delta(s - s')x(s')\}ds' = 0,$$

This integral is valid $\forall\ \Gamma \subset$ M. Thus $O_c(x(s')) = \delta(s - s')$. With this, the demostration of *iv.*, is completed. The identity in *v*, happens because $\mathcal{O}^{-1}(x - x') = -\Delta_F(x - x')$, considering the before property (simple conaequence of the property *iv*) [6].

The identity in *vi.*, happens in the phase space created by the cotangent space due to the image of the differential (21). Therefore, both members of integral identity will have to coincide in the intention given by $\Im_{Oc}$. Indeed, consider the integral

$$\int_{H} O_c(\phi(x))d\phi = \int_{\Omega(\Gamma)} (O_c) * \omega = \int_{H} d\Im_{Oc} = \Im_{Oc}, \tag{35}$$

On the other side, inside the quantum wave equation:

$$\begin{aligned} \int_{\Omega(\Gamma)} O(x - x')dx(s) &= \int_{\Omega(\Gamma)} (\Box_x + m^2 - i\varepsilon)\omega = \int_{\Omega} (\Box_x + m^2 - i\varepsilon)\phi(x)dx \\ &= \int_{H} (\Box_x + m^2 - i\varepsilon)d\phi = \int_{H} d\Im_{Oc}(\phi), \end{aligned} \tag{36}$$

Joining (35) with (36) we have *vi*.

3. Quantum intentionality and organized transformations

Considering the quantizations of our Lagrangian system describe in (11), (18) and (19) on $\mathbb{R}^n$, $n \geq 2$, coordinated by $\{x^j\}$, we describe terms of a graded commutative $C^{\infty}(M)$-algebra H, with generating elements

$$\{\partial_m x^a, \partial_m x^a{}_{l}, \partial_m x^a{}_{l1l2}, \ldots, \partial_m x^a{}_{l1\cdots lk}, \ldots\}, \tag{37}$$

and the bi-graded differential algebra H^*, of differential forms (the Chevalley–Eilenberg differential calculus) over H^0, as an $\mathbb{R}$-algebra [1-3]. One can think of generating elements (37) of H, as being *sui generis* coordinates of even and odd fields and their partial derivatives.

The graded commutative $\mathbb{R}$-algebra H^0, is provided with the even graded derivations (called total derivatives)

$$d_\lambda = \partial_\lambda + \sum_{0\leq|\Lambda|} \partial^a_{\lambda+\Lambda}\partial^\Lambda_a, \ d_\Lambda = d_{\lambda_1}\cdots d_{\lambda_k} \tag{38}$$

where $\Lambda = (\lambda_1 \ldots \lambda_k)$, $|\Lambda| = k$, and $\lambda+\Lambda = (\lambda, \lambda_1, \ldots, \lambda_k)$ are symmetric multi-indices. One can think of even elements

$$L = (x^l, \partial_\mu x^a)d^n x, \quad \delta L = d\partial_\mu x^a \wedge E_a d^n x = \sum_{0\leq|\Lambda|}(-1)^{|\Lambda|} d\partial_\mu x^a \wedge d_\Lambda(\partial_a{}^\Lambda L)d^n x, \tag{39}$$

where we observe that δL, is the 2-form given by ω_L, in the formula (20) with $n = 2$, and $\Lambda = \lambda_1\lambda_2$.

Now we consider the dual part of the space $(H, \Omega(\Gamma))$, that is to say, the space (H^*, L), be

We consider quantize this Lagrangian system in the framework of perturbative Euclidean QFT. We suppose that L, is a Lagrangian of Euclidean fields on $\Omega(\Gamma) \subseteq \mathbb{R}^n$. The key point is that the algebra of Euclidean quantum fields B_Φ, as like as H^0, is graded commutative. It is generated by elements $\phi^{\lambda\Lambda}{}_a$, $x \in \Omega(\Gamma)$. For any $x \in \Omega(\Gamma)$, there is a homomorphism belonging to space $H \rightarrow Hom(T(H), D)$ (with homomorphisms $Hom(T(H), D)$ given for $D\mathcal{G}$-algebra of cycles)

$$\Upsilon_x : \mathrm{I}^{\Lambda_1\cdots\Lambda_r}_{a_1\cdots a_r}\partial x^{a_1}_{\Lambda_1}\cdots\partial x^{a_r}_{\Lambda_r} \mapsto \mathrm{I}^{\Lambda_1\cdots\Lambda_r}_{a_1\cdots a_r}(x)\phi^{a_1}_{x\Lambda_1}\cdots\phi^{a_r}_{x\Lambda_r}, \qquad \mathrm{I}^{\Lambda_1\cdots\Lambda_r}_{a_1\cdots a_r} \in C^\infty(\Omega(\Gamma)) \tag{40}$$

Of the algebra H^0, of classical fields to the algebra B_Φ, which sends the basic elements $\partial x^a{}_\Lambda \in H^0$ to the elements $\phi^{\lambda\Lambda}{}_a \in B_\Phi$, and replaces coefficient functions $\mathfrak{I}$, of elements of H^0, with their values $\mathfrak{I}(x)$ (executions) at a point x. Then a state <, > of B_Φ, is given by symbolic functional integrals

$$<\phi^{a_1}_{x_1}\cdots\phi^{a_k}_{x_k}> \frac{1}{N}\int_H \phi^{a_1}_{x_1}\cdots\phi^{a_k}_{x_k}\exp\left\{-\int_{\Omega(\Gamma)} O_c(\phi^a_{x\Lambda})d^n x\right\}\prod_x[d\phi^a_x], \tag{41}$$

where this is an integral of type $\int_{H\times\Omega(\Gamma)} O_c(x(s))d\phi$, as give by the properties. When the intention expands to the whole space, infiltrating their information on the tangent spaces images of the cotangent bundle T^*M, (given by the imagen of $\phi(x)$, under $d\mathfrak{I}_{Oc}$).Then their intentionality will be the property of the field to spill or infiltrate their intention from a nano level of strings inside the quantum particles. Then from the energy states of the particles, and considering the intention spilled in them given by $O_c(\phi)$, we have the homomorphism (40) that establishes the action from M ($\cong \Omega(\Gamma)$), to ∂M, for their transformation through the action $\Im_T$, defined in (28) to any derivation given through their conscience operator (fiber (18)), like the graded derivation $\tilde{\partial}$, (considering the derivatives $O_c(x) = (O_c)^*\omega$, $D_F = d\phi$):

$$\tilde{\partial} : \phi^a_{x\Lambda} \mapsto (x, \partial^a_\mu x) \mapsto (O_c(x), \omega) \mapsto (O^*_c(x), \omega_L) = O^*_c(\phi^a_{x\Lambda})\omega, \tag{42}$$

of the algebra of quantum fields B_Φ. With an odd parameter α, let us consider the automorphism

$$\hat{U} = \exp\{\alpha\tilde{\partial}\} = \mathrm{Id} + \alpha\tilde{\partial},$$

Of the algebra B_Φ. This automorphism yields a new state $<,>$, of B_Φ, given by the equality

$$<\phi^{a_1}_{x_1} \cdots \phi^{a_k}_{x_k}> \frac{1}{N'}\int_H \hat{U}(\phi^{a_1}_{x_1})\cdots\hat{U}(\phi^{a_k}_{x_k})\exp \tag{43}$$

where the energy state has survived, since $d\hat{U}(\phi^a_x) = d\varphi^a_x$. That because the intention is the same. The intention has not changed.

What happens towards the interior of every particle? what is the field intention mechanism inside every particle?

To answer these questions we have to internalise the actions of field X, on the particles of the space M, and consider their spin. But for it, it is necessary to do the immersion of the Lagrangian ω, defined as the map

$$w : \mathrm{L} \to \mathrm{M}^{2n},$$ [4]

with rule of correspondence

$$Z_i \mapsto w(Z_i),$$

where the image of the 1-form ω, that the Lagrangian defines, $\omega(Z_i)$, is a symplectic form [8], and the variable Z_i, is constructed through the algebraic equations $W^a(Z_i) = 0$ [16].They describe the k-dimensional hypersurfaces denoted by S, such that $S \subset \mathcal{H}$, where $\mathcal{H}$ is the phase space defined in the section 2. The index $a = 1, \ldots, q$ runs over the number of polynomials $W^a(Z_i)$, in the variables Zi and i runs over the dimension of the ambient manifold which is assumed to be $\mathbb{C}^N$. If the space is a complete intersection, the constraints

---

[4] Having chosen M^{2n}, is to consider the two components of any point in the space $\mathbb{C}^N$, (that we are considering isomorfo to the ambient space of any quantum particle $x(s)$, in the space-time) to have the two components that characterise any quantum particle x(s), that is their spin (direction) and their energy state (density of energy or "living force of the particle"). $\mathbb{L}$, is the corresponding Lagrangian submanifold of the symplectic structure given by (M_{2n}, ω).

$W^a(Z_i)$ (there is exact solution to $W^a(Z_i) = 0$), are linearly independent and the differential form

$$\Theta^{(n-k)} = \in_{a_1 \cdots a_{N-k}} dW^{a_1} \wedge \cdots \wedge dW^{a_{N-K}}, \tag{44}$$

is not vanishing. In this case, q = N − *k* and the dimension of the surface is easily determined. For example, if the hypersurface is described by a single algebraic equation W(Z) = 0, the form (44) is given by $\Theta^{(1)} = dW$. On the other hand, if the hypersurface is not a complete intersection, then there exists a differential form

$$\Theta^{(n-k)} = T_{A[a_1 \cdots a_{N-k}]} dW^{a_1} \wedge \cdots \wedge dW^{a_{N-K}} \wedge \eta^{A,(N-k-q)}, \tag{45}$$

where $\eta^{A,(N-k-q)}$, is a set of $N - k - q$, forms defined such that $\Theta^{(N-k)}$, is non-vanishing on the constraints $W^a(Z_i) = 0$, and $T_{A,[a1 \ldots aN-k]}$, is a numerical tensor which is antisymmetric in the indices $a_1 \ldots a_q$. The construction of $\eta^{A,(N-k-q)}$, depends upon the precise form of the algebraic manifold (variety of the equations $W^a(Z_i) = 0$). In some cases a general form can be given, but in general it is not easy to find it and we did not find a general procedure for that computation.

To construct a global form on the space *S*, one can use a modification of the Griffiths residue method [16], by observing that given the global holomorphic form on the ambient space $\Omega^{(N)} = \in^{i_1 \cdots i_N} dZ_{i_1} \wedge \cdots \wedge dZ_{i_N}$, we can decompose the $\{Z_i\}$'s, into a set of coordinates $Y^a = W^a(Z)$, and the rest. By using the contraction with respect to *q*, vectors $\{\underline{Z}^a{}_i\}$, the top form for *S*, can be written as

$$\Omega^{(k)} = \frac{\iota_{\underline{Z}_1^a} \cdots \iota_{\underline{Z}_\theta^a} \Omega^{(N)}}{\iota_{\underline{Z}_1^a} \cdots \iota_{\underline{Z}_\theta^a} \Theta^{(N-k)}}, \tag{46}$$

which is independent from $\{\underline{Z}^a{}_i\}$, as can be easily proved by using the constraints $W^a(Z_i) = 0$. Notice that this form is nowhere-vanishing and non singular only the case of CY-space (Calabi-Yau manifold). The calabi-Yau manifols is a spin manifold and their existence in the our space M^{2n}, like product of this construction is the first evidence that a spin manifold is the spin of our space-time due to their holomorphicity [17]. The vectors $\{\underline{Z}^a{}_i\}$, play the role of gauge fixing parameters needed to choose a polarisation of the space *S*, into the ambient space.

For example, in the case of pure spinor we have: the ambient form $\Omega^{(16)} = \in_{\alpha_1 \cdots \alpha} d\lambda_{i_1} \wedge \cdots \wedge d\lambda_{i_6}$, and $\Theta^{(5)} = \lambda\gamma^m d\lambda\, \lambda\gamma^m d\lambda \lambda\gamma^m d\lambda d\lambda \gamma_{mnp} d\lambda$. From these data, we can get the holomorphic top form $\Omega^{(11)}$, by introducing 5, independent parameters $\underline{\lambda}$, and by using the formula (46).

The latter is independent from the choice of parameters $\underline{\lambda}$, (however, some care has to be devoted to the choice of the contour of integration and of the integrand: in the minimal formalism, the presence of delta function $\delta(\lambda)$, might introduce some singularities which prevent from proving the independence from $\underline{\lambda}$, as was pointed out in [18], [19]). Using $\Omega^{(k)} \wedge \Omega^{(k)}$, one can compute the correlation functions by integrating globally defined functions. When the space is Calabi-Yau, it also exists a globally-defined nowhere vanishing holomorphic form $\Omega_{hol}{}^{(k|0)}$, such that $\Omega_{hol}{}^{(k|0)} \wedge \Omega_{\underline{hol}}{}^{(\underline{k}|0)}$, is proportional to $\Omega^{(k)} \wedge \Omega^{(\underline{k})}$. The ratio of the two top forms is a globally defined function on the CY-space. In the case of the holomorphic measure $\Omega_{hol}{}^{(k|0)}$, the integration of holomorphic functions is related to the definition of a contour $\gamma \in S$, in the complex space

$$< \prod_{A} \mathrm{o}(Z_i, p_A) > \int_{\gamma \in S} \Omega^{(k,0)} \prod_{A} \mathrm{o}_0(Z_i, p_A), \tag{47}$$

where $\mathcal{O}(Z_i, p_A)$, are the vertex operators of the theory localized at the points p_A, of the Riemann surface and $\mathcal{O}_0(Z_i, p_A)$, is the zero-mode component of the vertex operators. Newly our conscience operator come given by the form $\Omega^{(k,0)}$.

Example **2.** All Calabi-Yau manifolds are *spin*. In hypothetical quantum process (from point of view QFT), to obtain a Calabi-Yau manifold is necessary add (*or sum*) strings in all directions. In the inverse imaginary process, all these strings define a direction or spin. The strings themselves are Lagrangian submanifolds whose Lagrangian action is a path integral.

In mathematics, an isotropic manifold is a manifold in which the geometry doesn't depend on directions. A simple example is the surface of a sphere. This directional independence grants us freedom to generate a quantum dimension process, since it does not import what direction falls ill through a string, the space is the same way affected and it presents the same aspect in any direction that is observed creating this way their isotropy.

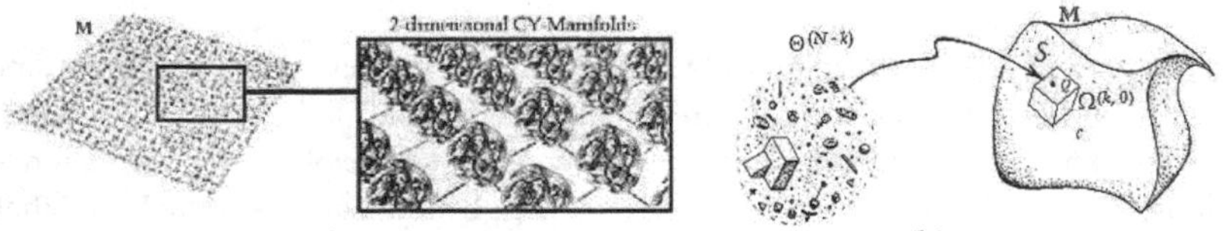

Figure 4. a) For instance, let us consider the hypersurface $S : \Sigma_i Z_i^2 = 0$, in $\mathbb{C}^N$. This equation can be put in the form using generalized coordinates $S(u_i) = wz$, where i, runs over $i = 1, 2, \ldots, N\text{-}2$, coordinates and w, z, are two combinations of the Z's. $S(u_i)$, is a polynomial of the coordinates u_i. For a given N, they are local CY-manifolds (*spin manifolds*) and there exist a globally-defined a organized transformation inside space $\mathbb{M}$ [20]. b) Intention inside particle Q.

The importance of this isotropy property in our spin manifold, helps us to establish that the transformations applied to the space that are directed to use (awakening) their nano-structure do it through an organized transformation that introduces the time as isotropic variable, creating a momentary timelessness in the space where the above mentioned transformation is created. Then the intentionality like a organized transformation is a co-action compose by field that act to realise the transformation of space and the field of the proper space that is transformed. Then the symplectic structure subjacent in M, receives sense.

A transformation T, it is said organized if in whole stage of execution of the transformation, isotropic images are obtained of the original manifold (object space of the transformation), under a finite number of endomorphisms of the underlying group to the manifold that co-acts with this transformation.

Likewise, if $\mathcal{T}$, is a transformation on the space M, whose subjacent group G, have endomorphisms $\sigma_1,\ldots,\sigma_n$, such that $\sigma_1\mathcal{T}(M),\ldots,\sigma_n\mathcal{T}(M)$, are isotropic then the infinite tensor product of isotropic submanifolds is a isotropic manifold, and is a organized transform equivalent to tensor product of spin representations $\sigma_1\mathcal{T}(M)\otimes\ldots\otimes\sigma_n\mathcal{T}(M)\otimes\ldots$[2, 5].

4. Quantum integral transforms: Elimination of distortions and quantum singularities

One of the quantum phenomena that can form or provoke the conditions of formation of singularities at this level is the propensity of a quantum system to develop scattering phenomena for the appearance of the anomalous states of energy, as antimatter energy or energy of particles of matter in the free state [21], which crowds (a big number of particles overlaps) due to the accumulation of the states of energy of the past or future (to see table 1) [21, 22],which on have different existence time and to having met their corresponding pairs of particles (particle/anti-particle pair) provoke bundles of energy that there form in the space time $M \times I_t$, singularities of certain weight (for mechanisms that can be explained inside the actions in $SU(3)$ and $SU(2)$ [11, 21]) due to its energy charge [21].

Studies in astrophysics and experiments in the CERN (Organisation Européenne pour la Recherche Nucléaire) establish that a similar mechanism although with substantial differences (known also like Schwinger mechanism) can explain the formation of a singularity such as the fundamental singularity (big-bang). This one establishes that the gravitational field turns into virtual pairs of particle- antiparticle of an environment of quantum gap in authentic pair's particle-antiparticle. If the black hole (singularity of the Universe) is done of matter (antimatter), it might repel violently to thousands of million antiparticles (particles), expelling them to the space in a second fraction, creating an event of ejection very similar to a Big-Bang. Nevertheless, in case of a singularity in the region of space - time of the particles in a quantum ambience is different in that aspect, since the small mass of a singularity $\ast$(s), might perform the order of the Planck mass, which is approximately 2×10^{-8} kg or $1{,}1 \times 10^{19}$ GeV. To this scale, the formulation of the thermodynamic theory of singularities of the space macroscopic time predicts that the quantum singularity ✱(s), could have an entropy of only 4πnats; and a Hawking temperature of needing quantum thermal energy comparable approximately to the mass of the finished singularity; and a Compton wavelength equivalent within a radius of Schwarzschild of the singularity in the Universe (this distance being equivalent to the Planck length). This is the point where the classic gravitational description of the object is not valid, being probably very important the quantum effects of the gravity. But there exists another mechanism or thermodynamic limit that is fundamental in the theory of the quantum dispersion and of the formation of quantum singularities.

The past and future in the quantum scattering phenomena

Z^+/Z^-	Particle $\phi(1)$	Anti-particle $\phi(-1)$
Input	$\bullet\longrightarrow\circ$ $\phi(1)$ Positive Future	$\bullet\longleftarrow\circ$ $-\phi(1)$ Negative Future
Output	$\circ\longrightarrow\bullet$ $\phi(1)$ Positive Past	$\circ\longleftarrow\bullet$ $-\phi(1)$ Negative Past

Table 1.

Proposition **1.** The energy of the singularity is born of the proper state of altered energy of the quantum space-time, although without a clear distinction of a suitable path of the particles (*without a normal sequence of superposition of (past and future) particles (a path does not exist*)). Then in this absence of paths, the singularity arises.

Proof. It is necessary to demonstrate that the past and present particles overlap without a normal sequential order in the causality, passing to the unconscious one ($O_c = 0$, since ${}^{L}\mathfrak{I}_0 = 0$ or $\omega_L = 0$) for fluctuations of energy that have a property of adherence to the transition of energy states forming energy bundles ([23]) that alter the normal behavior of the particles in the atoms. In effect, this happens when the quantum energy fragments and the photons exchange at electromagnetic level do not take in finished form (there is no exchange in the virtual field of the $SO(3)$). For which there is no path that the execution operator $\mathfrak{I}$, (see the section 2) could resolve through of a path integral between two photons (*virtual particles*). In this case the path integral does not exist. For the adherent effect, the virtual particles that do not manage to be exchanged accumulate forming the altered energy states (an excess very nearby of particles of the certain class (inclusive anti-particles of certain class) they add to themselves to the adherent space of photon (in this case the adherent photon is an excited photon $int\mathcal{E} \cup \delta(\mathcal{E})$, (where $\mathcal{E}$, is the influence of the singularity) with the infinitesimal nearby of points given by δ($\mathcal{E}$) [24])) which are the bundles of energy that defines the singularities. ■

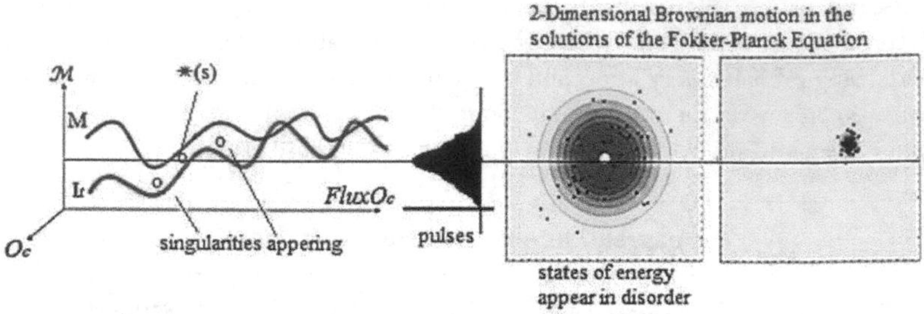

Figure 5. In a). is the elastic band of the space-time M, and b), result the pulse solutions of the Fokker-Planck Equation. In c), the singularity is being perturbed by states of different particles interacting and creating big scattering (*the red circles are perturbations created by the states of different particles creating scattering with a big level of particles pair annihilating. This produce defecting evolution in every sub-particles and increase of inflation in quantum level of the space-time*). In d), the singularity is formed.

Let $\hat{U}_0(t,s)$, the operator of evolution [6], of a particle $x(t)$, in the space of transition of the levels of conscience operator O_c, to all time $t \geq s$. Whose operator limits of s (*that is to say, coming to the process of understanding of a concept,* (border conditions of $\hat{U}_0(t,s)$)), satisfy

$$\lim_{t \to s^+} \hat{U}_0(t,s) = 1,[31] \tag{48}$$

But there are waves of certain level (*Table 1*), that act like moderators of wave length to the operator of evolution defined by ψ(s), satisfying that

$$\hat{U}_0(t,s)\left|\psi(s)\right> = \begin{cases} \left|\psi(s)\right>, & t \geq s \\ 0, & t < 0 \end{cases}, \tag{49}$$

having then that the singularity ✱(s), that is object of quantum transformation to along of the time, is that product obtained by the integral transform

$$✱(s) = \int_{X(C)} O_c(x(t))w(s,t)dt = (\mathrm{E}^+ - \mathrm{E}^-)\sqrt{\frac{m}{2\pi i n}} \int_{-\infty}^{\infty} \left\{ \int_{C} \mathrm{e}^{\frac{im(x-x')^2}{2\pi}} \hat{U}_0(s,t)dx \right\} dt, \tag{50}$$

where the singularity changes the time $t \geq s$, for the evolution operator $\hat{U}_0(t,s)$, [11], due to the evolution of the quantum system in the space-time. But by evolution operator [6, 21], this evolution comes given by the anomalous energy ($\mathrm{E}^- - \mathrm{E}^+$) [6] (*see figure* 3) , which establishes the energy load functions $w^+(s, t) - w^-(s, t)$, that define the energy load function $w(s, t)$, at time $t \geq s$, where $w(s, t) = (\mathrm{E}^- - \mathrm{E}^+)U_0(s, t)$ E^- and E^+, are amplitudes of the curves M = I$tx(s)$, It = M$x^{-1}(s)$, ∀ and M, I_t, functions of evolution curves of space-time *see figure* 3).

Then a corrective action is the inverse transform that transform the energy load function in energy useful to the process of re-establishment on the quantum space (*remember that it is necessary to release the bundle of energy captive*). How this inverse transformation realise?

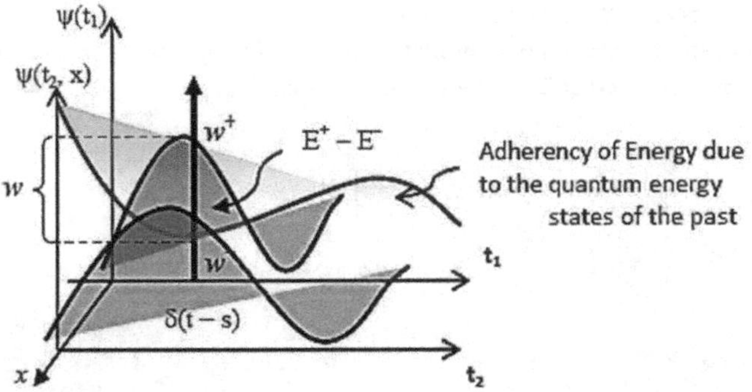

Figure 6. a). This is the graph that shows the formation of singularities of quantum type by the energy load (positives and negatives), rational amplitudes (extremes and defects), and bundle energy created and defined by $E^- - E^+$. The load is pre-determined by the impulse $\delta(t - s)$, which focalize the field action. The adherence zones are propitious to the formation of singularities with all conditions described. The wave function ψ, stays under constant regime in space evolution, which is $\psi = \psi(x, 0)$. In the second wave, the corresponding wave function is for other transition time (other evolution space), under the same conditions.

Lemma **1.** Let $\mathcal{M} = M \times I_t$, the unlimited space of the quantum space (Fock space [24]). A particle $x(t)$, that is focalized by a bad evolution given for the energy load function $w(t, s)$, comes given for

$$x(s) = \phi x(t)dx(t) = \int_{-\infty}^{+\infty} \delta(t - s)x(t)dt, \qquad (51)$$

Then to time $t = s$, begin the singularity.

Proof. We consider the function $w(t, s)$, like a Green function on the interval $t \geq s$. Given that this function is focalised for the emotional interpretation which is fed by the proper energy of the deep quantum energy (*since it produce an auto-disipant effect that deviates the evolution of every particles* [23]), then $O_c(x(t), x'(t))x(t) = \nabla^2 w(t, s) = \delta(t - s)$ [25]. By the nature of Green function of the weight function $w(t, s)$, we have

$$\int_M \phi(x')x(t)dx(t) = \int_{I_t}\left[\int_M \phi(x)\delta(x - x')dx(t)\right]x(t)\mu_t = \int_{-\infty}^{+\infty} O_c(x(t))x(t)\mu_t$$
$$= \int_{-\infty}^{+\infty} \nabla^2 w(s,t)x(t)\mu_t = \int_{-\infty}^{+\infty} \delta(t - s)x(t)dt,$$

Then all particle $x(t)$, in the space-time $M \times I_t$, affected by this regime to time $t = s$, and after, take the form (that is to say, to past and future particles)

$$\int_{-\infty}^{s} O_c(x(t))x(t)dt + \int_{s}^{+\infty} O_c(x(t))x(t)dt = 0 + \int_{s}^{+\infty} \delta(t-s)x(t)dt,$$

where the first integral is equal to cero, because there is no singularity before *s*, (the evolution happens after the time $t \geq s$ (see (49)). But this evolution is anomalous, since to all $t < s$, includes a captive energy not assimilated to $t = s$ (*this because it does not have a conscience operator at this moment (part of* (28) *defined by* $O^0{}_c$)). Then

$$\int_{t}^{+\infty} O_c(x)x(t)dt = \int_{t}^{+\infty} \delta(t-s)x(t)dt.$$

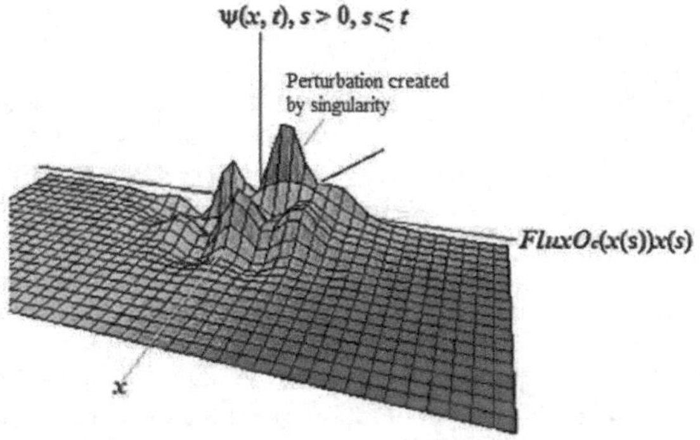

Figure 7. The surface represents the perturbation given by the existence of a singularity in the space-time of the quantum zone. The *computational model* obeys the solution of Planck-Fokker partial differential equation. The waves of perturbed state begins to $s \leq t$, such as it was predicted in the lemma 1. The surface include the waves established in figure 3, given by z = *Plot3D(exp(-1/2(x^2 + y^2))(cos(4x) + sin(2x) + 3sin(1/3y) + cos(5y)ln(2x)))*, through the *space-time 4.0 program*. Observe that is present a kernel of transformation for normal distribution given by *exp*$(-1/2(x^2 + y^2))$ that will appear in the transform that defines the singularity when a particle is not appropriately assimilate. The normal distribution kernel is the statistical weight that establishes the appearance of an abnormal evolution created by the existence of the singularity. That is to say, the singularity is detected by the anomalous effects that are glimpsed in the flux of the operator O_c, and that are observed in the surface bundle.

Theorem. **1. (F. Bulnes)**. Let consider a conscience operator with singularity $O_c(\ast)$, for the presence of an energy load $w(s, t)$. Then the *elimination* of the singularity $\ast(s)$, comes given for

$$x(t) = correction \quad + restoring \quad = \int_{X(C)} O_c(\ast(s))w(t,s)dt = dim\Lambda \int_{C} \left\{ \frac{1}{A} \prod_{j=1}^{\infty} \left\{ \int_{-\infty}^{+\infty} \phi(n_j)F(n_j) \right\} dx(t) \right\}, \quad (52)$$

where $dim\Lambda(\alpha)$, is the *Neumann dimension* corresponding to the *Weyl camera* of the roots α_j, [26, 27], used in the rotation process to eliminate the deviation [11], created by the singularity.

Proof. Consider an arbitrary irreducible diagram with nodes with *w*-parts (parts of diagrams with nodes of weight $w(t, s)$). Suppose that this points "nodes" with weight *w*, determines the singularity given by (50). In fact, by the theory of Van Hove on the singularities in the thermodynamic limit [23], each transition matrix of energy states has a correspondence with the product of Hermitian matrices of the corresponding evolution operators that to this case on a node en $t = s$, are given by $\upsilon_{\alpha}(s') = e^{is'h0}\upsilon_{\alpha}e^{-is'h^{0}}$, where we have used the *lemma 1*, to the arising of the quantum impurity in the space-time $\mathbb{M}$ (*quantum singularity*) in the image space of the conscience operator $\mathrm{T}\mathbb{M}^{*}$, located in $\mathbb{R}^3 \times \mathrm{I}_t$, in the point or node ($\cong \mathcal{G}$ [12, 23] (*w*-diagram)) $t = s$, corresponding of root space Φ_{α}. Then considering a irreducible diagram containing a $w(t, s)$-part, their contribution will be contained in $\mathbf{r} - \mathbf{R}$-space (which is the $E^{+} - E^{-}$-space (see figure 2)) [23] by the function $\delta(t - s)$:

$$x(t) = \int_{C}\left\{\prod_{j=1}^{\infty}\delta^{(3)}(\mathbf{r}^{j})e^{-iH_{0}t}\hat{U}(t,s)\right\}d^{3}r^{j}d^{3}R^{j}e^{iH_{0}s}n_{\mathbf{r}}(\mathbf{R},s^{j}), \tag{53}$$

to j*th*-particle in the interaction of a j*th*-thought to time s. The diagrams drawn here (figure 3) may be interpreted to represent the evolution of particles contributing the position density matrix $n_{\mathbf{r}}(\mathbf{R}, t)$ at $t = t$, in the corresponding path integral of correction. The true correction comes given by the evolution created in the quantum process of transformation of the functional $O_c(\circledast(s))$, where the information $\circledast(s)$, must be changed when $\lim_{t\to s} \pm iw(t,s) = \overline{n(s)}$ (is to say, when $M = I_t$). If we call $t_j = s$, (*j-th-time of evolution* in the thought process) then the integral (53), takes the form

$$e^{-i(t-t_j)H_0}\left\{1+\sum_{1}^{\infty}(-i)\lambda\int_{0}^{j\,t_1}dt_1\int_{0}^{t_2}dt_2\cdots\int_{0}^{t_{j-1}}dt_j\upsilon(t_1)\upsilon(t_2)\cdots\upsilon(t_j)\right\}=\prod_{j=1}^{\infty}\left\{\int_{-\infty}^{+\infty}\phi(n_j)F(n_j)\}dx(t)\right\}, \tag{54}$$

where the states ϕ_j, are established in the density matrix $\underline{n}(0)$, that to a vertex of $\mathcal{G}$, (*w* - diagram), arrange, those perturbations B (see figure 3), that they gave origin to the singularity, with the corresponding arrange of those positive perturbations A, that will realises the corrective action to transform the singularity signal $\circledast(s)$, of an adequate thought given by $x(t)$. Due to that, the information given by the product $A\underline{n}(0)B$, must be changed when $\lim_{t\to s} \pm iw(t,s) = \overline{n(s)}$, then a *w*-diagram must be change for $\Phi\alpha$–diagram [27]. But this live in the quantum field of the space-time $\mathrm{T}\mathbb{M}^{*}$, that is to say, in the corresponding zone of the executive operator $\mathfrak{I}$. Then in the material space-time (*Einstein universe*), the displacement of energy needs inside this transformation the application of an invariant given in the quantum space that guarantee that the new particle (*boson*) obtained let that correct. This is given by number $dim\Lambda(\alpha)$, since it depends on the roots system to the representation of the corresponding action group [27], that recover the recognition action. By the integral (4), the transformation due to the new conscience operator created in $\mathrm{T}\mathbb{M}^{*}$-zone obtained on whole the space-time is,

$$\int_{X(C)} O_{c}\left(\frac{x(t)}{dim\Lambda}\right)dx(t) = \int_{X(C)}\left\{\frac{1}{A}\prod_{j=1}^{\infty}\left\{\int_{-\infty}^{+\infty}\phi(n_j)F(n_j)\right\}dx(t)\right\},$$

that is the result wanted. ■

5. Re-composition and determination of the realities

We consider the space-time M, like space where $\mathbb{R}^d \times I_t$, is the macroscopic component of the space-time and we called F, the microscopic component of the space-time of ratio $10^{-33}cm$ (*length of a string* [21]). For previously described the quantum zone of the space-time $\mathcal{M}$, is connected with $\mathcal{N}$, which will called virtual zone of the space-time (zone of the space-time where the process and transformation of the virtual particles happen) are connected by possibilities causal space generated by certain class of photons and by the material particles interacting in the material space time, with permanent energy and the material particles re-combining their states they become in waves on having moved in $\mathbb{R}^d \times I_t$, on any path of Feynman. Likewise we can define the space of this double fibration of quantum processing as:

$$\mathcal{L} = \left\{ O_c\left(\phi, \partial_\mu \phi, x(t), t\right) \in C^2\left(\mathbb{R}^d \times I_t\right) \middle| \frac{\partial^2}{\partial t^2} - \nabla^2\left(O_c(\phi, \partial_\mu \phi, x(t), t)\right) = 0 \right\}, \tag{55}$$

with the states ϕ, of quantum field are in the quantum zone $\mathcal{M}$. Let $\mathcal{N}$, the ambi-space (*set of connection and field*) defined as:

$$\mathcal{N} = \left\{ (X, \nabla) \in M \times L \mid \nabla^{XY}_{X'Y'} \Psi + \Phi(X) = 0 \right\}, \tag{56}$$

where ∇, is the connection of virtual field X, with the quantum field Y, and Ψ, is the field whose action is always present to create perceptions in the quantum zone connected with Φ (2-form)[28]. Then we can create the correspondence given by the double fibration [29]:

(space of processing of quantum particles)

$$\mathcal{L}$$

$$\swarrow \quad \searrow^{\theta} \tag{57}$$

(virtual zone of the space-time)$\mathcal{N}$ $\mathcal{M}$(quantum zone of the space-time)

This double fibration conformed the interrelation between $\mathcal{M}$, and $\mathcal{N}$. $\forall\ x(t) \in \mathcal{M}$, give beginning to a complex submanifold (that represents the spaces where are the quantum hologram) that includes all these quantum images given by quantum holograms, why? Because this complex submanifolds, considering the causal structure given in the space-time by the light cones (see figure 6 a)) [26], of all trajectories that follow a particle in the space-time [29], they can write using (57) as:

$$\Theta_x = \theta\left[\pi^{-1}(x)\right], \tag{58}$$

of $\mathcal{N}$, such that $\Theta_x \cong \mathbb{P}^1 \times \mathbb{P}^1$, which by space-time properties to quantum level represent the space of all light rays that transit through x, conforming a hypersurface (projective surface) that is a light surface. This surface is called the sky x [30]. A sky in this context represents the set of light rays through x (*bosons*) that it comes of the virtual field.

If $\mathcal{M} \cong \mathbb{C}^4$, then $\mathbb{M} = \mathcal{M} \times Q_x$, is the complete universe (include the cosmogonist perception by the super-symmetry specialist [31]). But, what is there of our quantum universe with regard to our real universe (included the material part given by the atoms)?

The answer is the same, we have an universe of ten dimensions and $\mathbb{M} = \mathcal{N} \times \mathcal{M}$, where the quantum representation of the object $x(s)$, is the quantum space-time $\mathcal{M} = \mathbb{R}^3 \times It$, (*which is the Einstein cosmogonist perception*) then the cosmo-vision of the virtual particles is $\mathbb{C}^2 \times Q_x$, [21], then the execution operator $\mathfrak{I}$, that proceeds to connect virtual particles through the paths which have path integrals on double fibration, establishing the *material-quantum-virtual* connection required to a total reality:

$$\begin{array}{ccc} & \mathcal{L} & \\ {}^{\pi}\swarrow & & \searrow^{\theta} \\ \mathcal{N} & & \mathcal{M} \end{array} \quad \cup \quad \begin{array}{ccc} & \mathfrak{Sr} & \\ {}^{\rho}\swarrow & & \searrow^{\sigma} \\ \mathcal{M} & & C \end{array} \qquad (59)$$

where C, is the material part connected with the quantum zone of the space-time (space taken by atoms) $\mathcal{M}$. The corresponding path integral that connects virtual particles in the whole fibration is the integral of line type (5) defining feedback connection:

$$\mathrm{I}(\mathfrak{I}_{Q_x}(x(s))) = \oint_{\Gamma} O_c(\theta(\pi^{-1}(\sigma(\rho^{-1}(x))))\mu_s, \qquad (60)$$

always with the space $\{x(s) \in \mathcal{M} \mid \Theta_x \subset \mathcal{N}\}$, to the permanent field actions. Then the *reality state* is the obtained through the *integral of perception* (60), considering the fibre of the corresponding reality in the argument of the operator O_c, of the integrating from (60).

6. Applications to the nanosciences

6.1. Nanomedicine

The integral medicine into of the class of alternative medicine, fundament their methods of cure in to health and reactive the vital field X, of the human body B, the *regeneration* of the centers of energy of B, and the *corrections and restoration* of the flux of energy *Flux*, in and in each organ **B**, of the human body B, taking constant of gradient of their electromagnetic current, voltage and resistance, obtaining of this manner, the balance of each organ in sunstone with the other organs to characterize to B, like complete synergic system in equilibrium and harmony [10].

Now, the cure that is realized to nano-metric scale must be executed with a synergic action of constant field [33], equal to effect in each atom of our body to unison of *real conscience of cure* (*duality mind-body* [11]). Of this way, the conscience of B, is the obtained synergy by the atoms in this sense and that will come reflected in the reconstitution of the vital field X.

Then under this reinterpretation, the sickness is only an effect of the fragmentation of this *real conscience of cure* of *B*, that is deduced by disconnections and disparity of atoms [11]. The integral medicine helps to recover the *continuity of this conscience* through of the electronic memory of health of the proper body [11, 22] (see the figure 6 c)).

6.2. Quantology and neurosciences

Let $\mathfrak{M}$, the mind space and their organic component (material component) the brain space C. Also we consider the quantum component of the mind given by the space-time $\mathcal{M}$. Studies realised in statistical mechanics have revealed that the *Bose-Einstein statistics* stretches to accentuate the low energy levels. This reflects his closeness to the emission of a *virtual field,* where the virtual particles are not detected in a *virtual energy sea.* This allows to surmise that the radiation that takes place from the virtual field to the quantum field of $\mathfrak{M}$, is composed by *photons type bosons* (that is to say it obeys this Bose-Einstein statistics), since the quantum field interacts with the material particles that contains the material field of the mind which is anchored in the brain C, like material organ.

Theorem **(F. Bulnes).** [21] The *total Lagrangian of mental field* comes given by the superior action whose total conscience is

$$O_{total} = O_{QCD}\left(O_{EM}\right), \tag{61}$$

to one total action defined by the groups *SU*(3), (*quantum and virtual field*) and *SU*(2), (*material field*).

Proof. The Lagrangian of the theory is an invariant of Lorentz and invariant under local transformations of phase of the group *SU*(3), (*for the charge of color*) and has the following form [31]:

$$L_{TOTAL} = \left\{ \bar{q} i \gamma^{\mu} \partial_{\mu} q - \bar{q} m q - b \bar{q} \gamma^{\mu} T_a q \mathfrak{b}^{a}_{\mu\nu} - \frac{1}{4} \mathfrak{b}^{a}_{\mu\nu} \mathfrak{b}^{\mu\nu}_{a} \right\}, \tag{62}$$

This corresponds to the space of the mind $\mathfrak{M} = \mathcal{M} + C$, where O_{EM}, put in C, (*neurological studies have proved that the process of thought in the level at least visible is of electromagnetic type*) signals through charges in the synapses and neurons of C. Nevertheless these charges produce in one level deeper. The tensor $\mathfrak{b}^{a}_{\mu\nu}$, is anti-symmetric and represent a *bosonic field* created by the interaction of quarks q, $p = \bar{q}$, and $[p, q] = \bar{q}\, q - q\, \bar{q}$. Whose bosonic field has all the particles of spin 0, and the trace of tensor $\mathfrak{b}^{a}_{\mu\nu}$, has electromagnetic components conformed by the photons that are stable on the limit of the transformation $q(x(s)) \mapsto e^{-i\alpha_a(x(s))T_a}\mathfrak{b}(x(s))$, of the thought $x(s)$, where some $\alpha_a(x(s))T_a$, is the electromagnetic frequency k_a, of the term $ik_a v$ [7, 10 26], where in these cases the last term in the second member of (62), is $\frac{1}{4} F_{\mu\nu}F^{\mu\nu}$. Then the *SU*(2)-actions are included in the *SU*(3), actions and their points are electromagnetic particles transformed by these rules like photons (= thoughts in $\mathfrak{M}$) [28].■

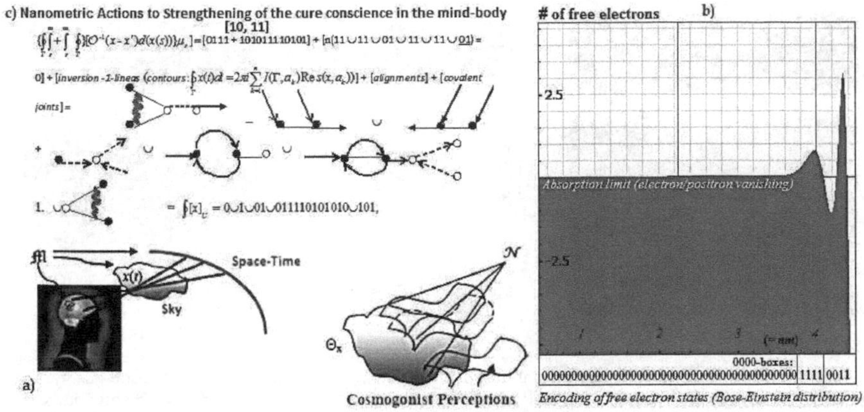

Figure 8. a). Concept of the topological space (*sky* [21, 30]) and the cosmogonist perceptions of the mind. The conscience operator of the different realities is given by (60). b) Modelling from boson-shape distribution in localizing of symptoms in the mind by cerebral signals with low interchange electrons [24, 25, 32, 33]. Bose-Einstein distribution was published in [22]. c) Total Action due Gelsem including the action de Bervul and *Ignamara*. The corresponding quantum intelligence code of this total action is given by 011110**1011011011101**10111110, [7]. The corresponding code given in black include de codes of *Gelsem* and *Bervul*. The code in orange is the code of *Gelsemium* equal to code of *Gelsem*. This quantum code was used to cure a patient with a digestive illness [11, 21, 22].

6.3. Electromagnetic vehicle with levitation magnetic conscience

Considering some applications of quantum electrodynamics in the design of flaying vehicles self-supported and their magnetic levitation, we find the magnetic conscience operator is defined for the transmitting of the diamagnetic property every particle of the ship structure.

This vehicle is controlled by one microchip that is programmed by conscience operators algebra of electromagnetic type that leads to the flow of Eddy currents, the iso-rotations and suspension of the special geometrical characteristics vehicle, generating also on the vehicle structure certain "magnetic conscience" that provokes all movements like succeeding the sidereal objects in the universe [26, 34]. This magnetic conscience is generating by the proper particles of the ship structure transmitted for the interaction of superconductor inside the reactor with the magnetic field generated by the rotating rings under the ship. By so doing, the Eddy's currents in the "skin effect" around the structure of the ship are given by the actions [35]

$$\Im_{\text{ship}} = \Im_{\text{M}} + \Im_{\text{rot}}, \tag{63}$$

and using the quantum $\mathfrak{E} \otimes \mathfrak{H}$-fields to create (magnetic conscious operator given by the integrals [36])

$$H(A, \Im_M) = \int_{ship} \left\{ L_M - \text{H}^2 / 8\pi \right\} d\text{V}, \tag{64}$$

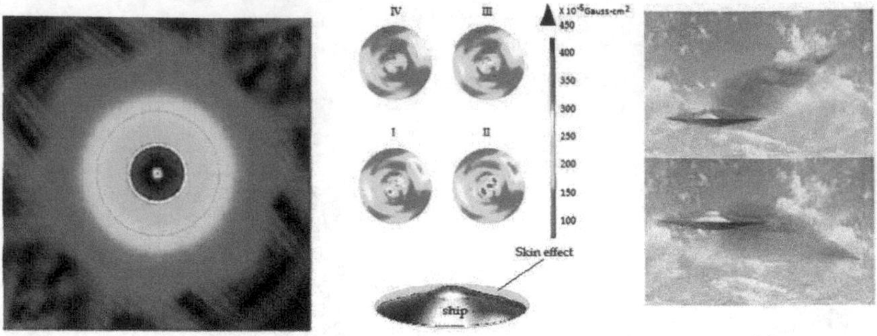

(a) (b) (c)

Figure 9. a). The green color represents the state of quantum particles in transition to obtain the anti-gravity states through the interaction of the $\mathfrak{E} \otimes \mathfrak{H}$-fields used on the structure of the vehicle. The blue flux represents Eddy's currents that interact with the lines of magnetic field to produce an diamagnetic effect in the top part of the vehicle. The red central ring is the magnetic field that generates the twistor surface (this computational simulation is published in [35, 36]). b) Top part of the flying plate showing the generation of the effect skin obtained in the interaction superconductor - magnetic field. The regions in light and obscure blue represent the flow of Eddy's currents, in this top part of the vehicle. The Skin effect is derived by the quantum interaction by the actions of $\mathfrak{E} \otimes \mathfrak{H}$ [26, 37]. c) The microscopic effects of the electromagnetic fields (quantum densities of field) created inside the algebra $\mathfrak{E} \otimes \mathfrak{H}$, create an effect of macro-particle of the vehicle [36] (the vehicle and their electromagnetic revetment behaves like a particle) where their displacement is realized in instantaneous form and their direction it is a macro-spin projected from the *magnetic conscious operator* (from $\Im_M$, from (63)) of the ship which defines their angular moment [35, 36].

Apendix

Technical notation

O_c – Is an operator that involves the Lagrangian but directing this Lagrangian in one specific fiber (direction) *prefixing tha Lagrangian action in one direction*. This is defined as the map: O_c : $T\mathbb{M} \rightarrow T\mathbb{M}^*$, with rule of correspondence $w \mapsto O_c(v)w$, where $w = L(v)$, with L, the classic Lagrangian. *This defines the quantum conscience*. If we locally restrict to O_c, that is to say, on the tangent space $T_x\mathbb{M} \times T_x\mathbb{M}$, $\forall\, x \in \mathbb{M}\ (\cong \Omega(\Gamma))$, we have that

$$O_c : T_x\mathbb{M} \times T_x\mathbb{M}\ (\underset{locally}{\cong}\ T\mathbb{M}) \longrightarrow T\mathbb{M}^*,$$

with rule of correspondence

$$(v, w) \longrightarrow O_c(v)w,$$

$O_c(v)w$, generalise the means of $O_c(v)v = L(v)$, $\forall\, v \in T_x\mathbb{M}$, $\forall\, x \in \Omega(\Gamma)$. Likewise, if $\Im$: $T\mathbb{M} \rightarrow \mathbb{R}$, with rule of correspondence $L(v) \mapsto \Im(L(v)) = O_c(v)v$, then the total action along the trajectory Γ, will be

$$\Im_\Gamma = \int_{\Omega(\Gamma)} O_c(v)v = \int_{\Omega(\Gamma)} L(v),$$

In the forms language, the conscience operator comes given by the map $\omega_L : TM \to TM^*$, with rule of correspondence given by (19). The quantum conscience shape a continuous flux of energy with an intention, involving a smooth map π (defined in the *example 1*). Then the conscience operator is related with the action $\Im$, and the trajectories γ_t, through of the following diagram:

$$\begin{array}{ccc} TM & \xrightarrow{O_C} & TM^* \\ \Im\downarrow & & \downarrow\pi \\ R & \xrightarrow[\gamma_t]{} & M \end{array}$$

$O_c(\ast)$ – Conscience Operator in the singularity $\ast$. This is a kernel of the quantum inverse transfom of path integrals to eliminate singularities. Their direct transform use the kernel $O_c(x(t))$.

$\mathcal{O}_0(Z_i, p_A)$ – Vertex operator of the theory given by the equations $W^a(Z_i) = 0$, localized at the points p_A, of the Riemann surface S.

$\mathcal{O}(x - x')$ – Is the functional operator $\mathcal{O}(x - x_j) = (\Box_x + m^2 - i\varepsilon)\delta^n(x - x_j)$. This operator involves to electronic propagator in a pulse impulse.

O_{QCD} – Quantum chromodynamics conscience operator. Their Lagrangian density using the quantum chromodynamics is $\mathcal{L}_{EM} = \Sigma_n (ihc\underline{\psi}_n \not{D} \psi_n - m_n c^2 \underline{\psi}_n \psi_n) - 1/4\, G^\alpha{}_{\mu\nu} G_\alpha{}^{\mu\nu}$, Where $\not{D}$, is the QCD gauge covariant derivative (in Feynman notation $\xi^\sigma D\sigma$), $n = 1, 2, \ldots 6$ counts the quark types, and is the gluon field strength tensor.

O_{EM} – Conscience operator defined through of the Lagrangian to quantum electromagnetic field (these like gauge fields). Their Lagrangian density is $\mathcal{L}_{EM} = ihc\underline{\psi}\not{D}\psi - mc^2\underline{\psi}\psi - (1/4\mu_0)F_{\mu\nu}F^{\mu\nu}$. where $F^{\mu\nu}$, is the electromagnetic tensor, D, is the gauge covariant derivative, and $\not{D}$, is Feynman notation for $\xi^\sigma D\sigma$.

O_{total} – Total quantum conscience operator. This is the composition of operators O_{EM} followed O_{QCD}.

$\Im_{O_c}$ – Action that involves a conscience operator O_c.

$\Omega^{(k,0)}$ – Differential form to complex hypersurfaces of dimension k. This form is analogous to the form ω_L, and involves the conscience operator $(O_c)^*$.

Θ_x – Fibers of the topological space Q_x, called *sky* conformed by the light rays through x (*bosons*) that it comes of the virtual field. This is a conscience operator when realises the reality transformation by the double fibration.

$\Im_{\partial M}$- Action that have codimension strata $n - k$. This action is due to the differential $d\Im(\phi)h$.

$\Im_{int_M}$- Action that have codimension strata k. This action is due by $\Im(\phi)$.

Author details

Francisco Bulnes

Department of Research in Mathematics and Engineering, TESCHA, Mexico

Acknowledgements

I am grateful with Carlos Sotero, Eng., for the help offered for the digital process of the images that were included in this chapter.

7. References

[1] Sobreiro R., editor. Quantum Gravity. Rijeka: InTech; 2012. http://www.intechopen.com/books/quantum-gravity- (accessed 20 January 2012).

[2] Bulnes F. Analysis of prospective and development of effective technologies through integral synergic operators of the mechanics. In: ISPJAE, Superior Education Ministry of Cuba (eds.) 14th Scientific Convention of Engineering and Arquitecture: proceedings of the 5th Cuban Congress of Mechanical Engineering, December 2-5, 2008, CCIA2008, 2-5 December, 2008, Havana, Cuba.

[3] Bulnes F. Theoretical Concepts of Quantum Mechanics. In: Mohammad Reza Pahlavani (ed.) Correction, Alignment, Restoration and Re-Composition of Quantum Mechanical Fields of Particles by Path Integrals and Their Applications. Rijeka: InTech; 2012. p Available from http://www.intechopen.com/books/theoretical-concepts-of-quantum-mechanics/correction-alignment-restoration-and-re-composition-of-fields-of-particles-by-path-integrals-and-the

[4] Marsden, JE.; Abraham, R. Manifolds, tensor analysis and applications. Massachusetts: Addison-Wesley; 1993.

[5] Lawson, HB.; Michelsohn, ML. Spin Geometry, Princeton University Press; 1989.

[6] Holstein, BR. Topics in Advanced Quantum Mechanics. CA, USA: Addison-Wesley Publishing Company; 1992.

[7] Feynman, RP; Leighton, RB; Sands, M. Electromagnetism and matter (Vol. II). USA: Addison-Wesley; 1964.

[8] Warner, FW. Foundations of Differential Manifolds and Lie Groups, New York: Springer, 1983.

[9] Sokolnikoff, IS. Tensor Analysis: Theory and Applications. New York: Wiley and Sons; 1951.

[10] Bulnes F; Bulnes H. F; Hernandez E; Maya J. Integral Medicine: New Methods of Organ-Regeneration by Cellular Encoding through Path Integrals applied to the Quantum Medicine. Journal of Nanotechnology in Engineering and Medicine ASME 2010; 030019(1) 7.

[11] Bulnes F; Bulnes H. F; Hernandez E; Maya J. Diagnosis and Spectral Encoding in Integral Medicine through Electronic Devices Designed and Developed by Path Integrals. Journal of Nanotechnology in Engineering and Medicine ASME 2011; 021009(2)10.

[12] Ionescu LM. Considerations on some algebraic properties of Feynman integrals. Surveys in Mathematics and Its Applications. Vol. 3 (2008). p79-110.

[13] Kontsevich M, Deformation quantization of Poisson manifolds I. In: Joseph A, Mignot F, Murat F, Prum B, Rentschler R.(eds.) Proceedings of the First European Congress of Mathematics, Vol. II, 6–10 July 1992, Paris, France. Progr. Math., 120, Birkhäuser, Basel, 1994.

[14] Kontsevich M. Feynman diagrams and low-dimensional topology. In: Joseph A, Mignot F, Murat F, Prum B, Rentschler R.(eds.) Proceedings of the first European congress of mathematics (ECM), Vol. II, 6–10 July 1992, Paris, France, Progr. Math., 120, Birkhäuser, Basel, 1994.

[15] Bulnes F. Cohomology of Cycles and Integral Topology. In: Bulnes, F. (ed.) Meeting of 27- 29 Autumn 2008, Mexico City, Appliedmath 4, IM-UNAM, Mexico, 2008. www.Appliedmath4.ipn.mx

[16] Griffiths P; Harris J. Principles of Algebraic Geometry. USA: Wiley-Interscience; 1994.

[17] Gross M; Huybrechts D; Joyce D. Calabi-Yau Manifolds and Related Geometries. Norway: Springer, 2001.

[18] Hoogeveen J; Skenderis K. Decoupling of unphysical states in the minimal pure spinor formalism I. JHEP 1001 (2010) 041. [arXiv:0906.3368].

[19] Berkovits N; Hoogeveen J; Skenderis K. Decoupling of unphysical states in the minimal pure spinor formalism II. JHEP 0909 (2009) 035. [arXiv:0906.3371]

[20] Kibbe TWB. Geometrization of Quantum Mechanics. Springer Online Journal Archives 1860-2000 1979; 65(2) 189-201.

[21] Bulnes F; Bulnes HF; Cote D. Symptom Quantum Theory: Loops and Nodes in Psychology and Nanometric Actions by Quantum Medicine on the Mind Mechanisms Programming Path Integrals. Journal of Smart Nanosystems in Engineering and Medicine 2012; 1(1) 97-121.

[22] Bulnes F; Bulnes HF. Quantum Medicine Actions: Programming Path Integrals on Integral Mono-Pharmacists for Strengthening and Arranging of the Mind on Body. Journal of Frontiers of Public Health 2012; accepted.

[23] Fujita S. Introduction to non-equilibrium quantum statistical mechanics. Malabar, Fla: W. Krieger Pub. Co. 1983.

[24] Simon B; Reed M. Mathematical methods for physics, Vol. I (functional analysis). New York: Academic Press, 1972.

[25] Eleftherios N; Economou E; Economou N. Green's Functions in Quantum Physics. Berlin Germany: Springer, 2006.

[26] Bulnes F, Doctoral course of mathematical electrodynamics. In: National Polytechnique Institute (ed.) Appliedmath3: Advanced Courses: Proceedings of the Applied Mathematics International Congress, Appliedmath3, 25-29 October SEPI-IPN, México, 2006.

[27] Bulnes F, Conferences of Mathematics: Seminar of Representation Theory of Reductive Lie Groups. Mexico: Compilation of Institute of Mathematics (ed.), UNAM Publications, 2000.

[28] Aharonov Y; Bohm D. Significance of electromagnetic potentials in quantum theory. Phys. Rev. 115 (1959) 485-491.

[29] Bulnes F; Shapiro M. On general theory of integral operators to analysis and geometry (Monograph in Mathematics). Mexico: SEPI-IPN, IMUMAM, 2007.

[30] LeBrun ER, Twistors, Ambitwistors and Conformal Gravity. In: Bailey TN, Baston RJ (ed.) Twistors in Mathematics and Physics, UK: Cambridge University; 1990. p71-86.

[31] Hughston LP; Shaw WT. Classical Strings in Ten Dimensions. Proceedings of the Royal Society of London. Series A, Mathematical and Physical Sciences, Vol 414. No. 1847 (December 8). UK, 1987.

[32] Dossey L. Space, Time and Medicine. Boston: Shambhala, USA, 1982.

[33] Truesdell C; Topin RA. The classical fields theories (in encyclopedia of physics, Vol. III/1). Berlin: Springer-Verlag. 1960.

[34] Bulnes F. Special dissertations of Maxwell equations. Mexico: unpublished. Only registered SEP, 1996.

[35] Bulnes F; Hernández E; Maya J. Design and Development of an Impeller Synergic System of Electromagnetic Type for Levitation, Suspension and Movement of Symmetrical Body. Imece2010: Fluid Flow, Heat Transfer and Thermal Systems Part A and B: Proceedings of 11th Symposium on Advances in Materials Processing Science and Manufacturing, 12-18 November2010, British Columbia, Canada, 2010.

[36] Bulnes F; Maya J; Martínez I. Design and Development of Impeller Synergic Systems of Electromagnetic Type to Levitation/Suspension Flight of Symmetrical Bodies. Journal of Electromagnetic Analysis and Applications 2012 1(4) 42-52. See in: http://www.scirp.org/journal/PaperInformation.aspx?paperID=17151

[37] Bulnes F. Foundations on possible technological applications of the mathematical electrodynamics. Masterful Conference in Section of Postgraduate Studies and Research (SEPI), National Polytechnic Institute, Federal District, Mexico 2007.

Permissions

The contributors of this book come from diverse backgrounds, making this book a truly international effort. This book will bring forth new frontiers with its revolutionizing research information and detailed analysis of the nascent developments around the world.

We would like to thank Professor Paul Bracken, for lending his expertise to make the book truly unique. He has played a crucial role in the development of this book. Without his invaluable contribution this book wouldn't have been possible. He has made vital efforts to compile up to date information on the varied aspects of this subject to make this book a valuable addition to the collection of many professionals and students.

This book was conceptualized with the vision of imparting up-to-date information and advanced data in this field. To ensure the same, a matchless editorial board was set up. Every individual on the board went through rigorous rounds of assessment to prove their worth. After which they invested a large part of their time researching and compiling the most relevant data for our readers. Conferences and sessions were held from time to time between the editorial board and the contributing authors to present the data in the most comprehensible form. The editorial team has worked tirelessly to provide valuable and valid information to help people across the globe.

Every chapter published in this book has been scrutinized by our experts. Their significance has been extensively debated. The topics covered herein carry significant findings which will fuel the growth of the discipline. They may even be implemented as practical applications or may be referred to as a beginning point for another development. Chapters in this book were first published by InTech; hereby published with permission under the Creative Commons Attribution License or equivalent.

The editorial board has been involved in producing this book since its inception. They have spent rigorous hours researching and exploring the diverse topics which have resulted in the successful publishing of this book. They have passed on their knowledge of decades through this book. To expedite this challenging task, the publisher supported the team at every step. A small team of assistant editors was also appointed to further simplify the editing procedure and attain best results for the readers.

Our editorial team has been hand-picked from every corner of the world. Their multi-ethnicity adds dynamic inputs to the discussions which result in innovative outcomes. These outcomes are then further discussed with the researchers and contributors who give their valuable feedback and opinion regarding the same. The feedback is then collaborated with the researches and they are edited in a comprehensive manner to aid the understanding of the subject.

Apart from the editorial board, the designing team has also invested a significant amount of their time in understanding the subject and creating the most relevant covers. They scrutinized every image to scout for the most suitable representation of the subject and create an appropriate cover for the book.

The publishing team has been involved in this book since its early stages. They were actively engaged in every process, be it collecting the data, connecting with the contributors or procuring relevant information. The team has been an ardent support to the editorial, designing and production team. Their endless efforts to recruit the best for this project, has resulted in the accomplishment of this book. They are a veteran in the field of academics and their pool of knowledge is as vast as their experience in printing. Their expertise and guidance has proved useful at every step. Their uncompromising quality standards have made this book an exceptional effort. Their encouragement from time to time has been an inspiration for everyone.

The publisher and the editorial board hope that this book will prove to be a valuable piece of knowledge for researchers, students, practitioners and scholars across the globe.

List of Contributors

Gabino Torres-Vega
Physics Department, Cinvestav, México

Francisco De Zela
Departamento de Ciencias, Sección Física, Pontificia Universidad Católica del Perú, Lima, Peru

Keita Sumiya, Hisakazu Uchiyama, Kazuhiro Kubo and Tokuzo Shimada
Department of Physics, School of Science and Technology, Meiji University, Japan

L. M. Arévalo Aguilar
Facultad de Ciencias Físico Matemáticas, Benemérita Universidad Autónoma de Puebla, Puebla, México

C. P. García Quijas
Departamento de F´sica, Universidad de Guadalajara, Guadalajara, Jalisco, México

Carlos Robledo-Sanchez
Facultad de Ciencias Fisico Matemáticas, Benemérita Universidad Autónoma de Puebla, Puebla, México

Jan J. Sławianowski and Vasyl Kovalchuk
Institute of Fundamental Technological Research, Polish Academy of Sciences, Warsaw, Poland

Miloš V. Lokajíˇcek, Vojtˇech Kundrát and Jiˇrí Procházka
Institute of Physics of the AS CR, Prague, Czech Republic

Kazuyuki Fujii
International College of Arts and Sciences, Yokohama City University, Yokohama, Japan

Valeriy I. Sbitnev
St.-Petersburg Nuclear Physics Institute, NRC Kurchatov Institute, Gatchina, Russia

Paul Bracken
Department of Mathematics, University of Texas, Edinburg, TX, USA

Francisco Bulnes
Department of Research in Mathematics and Engineering, TESCHA, Mexico

Printed in the USA
CPSIA information can be obtained
at www.ICGtesting.com
JSHW011430221024
72173JS00004B/745

9 781632 381606